T0093788

CMS/CAIMS Books in Mathematics

Volume 5

CMS/CAIMS Books in Mathematics is a collection of monographs and graduate-level textbooks published in cooperation jointly with the Canadian Mathematical Society- Societé mathématique du Canada and the Canadian Applied and Industrial Mathematics Society-Societé Canadienne de Mathématiques Appliquées et Industrielles. This series offers authors the joint advantage of publishing with two major mathematical societies and with a leading academic publishing company. The series is edited by Karl Dilcher, Frithjof Lutscher, Nilima Nigam, and Keith Taylor. The series publishes high-impact works across the breadth of mathematics and its applications. Books in this series will appeal to all mathematicians, students and established researchers. The series replaces the CMS Books in Mathematics series that successfully published over 45 volumes in 20 years.

CMS
SMC

CAIMS
SCMAI

Winfried Grassmann · Javad Tavakoli

Numerical Methods for Solving Discrete Event Systems

With Applications to Queueing Systems

 Springer

Winfried Grassmann
Department of Computer Science
University of Saskatchewan
Saskatoon, SK, Canada

Javad Tavakoli
Department of Mathematics
University of British Columbia
Kelowna, BC, Canada

ISSN 2730-650X   ISSN 2730-6518   (electronic)
CMS/CAIMS Books in Mathematics
ISBN 978-3-031-10081-9   ISBN 978-3-031-10082-6   (eBook)
https://doi.org/10.1007/978-3-031-10082-6

Mathematics Subject Classification: 60J27, 60J10, 60J22, 60K25, 65C40

This Springer imprint is published by the registered company Springer Nature Switzerland AG
The registered company address is: Gewerbestrasse 11, 6330 Cham, Switzerland

Preface

This book shows how to formulate and solve discrete event systems by numerical methods without using discrete event simulation. Discrete event systems are predominantly stochastic systems that are changed by discrete events. Examples of discrete event systems include queueing systems, where arrivals and departures are discrete events, and inventory models, where usages and replenishments are discrete events. Most systems we consider have several state variables, such as queues and inventories. We provide numerical methods to find transient and steady-state solutions for such systems.

The book is method rather than model oriented. In other words, the book helps researchers who already have a model they need to solve, and they are now looking for applicable methods to solve this model. The stress is on numerical tools as opposed to formulas. We provide algorithms that hopefully can be applied, possibly in a modified version, to solve the problems in question. Of course, for many researchers, the first approach to solve their problem is Monte Carlo simulation, but since there is an ample literature in this area, we do not cover it here. Instead, our algorithms are deterministic in the sense that no random numbers are used. We formulate our systems as Markov chains and solve them as such. This has certain advantages, especially when dealing with smaller systems, and in cases where high precision is required.

The use of Markov chains and their properties naturally leads to a better understanding of important concepts. It helps to gain insights into the concepts of recurrence and transience, and the concept of a statistical equilibrium. For this reason, this book may also serve people that merely want to have a better understanding of Markovian systems, and their use to solve discrete event systems.

A prime example of a discrete event system is a queueing system, and indeed we use many results from queueing theory in this book. However, we go beyond the standard models of queueing theory, such as one-server and multi-server queues. We are looking at situations where there are several queues, and where arrivals are not necessarily Poisson and service times are not necessarily exponential. We realize that dealing with such systems forces us to use numerical methods, which also allows us to avoid complex mathematics.

In traditional queueing theory, the emphasis is on mathematical derivations rather than on algorithms. This is understandable when considering the history of queueing theory and computing. When the classical results of queueing theory were derived, computers were slow or even non-existent, and computations were costly. Today, even a simple laptop can do millions of multiplications per second. This completely changes the situation. If you have no computers, you cannot use solution methods that require millions of additions and multiplications, but the derivation of formulas is not hampered by the lack of computing power. Today, even a simple laptop provides enough computing power to find transient and equilibrium solutions of non-trivial discrete event systems.

Though deterministic numerical methods gained enormously from the increased speed of the computers, simulation gained much more. The reason is the curse of dimensionality. Essentially, the numerical effort to find numerical solutions increases exponentially with the number of state variables needed to formulate the model. This severely limits the types of problems that can be solved by deterministic numerical methods. In simulation, the computational effort only increases linearly with the number of state variables, meaning that for large models, discrete event simulation is the method of choice. However, we found that for small models, deterministic methods have many advantages over simulations, including shorter execution times.

Concentrating on smaller models is not necessarily a disadvantage. There are many instances where models with very few state variables are appropriate, and for these models, solution methods based on probability theory are a good fit. Indeed, models with many state variables are often difficult to interpret because of their high dimensionality, which may hide essential features. For this reason, simple models are often more instructive.

Instead of using the discrete event paradigm, many people use stochastic Petri net [65] or stochastic activity nets [79]. Personally, we found these approaches both less general and more difficult to understand than discrete event systems, as they are described by Casandras and Lafortune [9]. Of course, discrete event systems are familiar to anyone using simulating queues or similar stochastic systems.

Here is the outline of this book. In Chapter 1, we introduce basic concepts which are needed to understand the remainder of this book. Chapters 2–5 show how to formulate discrete event systems, and how to convert them into Markov chains by using supplementary variables. Chapter 6 introduces computational complexity and discusses the different types of errors the modeler has to consider. In particular, the chapter deals with rounding errors, and it shows that rounding errors are magnified by subtraction. Fortunately, when dealing with probabilities, subtractions can often be avoided. Chapter 7 discusses transient solutions, with particular emphasis on the randomization method. Chapter 8 introduces the classifications of states, and it shows how to use eigenvalues to find transient solutions of discrete event systems. Both topics are required in order to estimate how fast a system converges toward its equilibrium solution. Chapter 9 describes how to find equilibrium solutions, both by direct and iterative methods. The direct method we prefer is an elimination method we call *state reduction method* or *state elimination method*, which is numerically very stable, an important property because the discrete event systems lead to Markov chains with many states, and rounding errors increase as the number of states increases. State elimination also has a probabilistic interpretation in terms of censoring, a method closely related to embedding. This leads to Chapter 10, which shows how to use censoring and embedding to reduce the state space. Chapter 11 covers systems that are almost or completely independent. It also covers product-form solutions where the state variables can be considered as independent, though only in systems in a statistical equilibrium. Chapter 12 deals with systems with an infinite state space, with emphasis on the matrix analytic method and related methods.

Since this book does not use advanced mathematical methods, it should be accessible to non-specialists who look for effective methods to solve the stochastic systems they encounter in their research. It should also be useful to graduate students for their projects and theses. We believe that this book is also useful as a text/reference for any typical stochastic process and related topic courses.

Aarau, Switzerland   Winfried Grassmann
Kelowna, BC, Canada   Javad Tavakoli

Contents

Chapter 1
Basic Concepts and Definitions

This book is about stochastic systems, but instead of simulation, as in discrete event simulation, we use Markov chains to analyze these systems. This chapter provides the basic mathematical tools for this endeavor. We thus describe what a system is, and how to work with Markov chains. Since our systems are stochastic, we also review random variables and their distributions. Many systems covered in this book are in fact queueing systems, and we therefore discuss a notation to describe such systems, which is due to Kendall. We will also cover the so-called Little's law, which provides a relation between expected waiting times and the expected number of elements in a given subsystem. To simplify our notations, we frequently use sets, which are covered in the last section of this chapter.

1.1 The Definition of a Discrete Event System

There are many different types of systems. There are production systems, computer systems, communication systems, and queueing systems. These are the types of systems we are looking at. These systems are *dynamic* in the sense that they change trough time. The systems we are looking at are also stochastic. Many systems change only in discrete steps, and these steps are caused by *events*. Such systems are called *discrete event systems*. In a queueing system, events include arrivals, moves from one queue to another one, and departures. In production systems, the events may be the arrival of raw material, completion of a task at a particular work station, or breakdown of machinery. To analyze such systems properly, they must be formulated in mathematical terms. In particular, we have to define the *state* of the system, which describes the present situation the system is in, and the *events*, which change the state of the system.

© The Author(s), under exclusive license to Springer Nature Switzerland AG 2022
W. Grassmann and J. Tavakoli, *Numerical Methods for Solving Discrete Event Systems*,
CMS/CAIMS Books in Mathematics 5, https://doi.org/10.1007/978-3-031-10082-6_1

1.1.1 State Variables

To be mathematically tractable, the state of the system must be represented by a number of variables. Some of these variables can easily be recognized because they are easily visible or physically measurable, such as queue length or inventory levels, and they will be referred to as *physical variables*. We assume that there are d physical variables, denoted by $X_1, X_2, \ldots, X_d$. These variables change over time, and at time t, they assume the values $X_k(t)$, $k = 1, 2, \ldots, d$. The vector $X = [X_1, X_2, \ldots, X_d]$ describes the *physical state*. Typically, the physical variables at time t are stochastic, that is, one cannot predict the $X_k(t)$ themselves, only their distribution can be found. Consequently, for each k, $\{X_k(t), t \geq 0\}$ forms a stochastic process. These processes are interdependent, and what is of interest is their joint distribution at some time in the future. This is what we want to calculate.

Let us suppose we are now at time 0, and our aim is to find the joint distribution of the physical variables for different times t. The first question is: What do we have to know in order to predict the joint distribution of the state variables in the future? Is it enough to know the physical variables at time 0, or do we need to know more? Often, we do. Frequently, variables other than the physical ones influence the resulting joint distributions of the $X_k(t)$. These variables will be called *supplementary variables*. Let $Y_1, Y_2, \ldots, Y_g$ be the supplementary variables, and let $Y_1(t), Y_2(t), \ldots Y_g(t)$ be the value of these variables at time t. For convenience, we define $Y = [Y_1, Y_2, \ldots, Y_g]$. $X(t)$ and $Y(t)$ are then the vectors of the values the entries assume at time t.

The state is defined as the information one needs at time 0 to find the distributions of $X(t)$ and $Y(t)$ for any $t > 0$. We assume that the state is given by the vectors X and Y. The variables used to define the state are called state variables. Some authors require that the set of state variables must be as small as possible, but we will not insist on this. There are often other variables of interest, so-called auxiliary variables. For instance, in a production system, the number of products at the different stations may all be state variables. Their sum, that is the total number of products in the system, is an auxiliary variable. The state variables, and their development through time form what is called a system. If the system is subject to random effects, then we have a stochastic system.

For our later discussion, we define S_X to be the set of all possible values X can assume, and S_Y similarly the set of all values Y can assume. S_X and S_Y will be called the physical state space and the supplementary state space, respectively. The state space S is then the union of S_X and S_Y. In the next chapters, we assume that the number of elements in S_X is finite, but later, we allow this number to be countably infinite. Typically, the elements of S_X will be integers. Some Y_k may be real, which means that the supplementary space may be continuous and therefore infinite. However, infinite spaces are difficult to deal with numerically. Therefore, we will search for methods to represent, or at least approximate, the infinite state space in ways that allow us to use numerical methods to find the results of interest.

1.1.2 Events

A system is a *discrete event system* if $X(t)$ can only change in discrete steps, typically in steps of size i where i is an integer. In other words, the changes of physical variables are discontinuous. This is certainly true for queueing systems, and for many other systems as well. The changes of the systems are caused by events. The events can be of different types, say type Ev_1, $Ev_2 \ldots$, Ev_e, where e is the number of event types. In queueing systems, event types include arrivals, in inventory system, deliveries, and so on. These event types are created by stochastic processes, such as arrival processes. In fact, with each event type, we will associate an event process. We do not allow any change of the physical state other than through events. This can easily be accomplished by associating any change of the system with an event type.

We distinguish between *determinist* and *stochastic* events. An event is deterministic if for a given state, the effect of the event is certain. If an event occurs at time t, and $X(t)$ is the physical state at this time, then the state at t^+ is a function of $X(t)$. Here, t^+ stands for the time right after t. Hence, if Ev_k happens at t, $X(t^+) = f_k(X(t))$. Here, $f_k(X)$ will be referred to as *event function*. We assume that $f_k(X)$ depends only on k and X, and on nothing else. An example of a deterministic event is the arrival to a one-sever queue. In such a system, the physical state is given by a single variable X, the length of the waiting line. If Ev_1 denotes arrivals, then $f_1(X) = X + 1$.

If Ev_k is a stochastic event, then there would be different outcomes, each associated with some probability. Consequently, if there are ℓ outcomes, there are ℓ event functions $f_{k,1}, f_{k,2}, \ldots, f_{k,\ell}$, with the outcome i having a probability p_i. For example, consider a queueing system with two different lines. An arriving customer may join line 1 with probability p, and line 2 with probability $1 - p$. If the physical state is given by $[X_1, X_2]$, where X_i, $i = 1, 2$ is the length of line i, and Ev_1 denotes arrivals, then $f_{1,1} = [X_1 + 1, X_2]$, and $f_{1,2} = [X_1, X_2 + 1]$.

Some events are not allowed in certain states. In particular, in a queueing system, there cannot be a departure from any empty line. For this reason, event Ev_k may have an *event condition* $C_k(X)$, and Ev_k cannot happen unless $C_k(X)$ is true in state X. For instance, in a one-server queue, with X being the only physical state variable, $C_2(X) = X > 0$, where Ev_2 represents departures. Hence, there are no departures if $C_2(X)$ is false.

As an example of a discrete event system, consider a system with three waiting lines. To represent these lines, three physical state variables are used, such as X_1, X_2, and X_3, and possibly some supplementary variables, representing the states of the processes generating the events. Possible events would be arrivals to line one, a move from line one to line two, and a move from line two to line three. The corresponding event functions would be $f_1(X_1, X_2, X_3) = [X_1 + 1, X_2, X_3]$, $f_2(X_1, X_2, X_3) = [X_1 - 1, X_2 + 1, X_3]$, and $f_3(X_1, X_2, X_3) = [X_1, X_2 - 1, X_3 + 1]$. Other events are also possible. In a different system, there may be an event, such as a repair, that takes two units from line one and one unit from line two. This event would have the event function $f_3(X_1, X_2, X_3) = [X_1 - 2, X_2 - 1, X_3]$.

Discrete event systems can either be *terminating* or *non-terminating*. A system can terminate after a certain amount of time, say each production run ends at the end of the working day, or after the system has entered a certain state, such as all orders have been completed. The main emphasis in literature is on non-terminating systems. Such systems may converge to a stochastic equilibrium in the sense that the distributions of $X(t)$ and $Y(t)$ converge toward an equilibrium distribution.

Discrete event systems may either be discrete-time systems or continuous-time systems. If t represents time, then discrete-time systems are only defined at times $t = 0, 1, 2, \ldots$, that is, $X(t)$ and $Y(t)$ exist only for $t = 0, 1, 2, \ldots$. Continuous-time systems, on the other hand, exist for any time t within a given interval.

We frequently talk about *stochastic processes*. In literature, a stochastic process is defined as a sequence of random variables [92]. For instance, $X_1(t)$, $t = 1, 2, \ldots$ would be a stochastic process. However, here we think of a process as a variable that evolves through time. The time dimension implies that actions of the past can affect the future, but no action in the future can change what happened in the past. This means that the past may be unknown, but it is not random. Also, the randomness of the process increases, at least initially, as we go further into the future. For instance, if $X_1(t)$ is a stochastic process, and if we know the state at time 0, then the variance of $X(t)$ tends to increase with t, at least initially. It may, however, converge to a steady state.

The definition of a state implies that if we know the state at time τ, the probabilities of all possible states the system can be in at time $\tau + t$, $t > 0$ are fully determined. Any information of the state before τ is irrelevant because otherwise, the state would not include all information needed to determine the distributions of $X(\tau + t)$ and $Y(\tau + t)$. Systems are multidimensional processes, and processes in which the past is irrelevant are called *Markov chains*. It follows that discrete event systems, as they are defined here, can be considered as Markov chains and solved as such.

1.2 Markov Chains

Since discrete event systems are Markov chains, we will now review the basic concepts and formulas relating to Markov chains. In Markov chains, we have one variable, say X, which changes over time. We also assume that X can only assume the values between 1 and N. There are two types of Markov chains: discrete-time Markov chains (DTMC) and continuous-time Markov chains (CTMC). In a discrete-time Markov chain, $X(t)$, the value of X at time t, can only change at times 1, 2, 3, $\ldots$, whereas in a continuous-time Markov chain, it can change at any $t \geq 0$. The essential characteristic of a Markov chain is the fact that once $X(t)$ is known, the distributions of $X(t + h)$ for any $h > 0$ are fully determined.

We stated above that discrete event systems are Markov chains once supplementary variables are added. Of course, the Markov chains representing discrete event systems are multidimensional, whereas here we have one-dimensional Markov chains. To convert multidimensional systems into systems with only one dimension, we give each state vector a unique number. In effect, if the size of the state space S is N, then we give each state vector a unique number between 1 and N. This is also necessary because the states have to be stored in the computer memory in a sequential way. The enumeration can be quite complicated, as will be shown in Chapter 3. We also note that the number of states may be huge, and special methods are required to handle such large numbers.

1.2.1 Discrete-time Markov Chains (DTMCs)

DTMCs as well as CTMCs are characterized by their *transition matrix*. In the case of DTMCs, the entries of the transition matrix are the probabilities $p_{i,j}$, which are defined as

$$p_{i,j} = P\{X(t + 1) = j \mid X(t) = i\}.$$

Here, i will be referred to as the original state and j as the target state. The transition matrix is now $P = [p_{i,j}]$. Let $\pi_j(t)$ be the probability that $X(t) = j$, with $\pi_\ell(0)$, $\ell = 1, 2, \ldots, N$ given. In the DTMC, one can calculate the $\pi_j(t)$ recursively as

$$\pi_j(t + 1) = \sum_{i=1}^{N} \pi_i(t) p_{i,j}, \ t = 0, 1, \ldots. \tag{1.1}$$

This relation can also be expressed in matrix form, using the row vector $\pi(t) = [\pi_i(t), i = 1, 2, \ldots, N]$

$$\pi(t + 1) = \pi(t)P. \tag{1.2}$$

This formula allows us to find $\pi_i(t)$ for any time t.

In some cases, we are also interested in a given interval, say from 0 to m, and we want to determine how often state i was visited. For instance, if $X(t)$ is the amount on hand in an inventory system, one may want to know for how many t $X(t)$ assumed the value i. In particular, one may want to know how often there was no stock on hand, which is true when $X(t) = 0$. Generally, the expected number of visits to state i is given as

$$E(T_i) = \sum_{t=0}^{m} \pi_i(t).$$

Here, T_i is the actual number of visits to state i. To justify this formula, let $I(i, t) = 1$ if $X(t) = i$, and 0 otherwise. Then,

$$T_i = \sum_{t=0}^{m} I(i, t),$$

and the result follows because $E(I(i, t)) = \pi_i(t)$.

In order to deal with processes that terminate because of internal transitions, one uses what is called *absorbing states* or *trapping states*. An absorbing state is a state that cannot be left. Mathematically, a state i is absorbing if $p_{i,i} = 1$, $p_{i,j} = 0$, $j \neq i$. It is also possible to have several absorbing states. Often, one is interested in the expected time the process is in state i, $1 \leq i \leq N$ before absorption, and, if there are several absorbing states, in which state the process ends. Markov chains with absorbing states can be used, for instance, to find the probability distribution of the time an item is in a productions system, starting from the moment it enters the system, until the moment it leaves the system.

The transition matrices of DTMCs are stochastic matrices. A matrix is said to be stochastic if it satisfies the following condition:

1. $p_{i,j} \geq 0$;
2. $\sum_{j=1}^{N} p_{i,j} = 1$, $i = 1, 2, \ldots, N$.

If the rows and columns of all absorbing states are deleted, the resulting matrix is *substochastic*. Generally, a matrix is substochastic if it satisfies

1. $p_{i,j} \geq 0$;
2. $\sum_{j=1}^{N} p_{i,j} \leq 1$, $i = 1, 2, \ldots, N$.

According to this definition, stochastic matrices are also substochastic. To indicate that at least one row-sum is less than 1, we use the term *strictly substochastic*. Note that if P in equation (1.2) is substochastic, then (1.2) can still be used to find the probabilities $\pi_i(t)$ of the non-absorbing states.

Every strictly substochastic matrix can be converted into a stochastic matrix. One merely adds one column, say column $N + 1$, with $p_{i,N+1} = 1 - \sum_{j=1}^{N} p_{i,j}$, as well as row $N + 1$, with $p_{N+1,j} = 0$, except that $p_{N+1,N+1} = 1$. Since substochastic matrices can be converted to stochastic matrices, we do not need to derive special formulas for substochastic matrices.

Consider now the non-terminating processes. In this situation, $\pi(t)$ may converge to a vector π, the equilibrium vector. This vector represents the equilibrium distribution of X. Its entries π_j are determined as the solution of the following system of equations, which can be derived from equation (1.1) by replacing all $\pi_j(t)$, $j = 1, 2, \ldots, N$ by π_j:

$$\pi_j = \sum_{i=1}^{N} \pi_i p_{i,j}, \; j = 1, \ldots, N. \tag{1.3}$$

In matrix form, (1.3) yields

$$\pi = \pi P.$$

Note that if this system has a solution $[\pi_1, \pi_2, \ldots, \pi_N]$, and c is a constant, then $[c\pi_1, c\pi_2, \ldots, c\pi_N]$ is also a solution. The solution is thus not unique, and we have to add the condition that the sum of probabilities must add up to 1, that is,

$$\sum_{i=1}^{N} \pi_i = 1.$$

This equation will be referred as *norming condition*. One can also consider unnormed solutions of (1.3), that is, solutions leading to values π_i that do not satisfy the norming condition, yet they satisfy (1.3). If $\tilde{\pi}_i$ is such a solution, the normed solution can be found as

$$\pi_i = \frac{\tilde{\pi}_i}{\sum_{j=1}^{N} \tilde{\pi}_j}.$$

In most situations, we set $\tilde{\pi}_1 = 1$, and then, we normally have enough equations to find $\tilde{\pi}_i$ for all $i > 1$.

It is important to note that only the distribution, but not the process, converges to an equilibrium. For this reason, one frequently uses the term *statistical equilibrium*.

To find equilibrium solutions, one can use software tools. Indeed, spreadsheets, like Excel$^©$, allow one to solve them easily, and so does MATLAB, Maple, or Mathematica. However, we will introduce very stable, but easily programmable methods in Chapter 9. For the moment, we convert our problem to the solution of linear equations as follows. Let

$$\tilde{\pi}^- = [\tilde{\pi}_2, \tilde{\pi}_3, \ldots, \tilde{\pi}_N]$$
$$u = [p_{1,2}, p_{1,3}, \ldots, p_{1,N}]$$
$$s = [p_{2,1}, p_{3,1}, \ldots, p_{N,1}]$$
$$P^- = [p_{i,j}, 2 \le i, j \le N].$$

In all these formulas, we omit the first row and the first column. This allows us to partition $\tilde{\pi}$ and P, and $\tilde{\pi} = \pi P$ becomes

$$[1, \tilde{\pi}^-] = [1, \tilde{\pi}^-] \begin{bmatrix} p_{0,0} & u \\ s & P^- \end{bmatrix},$$

or

$$[1, \tilde{\pi}^-] = [p_{0,0} + \tilde{\pi}^- s, u + \tilde{\pi}^- P^-].$$

Consequently

$$\tilde{\pi}^- = u + \tilde{\pi}^- P^-.$$

Solving this equation for $\tilde{\pi}^-$ yields

$$\tilde{\pi}^- = u(I - P^-)^{-1}. \tag{1.4}$$

When using MATLAB, with vector P and u, tildepi ($\tilde{\pi}$) can be determined as follows:

tildepi = [1, P(1, 2:N)/(eye(N−1) − P(2:N, 2:N))].

Here, eye(N−1) is the identity matrix of size N−1. When executing this statement, MATLAB actually uses a factorization approach, which is more efficient than inverting the matrix $I - P$.

We now discuss a result that applies to any distribution, including the equilibrium distribution. If there is a random variable X, and if we know its distribution $F_X(x)$, then as soon as we obtain additional information about X, this distribution is no longer valid. For instance, if we know that $X > a$, where a is given, then we must update the distribution of X. Similarly, if there is any event that affects X of the Markov chain $\{X(t) \; t > 0\}$, then the equilibrium distribution must be updated.

For instance, in a queueing system, if one knows that an arrival has occurred, then the system ceases to be in its statistical equilibrium, even if it was in a statistical equilibrium before the arrival.

1.2.2 Continuous-time Markov Chains (CTMCs)

The transition matrix of CTMCs is denoted by $Q = [q_{i,j}]$, where

$$q_{i,j} = \begin{cases} \lim_{h \to 0} \frac{P\{X(t+h)=j\}|X(t)=i\}}{h}, & i \neq j \\ -\sum_{k=1,k \neq j}^{N} q_{i,k}, & i = j. \end{cases}$$

For the CTMCs, we have, if $\pi'_j(t)$ is the derivative of π_j with respect to t:

$$\pi'_j(t) = \sum_{i=1}^{N} \pi_i(t)q_{i,j}, \ t > 0. \tag{1.5}$$

In many cases, it is convenient to write

$$\pi'_j(t) = \sum_{i=1,i \neq j}^{N} \pi_i(t)q_{i,j} - \pi_j(t) \sum_{k=1,k \neq j}^{N} p_{j,k}.$$

Since the first term on the right is the flow into state j, whereas the second term is the flow out of state j, this equation indicates that the increase of $\pi_j(t)$ is given by the flow into state j, minus the flow out of state j.

Equation (1.5) can be written in matrix form. If $\pi(t) = [\pi_1, \pi_2, \ldots, \pi_N]$ and $Q = [q_{i,j}]$, then equation (1.5) becomes

$$\pi'(t) = \pi(t)Q. \tag{1.6}$$

As in the discrete case, we have terminating and non-terminating CTMCs. If the process runs for a fixed time τ, then one may be interested in $\pi_i(\tau)$. We postpone the discussion of calculating $\pi_i(\tau)$ to Chapter 7. Of interest is also the expected time the process is in state i. This expected time is given as

$$E(T \mid i) = \int_0^T \pi_i(t)dt.$$

As in the discrete case, one can use absorbing states to model processes that are terminated by internal conditions. The rows corresponding to absorbing states only contain zeros, that is, if i is an absorbing state, $q_{i,j} = 0$ for all j. For this reason, the rows and columns of absorbing states are often omitted.

The transition matrices of CTMCs are called *infinitesimal generators* or *generator matrices*. We say that a square matrix $Q = [q_{i,j}, \ i, j = 1, 2, \ldots, N]$ is a generator matrix if

1. $q_{i,j} \geq 0, i \neq j$
2. $\sum_{j=1}^{N} q_{i,j} = 0, i = 1, 2, \ldots, N$.

If $\sum_{j=1}^{N} q_{i,j} \leq 0$, we say that the matrix is a *subgenerator matrix*. A matrix is strictly a subgenerator if for at least one i, $\sum_{j=1}^{N} q_{i,j} < 0$. To convert a subgenerator into a generator, one can always add an absorbing state, just like in the case of stochastic matrices. Subgenerators are useful to find expected times to absorption, such as the time needed from the moment an entity enters the system, until it leaves the system. Equation (1.5) also holds for subgenerators.

Consider now the non-terminating processes. Like DTMCs, CTMCs frequently converge toward an equilibrium distribution $\pi = [\pi_i]$, which is given as

$$0 = \sum_{i=1}^{N} \pi_i q_{i,j}, \ j = 1, 2, \ldots, N \tag{1.7}$$

or

$$0 = \sum_{i=1, i \neq j}^{N} \pi_i q_{i,j} - \pi_j \sum_{k=1, k \neq j}^{N} q_{k,j}.$$

This equation expresses that in equilibrium, the flow into state j must be equal to the flow out of state j. Only if this holds for all j, then we can conclude that an equilibrium distribution is reached. As in DTMC, we also need

$$\sum_{i=1}^{N} \pi_i = 1.$$

In matrix form, (1.7) becomes

$$0 = \pi Q.$$

To find π, one can use an equation similar to (1.4). One defines $\tilde{\pi}^-$ as before and uses Q^-, which is Q with the first row and column deleted, $r = [q_{1,j}, j = 2, 3, \ldots, n]$ and solves

$$\tilde{\pi}^- = r^{'}(-Q^-)^{-1}. \tag{1.8}$$

In the rest of this book, we refer to equations (1.1) and (1.5) as transient equations, and to equations (1.3) and (1.7) as equilibrium or steady-state equations. In literature, one frequently uses the so-called *Chapman–Kolmogorov* equation. They are specified as

$$p_{i,j}(t) = \sum_{k=1}^{N} p_{i,k}(\tau) p_{k,j}(t - \tau), \ \tau > 0.$$

Here, $p_{i,j}(t)$ is the probability that given the system starts in state i at time 0, the system is in state j at time t.

1.3 Random Variables and Their Distributions

The state variables used in discrete event systems are random variables, and so are the times between events, such as the times between arrivals. This is the reason we now review random variables.

Generally, if X is a random variable, its *distribution* $F_X(x)$ is defined as

$$F_X(x) = \mathrm{P}\{X \leq x\}.$$

If X can only assume the values $0, 1, 2, 3, \ldots$, then we define the *probability mass function* as

$$p_i = \mathrm{P}\{X = i\}, \ i = 0, 1, 2, \ldots.$$

In this case,

$$F_X(j) = \sum_{i=0}^{j} p_i.$$

In the case of continuous distributions, one frequently uses the density, which is given as the derivative of $F_X(x)$ with respect to x:

$$f_X(x) = F_X'(x).$$

The higher the density of a certain value x, the higher the probability that X is close to x.

Discrete distributions are important because the systems we discuss have discrete state spaces. Also, empirical distributions are frequently presented as frequency distributions, which lead to discrete distributions. For numerical purposes, we are often forced to discretize continuous distributions, that is, we effectively deal with discrete distributions.

In addition to being discrete, we frequently require that the distributions have finite support. This means that if X is a random variable, then X can assume only a finite number of values. Typically the values run from 0 to N. Moreover, unless stated otherwise, we always assume that $X \geq 0$.

1.3.1 Expectation and Variance

Given a random variable X, there are two important measures to characterize its distribution: the *expectation* $\mathrm{E}(X)$ and the *variance* $\mathrm{Var}(X)$. First, we define the expectation. For discrete distributions, we have

$$E(X) = \sum_{i=0}^{\infty} i p_i. \qquad (1.9)$$

In the case of continuous distribution, $E(X)$ is defined as

$$E(X) = \int_{0}^{\infty} x f_X(x) dx. \qquad (1.10)$$

Instead of $E(X)$ we frequently write μ_X.

Consider now the squared deviation of X, defined as $(X - \mu_X)^2$. Clearly, the squared deviation is also a random variable, and it has an expectation. The expected squared deviation is called *variance*, and it is given as

$$\text{Var}(X) = E((X - \mu_X)^2).$$

In the case of a discrete distribution, one has

$$\text{Var}(X) = \sum_{i=0}^{\infty} (i - \mu_X)^2 p_i. \qquad (1.11)$$

Instead of $\text{Var}(X)$, we also write σ_X^2. If the random variable in question is continuous, the expectation is measured in the same units as X, but the variance is not. For this reason, one uses the *standard deviation* σ_X, which is defined as

$$\sigma_X = \sqrt{\text{Var}(X)}.$$

We now define the *coefficient of variation* (CV) as

$$\text{CV} = \frac{\sigma}{E(X)}.$$

The CV is dimensionless, that is, it does not depend on the unit used for measuring X. Hence, if X is, for example, an inter-arrival time, it does not matter whether X is measured in seconds, minutes, or hours: the CV remains unchanged.

We have for any distribution:

$$\text{Var}(X) = E\left((X - \mu_X)^2\right) = E(X^2) - E^2(X).$$

To demonstrate this, use $(X - \mu_X)^2 = X^2 - 2\mu_X E(X) + \mu_X^2$, which is $X^2 - E^2(X)$. The result follows by taking the expectation of $X^2 - E^2(X)$.

Two random variables X_1 and X_2 can be *dependent*, and the amount of dependency can be measured by their *covariance*, which is defined as

$$\text{Cov}(X_1, X_2) = E(X_1 - \mu_1)(X_2 - \mu_2) = E(X_1 X_2) - \mu_1 \mu_2.$$

Here, $\mu_i = E(X_i), i = 1, 2$. Note that $\text{Cov}(X_i, X_i) = \text{Var}(X_i), i = 1, 2$. If $\text{Cov}(X_1, X_2) > 0$, X_1 and X_2 are said to be *positively correlated*, which means that if X_1 is large, X_2

can be expected to be large as well. On the other hand, if $\text{Cov}(X_1, X_2) < 0$, X_1 then X_2 are *negatively correlated*.

The discrete random X_1 and X_2 are *independent* if for all i and j,

$$P\{X_1 = i \text{ and } X_2 = j\} = P\{X_1 = i\}P\{X_2 = j\}.$$

Hence, the *joint distribution* $P\{X_1 = i \text{ and } X_2 = j\}$, $i, j \geq 0$ of the independent random variables X_1 and X_2 is the product of the distributions of X_1 and X_2. In the case of continuous random variables, the density of the joint distribution is the product of their densities. If X_1 and X_2 are independent random variables, then $\text{Cov}(X_1, X_2) = 0$. The reverse, however, is not necessarily true.

The *correlation coefficient* ρ is defined as

$$\rho = \frac{\text{Cov}(X_1, X_2)}{\sigma_1 \sigma_2}.$$

Here, σ_i is the standard deviation of X_i, $i = 1, 2$. The correlation coefficient is not defined if either $E(X_1)$ or $E(X_2)$ are constants. Otherwise, ρ must always be between -1 and 1.

Another relation for discrete distributions is

$$E(X) = \sum_{i=0}^{\infty} i p_i = \sum_{j=0}^{\infty} \sum_{i=j+1}^{\infty} p_i = \sum_{k=1}^{\infty} \sum_{i=k}^{\infty} p_i. \tag{1.12}$$

To prove this equation, replace i by $\sum_{j=0}^{i-1} 1$ to obtain

$$\sum_{i=0}^{\infty} i p_i = \sum_{i=0}^{\infty} p_i \sum_{j=0}^{i-1} 1 = \sum_{j=0}^{\infty} 1 \sum_{i=j+1}^{\infty} p_i,$$

proving $E(X) = \sum_{j=0}^{\infty} \sum_{i=j+1}^{\infty} p_i$. $E(X) = \sum_{k=1}^{\infty} \sum_{i=k}^{\infty} p_i$ follows by setting $k = j + 1$. Equation (1.12) will be used repeatedly in later chapters.

1.3.2 Sums of Random Variables

We frequently need the sum of random variables. We first assume that there are only two random variables: X_1 and X_2. If they are independent, then the probability of $Z = X_1 + X_2$ can easily be calculated. If X_1 and X_2 are both non-negative integer random variables, then one has

$$P\{Z = i\} = \sum_{k=0}^{i} P\{X_1 = k\}P\{X_2 = i - k\}.$$

If X_1 and X_2 are both continuous, the sum is replaced by an integral, that is,

$$f_Z(x) = \int_0^x f_X(y) f_{X_2}(x-y) dy.$$

If there is a sum of m independent random variables, then one first calculates the distribution of $Z_2 = X_1 + X_2$. After that one calculates the distribution of $Z_3 = Z_2 + X_3$, and so on, until finally, one calculates the distribution of $Z_m = Z_{m-1} + X_m$.

If X_i, $i = 1, 2, \ldots, m$ are m random variables, then one may be interested in the expectation of their sum. Let

$$Y = \sum_{i=1}^m X_i.$$

In this case, one has

$$E(Y) = E\left(\sum_{i=1}^m X_i\right) = \sum_{i=1}^m E(X_i). \tag{1.13}$$

In other words, the sum and the expectation symbols can be interchanged. This is even true if the X_i are dependent random variables.

The variance of a sum is given as

$$\mathrm{Var}(Z) = \sum_{i=1}^m \sum_{j=1}^m \mathrm{Cov}(X_i, X_j)$$

$$= 2\sum_{i=1}^m \sum_{j=1}^{i-1} \mathrm{Cov}(X_i, X_j) + \sum_{i=1}^m \mathrm{Var}(X_i).$$

If the random variables are independent, their covariance is 0, and we obtain

$$\mathrm{Var}(Z) = \mathrm{Var}\left(\sum_{i=1}^m X_i\right) = \sum_{i=1}^m \mathrm{Var}(X_i). \tag{1.14}$$

1.3.3 Some Distributions

In this section, we review some distributions, starting with distributions for discrete random variables, also called *discrete distributions*. After that, we introduce some continuous distributions.

1.3.3.1 Discrete Distributions

The simplest discrete distribution is the Bernoulli distribution, which has only two values, 0 and 1. If X is a Bernoulli distributed random variable, then we define p as the probability that $X = 1$. $1 - p$ is then the probability that $X = 0$. Therefore, the expectation is

$$E(X) = 0P\{X = 0\} + 1P\{X = 1\} = p.$$

Consequently, $E(X) = p$. Clearly, $X^2 = X$ if X can only assume the values 0 and 1. $E(X^2)$ is therefore also p. This implies that the variance is

$$\text{Var}(X) = p - p^2 = p(1 - p).$$

Trials with two possible outcomes are called *Bernoulli trials*. The outcomes are called *success* and *failure*. The probability of a success is p, and probability of failure is therefore $1 - p$. A sequence of independent Bernoulli trials is called a *Bernoulli process*. For instance, tossing a coin repeatedly results in a Bernoulli process.

Next, consider the binomial distribution. The binomial distribution has two parameters: n and p, and it is given as

$$b(k; n, p) = \frac{n!}{k!(n-k)!} p^k (1-p)^{n-k}.$$

In can be shown that the number of successes in n independent Bernoulli trials has a binomial distribution.

One can interpret the binomial distribution as the sum of n independent random variables X_i, each having a Bernoulli distribution, that is,

$$X = \sum_{i=1}^{n} X_i$$

has a binomial distribution. Equations (1.13) and (1.14) now imply

$$E(X) = np, \ \text{Var}(X) = np(1 - p).$$

Next, consider the geometric distribution. There are two definitions of the geometric distribution, which only differ slightly. They both have one parameter, p. The probabilities that the random variable in question is n for the two versions of the geometric distribution are

$$\text{Geom}(k; p) = (1 - p)^k p, \ k = 0, 1, \ldots \tag{1.15}$$
$$\text{Geom}_1(k; p) = (1 - p)^{k-1} p, \ k = 1, 2, \ldots . \tag{1.16}$$

Consequently, if X has the distribution given by equation (1.15), then $X + 1$ has the distribution given by equation (1.16). Both distributions have their roots in Bernoulli processes with parameter p. The distribution Geom gives the number of failures preceding the first success, whereas Geom_1 gives the time of the first success. To prove this, note that the probability of having k failures before the first success is $(1 - p)^k$, and to have a success following k failures is therefore $(1 - p)^k p$.

The expectations of the two geometric distributions are

$$E(X) = \begin{cases} \frac{1-p}{p} & \text{Geom} \\ \frac{1}{p} & \text{Geom}_1. \end{cases}$$

In both versions, we have

$$\text{Var}(X) = \frac{1-p}{p^2}, \; \cdot$$

In contrast to the binomial distribution, the geometric distribution has a range from 0 to infinity.

In the geometric distribution, the cumulative distribution can be obtained directly. The probability of having k failures before the first success is obviously $(1 - p)^k$. If X is given by (1.15), it follows that

$$P\{X \geq k\} = (1 - p)^k, \; k \geq 0.$$

The number of failures between two successes in a Bernoulli process has the distribution Geom. To see this, just start counting right after a success. In a DTMC, with $p_{i,i} > 0$, we can consider staying in state i as failure, meaning that the time the process remains in state i has a geometric distribution.

Our next discrete distribution is the Poisson distribution, which has one parameter, λ. It is given as

$$p(i; \lambda) = \frac{\lambda^i}{i!} e^{-\lambda}, \; i \leq 0. \tag{1.17}$$

The Poisson distributions can be obtained from the binomial distribution by decreasing p while increasing n such that np does not change. We set $np = \lambda$. With $p = \frac{\lambda}{n}$, the binomial distribution becomes

$$\frac{n!}{i!(n-i)!} \left(\frac{\lambda}{n}\right)^i \left(1 - \frac{\lambda}{n}\right)^{n-i} = \frac{n!}{i!(n-i)!n^i} \left(1 - \frac{\lambda}{n}\right)^{n-i} \lambda^i. \tag{1.18}$$

If we let n going to infinity, then

$$\lim_{n \to \infty} \frac{n!}{(n-i)!n^i} = \lim_{n \to \infty} \frac{n(n-1)\dots(n-i+1)}{n^i} = 1.$$

Moreover, it is well known that

$$e^x = \lim_{n \to \infty} \left(1 + \frac{x}{n}\right)^n.$$

Consequently, one has for the second factor

$$\lim_{n \to \infty} \left(1 - \frac{\lambda}{n}\right)^{n-i} = e^{-\lambda},$$

and (1.18) becomes (1.17) as claimed.

By using $np = \lambda$, one finds the expectation and the variance of the Poisson distribution from the one of the binomial distribution as

$$E(X) = \lambda, \ \text{Var}(X) = \lambda.$$

We now come to what we call the *explicit distribution*. In this case, the p_i are just given for $i = 1, 2, \ldots, n$. This happens, for instance, when the distribution is obtained from an empirical frequency distribution. Also, when discretizing the continuous distribution $f_X(x)$, one divides the unit into n equal parts, each having a length of Δ. The probability p_i then becomes $cf_X(i\Delta)$, where c is chosen such that $\sum_i p_i = 1$. In explicit distributions, expectation and variance can be obtained from equations (1.9) and (1.11).

In some frequency distributions, the last class is open ended. To account for this, suppose that for $i \leq g$, the p_i are given explicitly, but for $i > g$, only $P = P\{X > g\}$ is given. We then assume that for $i > g$, $p_i = p_g q^{i-g}$. We have

$$p_g + P = p_g + p_{g+1} + p_{g+2} + \cdots = p_g + p_g q + p_g q^2 + \cdots = \frac{p_g}{1 - q}, \qquad (1.19)$$

or

$$p_g + P = \frac{p_g}{1 - q}.$$

Solving this equation for q yields

$$q = \frac{P}{p_g + P}.$$

We will find expectation and variance of this distribution later.

1.3.3.2 Continuous Distributions

We will discuss only two continuous distributions: the exponential distribution and the normal distribution. The phase-type distributions, another type of continuous distributions, will be described in Chapter 5.

The exponential distribution has only one parameter, namely, λ. It is defined by

$$f_X(x) = \lambda e^{-\lambda x}, \ x \geq 0. \qquad (1.20)$$

The exponential distribution can be interpreted as the limit of the geometric distribution. Specifically, if $\lambda = np$, and X_G has a geometric distribution, then as $n \to \infty$, $X = X_G/n$ has an exponential distribution. Indeed, we have, after substituting λ/n for p:

$$P\{X > t\} = P\{X_G > nt\} = (1 - p)^{nt} = \left(1 - \frac{\lambda}{n}\right)^{nt}.$$

The right side of this equation converges to $e^{-\lambda t}$, which proves our claim. A similar calculation for the mean and variance yields

$$E(X) = E(X_G/n) = \frac{1}{n}E(X_G) = \frac{1}{n}\frac{1}{p} = \frac{1}{\lambda}$$

$$Var(X) = Var(X_G/n) = \frac{1}{n^2}\frac{1-p}{p^2} = \frac{1}{\lambda^2} \text{ as } n \to \infty.$$

The final distribution we discuss is the normal distribution. It has two parameters, μ and σ, which are, respectively, its expectations and its standard deviation. Its density is given as

$$f_X(x) = \frac{1}{\sigma\sqrt{2\pi}}e^{-\frac{(x-\mu)^2}{2\sigma^2}}.$$

Any distribution representing the sum of independent random variables with finite expectation and variance converges to a normal distribution as the number of summands increases, provided the contribution of the variances of every term is negligible when compared to the variance of the total. Since the binomial distribution can be interpreted as the sum of Bernoulli random variables, it converges toward the normal distribution. The Poisson distribution also converges to the normal distribution as λ goes to infinity.

1.3.4 Generating Functions

In this section, we first discuss the *probability generating function* or z-transform, which will be used to find the expectation and the variance for discrete-time distributions. Next, we use the *moment generating function* which is applied to find expectations and variances in continuous distributions. Without proof, we also note that given a generating function, the corresponding distribution is uniquely determined.

We assume that the random variables in question range from 0 to infinity. If the discrete random variable has a finite range, say from 0 to g, we set $p_i = 0$ for $i > g$, and convert the finite range into an infinite range. A similar device can be used to extend the range of a continuous random variable.

1.3.4.1 Probability Generating Function

If $p_0, p_1, p_2, \ldots$ defines the probability distribution of X, where X can only assume the values $0,1,2, \ldots$, then its probability generating function is defined as

$$G_X(z) = \sum_{i=0}^{\infty} p_i z^i = E(z^X).$$

The probability generating function of a random variable can be used to find its expectation and variance. To find these measures, one uses $G'(z)$, the first derivative of $G_X(z)$ with respect to z, and $G''(z)$, its second derivative:

$$G'_X(z) = \sum_{i=1}^{\infty} i p_i z^{i-1} = E(X z^{X-1}),$$

$$G''_X(z) = \sum_{i=2}^{\infty} i(i-1) p_i z^{i-2} = E((X-1) X z^{X-2}).$$

If we set $z = 1$, this yields

$$G_X(1) = \sum_{i=0}^{\infty} p_i$$

$$G'_X(1) = \sum_{i=1}^{\infty} i p_i = E(X),$$

$$G''_X(1) = \sum_{i=2}^{\infty} i(i-1) p_i = E((X-1) X).$$

Consequently, $G_X(1)$ should be 1, $E(X) = G'_X(1)$. Also, since $Var(X) = E(X^2) - E^2(X)$, we have

$$Var(X) = E(X(X-1)) + E(X) - E^2(X) = G''_X(1) + E(X) - E^2(X).$$

For the Poisson distribution $p(n; \mu)$, we get

$$G_X(z) = \sum_{n=0}^{\infty} \frac{\mu^n}{n!} e^{-\mu} z^n = e^{-\mu(1-z)}.$$

Clearly, $G_X(1) = e^0 = 1$ as required. The derivatives of $e^{\mu(1-z)}$ are

$$G'_X(z) = \mu e^{-\mu(1-z)}$$
$$G''_X(z) = \mu^2 e^{-\mu(1-z)}.$$

The expectation is therefore $G'_X(1) = \mu$, and the variance is

$$Var(X) = G''_X(1) + E(X) - E^2(X) = \mu^2 + \mu - \mu^2 = \mu.$$

This shows that both expectation and the variance are μ.

Consider now the geometric distribution $Geom_1$, which is given as

$$p_i = (1-p)^{i-1} p, \ i \geq 1.$$

Note that $X \geq 1$ in this case. Its generating function is

$$G(z) = \frac{pz}{1 - (1-p)z}.$$

A simple check verifies that $G(1) = 1$. To find its expectation and variance, we use a method that avoids differentiating a fraction, a method we will also use in later chapters.

$$G_X(z)(1 - (1 - p)z) = pz.$$

We differentiate both sides of this equation to obtain

$$G'_X(z)(1 - (1 - p)z) - G(z)(1 - p) = p$$
$$G''_X(z)(1 - (1 - p)z) - 2G'_X(z)(1 - p) = 0.$$

Setting $z = 1$ now yields

$$G'_X(1)p - (1 - p) = p$$
$$G_X(1)''p - 2G'_X(1)(1 - p) = 0.$$

From this, one obtains $G'_X(1) = E(X) = \frac{1}{p}$, and $G_X(1)'' = 2\frac{1-p}{p^2}$. Consequently

$$\text{Var}(X) = 2\frac{1 - p}{p^2} + \frac{1}{p} - \frac{1}{p^2} = \frac{1 - p}{p^2}.$$

Let us now consider the distribution with a geometric tail as indicated in (1.19). One finds

$$G_X(z) = \sum_{i=0}^{g-1} p_i z^i + \frac{z^g p_g}{1 - qz}.$$

To simplify differentiation, we multiply both sides by $1 - qz$ to obtain

$$G_X(z)(1 - qz) = \sum_{i=0}^{g-1} p_i z^i (1 - qz) + z^g p_g. \tag{1.21}$$

Differentiating both sides yields

$$G'_X(z)(1 - qz) - G_X(z)q = \sum_{i=1}^{g-1} i p_i z^{i-1}(1 - qz) - \sum_{i=0}^{g-1} p_i z^i q + g p_g z^{g-1}$$

$$G''_X(z)(1 - qz) - 2G'_X(z)q = \sum_{i=2}^{g-1} i(i - 1)p_i z^{i-2}(1 - qz) - 2\sum_{i=1}^{g-1} i p_i z^{i-1} q$$
$$+ g(g - 1)p_g z^{g-2}.$$

We set $z = 1$ to obtain

$$G'_X(1)(1-q) - G_X(1)q = \sum_{i=1}^{g-1} i p_i (1-q) - \sum_{i=0}^{g-1} p_i q + g p_g \tag{1.22}$$

$$G''_X(1)(1-q) - 2G'_X(1)q = \sum_{i=2}^{g-1} i(i-1)p_i(1-q) - 2\sum_{i=0}^{g-1} i p_i q$$
$$+ g(g-1)p_g. \tag{1.23}$$

To simplify equation (1.22), we use $G(1) = 1$ and

$$\sum_{i=1}^{g-1} p_i = 1 - \frac{p_g}{1-q}$$

to obtain

$$G'_X(1)(1-q) = \sum_{i=1}^{g-1} i p_i (1-q) - q\left(1 - \frac{p_g}{1-q}\right) + q + g p_g.$$

We divide by $1 - q$ and simplify to obtain

$$E(X) = G'_X(1) = \sum_{i=0}^{g-1} i p_i + p_g \left(\frac{q}{(1-q)^2} + \frac{g}{1-q}\right). \tag{1.24}$$

To find $G''_X(1)$ from (1.23) we use $G'_X(1) = E(X)$ and the following relation, which can easily be derived from equation (1.24)

$$\sum_{i=0}^{g-1} i p_i = E(X) - p_g \left(\frac{q}{(1-q)^2} + \frac{g}{1-q}\right).$$

With these values, (1.23) yields

$$G''_X(1)(1-q) = 2E(X)q + \sum_{i=2}^{g-1} i(i-1)p_i(1-q)$$
$$- 2q\left(E(X) - p_g\left(\frac{q}{(1-q)^2} + \frac{g}{1-q}\right)\right) + g(g-1)p_g.$$

We conclude

$$G''(1) = \sum_{i=0}^{g-1} i(i-1)p_i + p_g\left(\frac{2q}{(1-q)^3} + \frac{2g}{(1-q)^2} + \frac{g(g-1)}{1-q}\right).$$

The variance can now be obtained as usual as

$$\text{Var}(X) = G''(1) + E(X) - E^2(X).$$

With this, we have found the expectations and variances of all discrete distributions discussed in this section.

If X_1 and X_2 are two independent discrete random variables with z-transforms $G_1(z)$ and $G_2(z)$, respectively, then $G_S(z)$, the transform of $S = X_1 + X_2$, becomes

$$G_S(z) = G_1(z)G_2(z).$$

For instance, in the case of the Poisson distribution, the sum of $S = X_1 + X_2$, with $G_1(z) = e^{\mu_1(1-z)}$, and $G_2 = e^{-\mu_2(1-z)}$, one finds

$$G_S(z) = e^{-\mu_1(1-z)}e^{\mu_2(1-z)} = e^{-(\mu_1+\mu_2)(1-z)}.$$

The result S has also a Poisson distribution, with parameter $\mu_1 + \mu_2$.

1.3.4.2 Moment Generating Function

For continuous distributions, one uses the moment generating function, which, for the density $f_T(t)$, is defined as follows:

$$M(s) = \int_0^\infty e^{st} f_T(t)dt = \mathrm{E}(e^{sT}).$$

The nth derivative of $M(s)$ with respect to s can be found to be

$$M^{(n)}(s) = \int_0^\infty t^n e^{st} f_T(t)dt = \mathrm{E}(T^n e^{sT}).$$

It follows that $M^{(n)}(0) = \mathrm{E}(T^n)$, which is the nth moment of $f_T(t)$. Consequently

$$\mathrm{E}(T) = M'(0), \quad \mathrm{Var}(T) = M''(0) - \mathrm{E}^2(T).$$

Also, $M(0) = \int_0^\infty f_T(t)dt$, which must be 1.

The moment generating function of the exponential distribution can be found to be

$$M_T(s) = \int_0^\infty e^{st} \mu e^{-\mu t} dt = \int_0^\infty \mu e^{t(s-\mu)} dt = \frac{\mu}{\mu - s}.$$

Clearly, $M(0) = 1$ as required. The derivatives of $M(s) = \frac{\mu}{\mu-s}$ are

$$M^{(n)}(s) = \frac{n!\mu}{(\mu - s)^{(n+1)}}.$$

Consequently,

$$\mathrm{E}(T) = \frac{1}{\mu}, \quad \mathrm{Var}(T) = \frac{2}{\mu^2} - \mathrm{E}^2(T) = \frac{1}{\mu^2}.$$

The standard deviation is equal to $\sqrt{\text{Var}(T)}$, or $\frac{1}{\mu}$. Note that the standard deviation of the exponential distribution is equal to the expectation. The coefficient of variation (CV) of the exponential distribution is therefore 1.

In the case that X_1 and X_2 are continuous random variables with moment generating functions $M_1(s)$ and $M_2(s)$, the moment generating function of $S = X_1 + X_2$ is given as

$$M_S(s) = M_1(s)M_2(s).$$

For instance, if X_1 and X_2 both are exponential random variables with common expectations $1/\mu$, the moment generating function of their sum is

$$M_S(s) = \left(\frac{\mu}{\mu - s}\right)^2.$$

1.4 The Kendall Notation

One of the most important applications of discrete event systems are queueing systems, in particular, one and multi-server queues. To describe these, there is a classical notation, given by Kendall [58]. Essentially, Kendall abbreviated queueing systems with a single queue by three letters, separated by a slash, such as $X/Y/Z$. Here, X stands for the arrival process, Y for the service process, and Z for the number of servers. If the inter-arrival times are exponential, the letter M is used for Markovian. In fact, if the times between arrivals are exponential, the arrival process is Markovian in the sense that the past, the time since the previous arrival, plays no role. If the distributions of the inter-arrival times are independent, but not otherwise specified, the letters GI for General Independent are used. Some people also use G if the inter-arrival times are not necessarily independent. Similarly, if the service times are exponential, one uses the letter M, and if the service times are not specified, the letter G is used. The use here is somewhat inconsistent because for service times, the abbreviation GI is not normally used, even when the service times are independent.

Using this notation, the $M/M/1$ queue represents a queue with Poisson arrivals, a Poisson service process, and 1 server. The $M/M/c$ queue, on the other hand, is a queue with the same inter-arrival and service times as the $M/M/1$ queue, but it has c servers, where $c \geq 1$. We will also discuss the $GI/G/1$ queue, which is a queue where inter-arrival times are independent, and so are the service times and there is one server.

The notation of Kendall has frequently been extended. First, abbreviations for additional distributions have been introduced. We mention here, in particular, the abbreviation for geometric inter-event times, where the abbreviation $Geom$ is used. Later, we will discuss phase-type distributions, which are abbreviated by the letter PH. Consequently, the $PH/PH/1$ queue represents a queue where inter-arrival times are phase type, and so are service time, and there is one server.

In many cases, no arrivals are admitted once the number in the system reaches N. To account for this, one adds an N to the original Kendall notation, such as

$M/M/1/N$. We will make use of this notation as well. Some authors also extended the Kendall notation to situations involving several queues. However, we will not use such extensions.

We mention here one further notational convention: In this text, the word "queue" refers to everyone except the element served, whereas the term "line" includes the element receiving service.

1.5 Little's Law

One of the most famous laws in queueing theory is *Little's law*, which says that in a queueing system in statistical equilibrium, the expected waiting time W in a queue is equal to L, the expected number in a queue, divided by λ, the expected number of arrivals, that is,

$$W = L/\lambda.$$

Many articles have been written about Little's law, see, for example, [17] and [88], and it has many applications. Indeed, if there is any collection or set, where the collection can be a queue, customers in a shopping center, jobs in a production system, and there is a throughput λ, then the expected time an element spends in the collection is equal to the expected number in the set, divided by λ.

We will not give a rigorous proof of this law here because such a proof is available in [88], but it is useful to comprehend the main idea, which is as follows. If the expected number in a set is L, and the system is observed for a long time T, then the total wait spent by all members of the set together is LT. If the system is initially empty, then the time LT is shared by all arrivals to the set, and since the expected number of arrivals during a time T is λT, the average wait $\overline{W}$ is $(LT)/(\lambda T) = W?\lambda$. Consequently, if the average is meaningful, then Little's law follows. Little's law is not valid if a single waiting time can dominate the average. For an example of this see [88]. To avoid this problem, it is normally sufficient to require that the variance of the waiting time is finite. Little's law is also invalid if one considers several sets in parallel, as the following simple example demonstrates. Suppose that there are two situations, and each one can happen with probability 0.5. In the first situation, $L_1 = 10$, $\lambda_1 = 2$, and according to Little's law, $W_1 = 5$. In the second situation, $L_2 = 20$, $\lambda_2 = 8$, and $W_2 = 2.5$, again using Little's law. If you take expectations over the two situations, then $L = 0.5L_1 + 0.5L_2 = 15$, $\lambda = 0.5\lambda_1 + 0.5\lambda_2 = 5$, and $0.5W_1 + 0.5W_2 = 3.75$. On the other hand, when using Little's law on the combined system, one finds $L/\lambda = 15/5$ which is 3 and not 3.5. Hence, Little's law fails. It is correct for each single sample functions, but not necessarily for a collection of sample functions.

Little's law can be used to find the expected number of busy servers in a queueing system. One has, if λ is the arrival rate and $E(S)$ is the expected service time:

$$E(\text{busy servers}) = E(S)\lambda. \tag{1.25}$$

To understand this formula, note the number of busy servers is equal to the number of customers being served, and the time these customers spend with the server is their time they are part of the group being served.

If there is only one server, then the expected number of busy servers is equal to $1 - P_0$, the probability that the system is not idle. If $\mu = 1/E(S)$ is the service rate, $1 - P_0$ is thus $\frac{\lambda}{\mu}$, and

$$P_0 = 1 - E(S)\lambda = 1 - \frac{\lambda}{\mu}. \tag{1.26}$$

This, too, is a useful relation.

1.6 Sets and Sums

We will frequently use sets and sums. A set is a collection of entities, such as integers, events, or states. For instance, $\{3, 5, 8\}$ is the set of the integers 3, 5, and 8, and $\{Ev_1, Ev_3\}$ is the set of consisting events Ev_1 and Ev_3. When x is an element of set A, we write $x \in A$. For instance $2 \in \{1, 2, 3\}$, but 4 is not in A, that is, $4 \notin A$. Sets are frequently defined by conditions, such as $\{i \mid i \text{ even }\}$, which denotes the even integers.

We use $\aleph$ for the set of natural numbers, that is, $\aleph = \{0, 1, 2, \ldots\}$ and $\Re$ for the set of real numbers. Also $\Re^+$ stands for the positive real numbers. The *Cartesian product* of two sets A and B, denoted by $A \times B$, is the set of all pairs, with the first element taken from set A, and the second one from set B. For instance, if $A = \{1, 2\}$ and $B = \{c, d\}$ then $A \times B = \{(1, c), (1, d), (2, c), (2, d)\}$. The Cartesian product $\aleph \times \aleph = \aleph^2$ is thus the set of all pairs of natural numbers. Similarly, $\aleph^n$ is the set of all n-tuples of natural numbers. The set with no element in it is the *empty set* and denoted by $\emptyset$ or by $\{\}$. Thus, the set $\{x \mid x > 4 \text{ and } x < 2\}$ contains no elements, it is empty.

Set A is *subset* of set B if all elements of A are also elements of B, and, in this case, we write $A \subseteq B$. The set of all subsets of set A is called the *power set* of A, denoted by 2^A. If $A = \{a, b, c\}$, then

$$2^A = \{\{\}, \{a\}, \{b\}, \{c\}, \{a, b\}, \{a, c\}, \{b, c\}, \{a, b, c\}\}.$$

Each set B contained in the powerset 2^A can be matched with a binary number as follows: Each element of A is associated with a digit of the binary number, and if this digit is 1, then the element is also in set B, and if the digit is zero, it is not in set B.

If A and B are two sets, then $A \cup B$ is the *union* of A and B, and $A \cap B$ is the *intersection* of A and B. The union $A \cup B$ contains all elements that are either elements of A or elements of B, and the intersection $A \cap B$ is the set of elements that are both in A and B. Thus, if $x \in A \cup B$, then $x \in A$ or $x \in B$, and if $x \in A \cap B$, then $x \in A$ and $x \in B$. The set difference $A - B$ contains all elements of set A that are not elements of set B

Sometimes, we assume that there is a universal set, which includes all elements in question. For instance, when formulating discrete event systems, the set of all events that can occur in this system forms a universal set. We denote the universal set by Ω.

The notion of a sum should be familiar to all readers. Thus, $\sum_{i=1}^{n} a_i$ is an abbreviation of $a_1 + a_2 + \cdots a_n$. To specify the terms included in a sum, we sometimes use the set notation. For instance, if $A = \{1, 3, 5\}$, then $\sum_{i \in A} a_i$ is an abbreviation for $a_1 + a_3 + a_5$. Generally, $\sum_{i=n}^{m} a_i$ is understood as an abbreviation of $\sum_{i \in A} a_i$, with $A = \{n, n+1, \ldots, m\}$. This implies that if $m < n$, A is empty, and then the sum is zero. The notation $\prod_{i=1}^{n} a_i$ is used for the product of the a_i, where i runs from 1 to n. Similarly, $\prod_{i \in A} a_i$ is the product of the c_i satisfying $i \in A$. Note that if $A = \emptyset$, then $\prod_{i \in A} a_i = 1$.

Problems

1.1 Consider the DTMC $X(t)$, with a transition matrix as follows:

$$P = \begin{bmatrix} 0.4 & 0.2 & 0.4 \\ 0.3 & 0.6 & 0.1 \\ 0.0 & 0.5 & 0.5 \end{bmatrix}.$$

Here, $X(t)$ can only assume the values 1, 2, and 3. Also, $X(0) = 1$.

1. Find the distribution of $X(1)$, $X(2)$, and $X(3)$.
2. Find $E(X(i))$ and $Var(X(i))$, $i = 2, 3$.
3. Find the joint distribution of $X(2)$ and $X(3)$.
4. Find the distribution of $X(2) + X(3)$.
5. Find $Var(X(2) + X(3))$ from the distribution of $X(1) + X(2)$.
6. Is the formula $Var(X(2) + X(3)) = Var(X(2)) + Var(X(3))$ correct? If it is wrong, provide the correct formula.

1.2 Prove that in a Markov chain defined on the set $\{1, 2, \ldots, N\}$, with $\sum_{j=1}^{N} p_{i,j} = 1$ and $\sum_{i=1}^{N} p_{i,j} = 1$, the equilibrium probabilities must all be equal to $\frac{1}{N}$.

1.3 Find the generating function $P(z) = \sum_{i=0}^{\infty} p_i z^i$, where the p_i are defined by the recursion $p_1 = 0.2$, $p_i = 0.5 p_{i-1} + 0.2 p_{i-2}$, $i > 1$. Here, p_0 is unknown, and must be found by the condition that $P(1) = 1$. Also find $\sum_{i=1}^{\infty} i p_i$.

1.4 The truncated geometric distribution with parameter p and N is defined as follows: $p_{i+1} = p_i p$, $i < N$, $p_i = 0$, $i > N$, with p_0 determined in such a way that $\sum_{i=0}^{N} p_i = 1$. Find p_0, the generating function of the truncated geometric distribution, its mean and its variance.

1.5 Consider the matrix

$$P = \begin{bmatrix} 0.5 & 0.2 & 0.3 & 0 & 0 & 0 \\ 0.6 & 0 & 0 & 0.4 & 0 & 0 \\ 0.6 & 0 & 0 & 0 & 0.4 & 0 \\ 0 & 0.6 & 0 & 0 & 0 & 0.4 \\ 0 & 0 & 0.6 & 0 & 0 & 0.4 \\ 0 & 0 & 0 & 0.6 & 0 & 0.4 \end{bmatrix}.$$

Use a system that allows you to multiply matrices, such as "MMULT(array 1, array 2)" in Excel$^{©}$ or in OpenOffice to find P^n for $n = 2, 3, 4, 5$. What do you observe regarding the density of the resulting matrices. Explain your observation.

1.6 Find the generating function of the Bernoulli distribution, then show that the binomial distribution has the same generating function as the sum of n independent, identically distributed Bernoulli distributions.

Chapter 2
Systems with Events Generated by Poisson or by Binomial Processes

This chapter is mainly devoted to systems where all events are generated by Poisson processes. Hence, arrivals are Poisson, and so are moves from one waiting line to another, as well as departures. However, the rate of the process may depend on the state of the system. We will also discuss systems where the events are generated by binomial processes.

Both the Poisson process and the binomial process are Markov chains. Both are counting processes, that is, they count the number of events, such as arrivals, that occur within a given time interval. However, whereas the Poisson process is a CTMC, the binomial process is a DTMC. The Poisson process generates events at a given rate, say a rate λ, and the binomial process generates at each time t, $t = 1, 2, \ldots$ an event with probability p.

Poisson event systems are systems where events are generated by Poisson processes whereas binomial event systems are systems where events are generated by binomial processes. For the effective formulation of Poisson event systems, we have to modify our definition of the Poisson process slightly and allow rates that depend on the state of the system, and a similar extension applies to the binomial process. Note that in a Poisson process, the times between events of a given type are exponential. If these times are not exponential, then the system is not a Poisson event system. In particular, if the times between events are phase type, they are not Poisson event systems. A similar convention applies to binomial event systems, where the times between events of a given type are geometric.

Binomial event systems are in many respects similar to Poisson event systems. In particular, their formulation is almost identical. However, since binomial event systems are discrete in time, the possibility of several events occurring at exactly the same time cannot be neglected, and this makes the generation of transition matrices considerably more difficult.

Whenever we deal with a discrete event system, we must perform three steps. In the first step, the system must be formulated. The second step uses this formulation to create a transition matrix. This is followed by the third step, the calculation of transient and steady-state solutions. This chapter concentrates on the first two steps, the formulation and the generation of the transition matrix. We also present

© The Author(s), under exclusive license to Springer Nature Switzerland AG 2022   27
W. Grassmann and J. Tavakoli, *Numerical Methods for Solving Discrete Event Systems*,
CMS/CAIMS Books in Mathematics 5, https://doi.org/10.1007/978-3-031-10082-6_2

equilibrium solutions for such systems, but effective methods for finding transient and equilibrium solutions are postponed to Chapters 7 and 9.

The three-step approach we use deviates somewhat from the classical queueing approach. In the classical approach, after describing the system, one derives transient and equilibrium equations, which are solved explicitly for the equilibrium probabilities, and numerical methods are considered only if this is not possible,. We will describe the classical method as well, and for some simple cases, we will use it. Unfortunately, in many cases of interest, numerical methods are the only way to obtain transient and equilibrium results, and then, no mathematical formulation of the transient and equilibrium equations is needed.

The outline of this chapter is as follows: we will first discuss the binomial and the Poisson processes, and then we will show how to formulate Poisson event systems in a concise form. A number of examples then demonstrate our method for formulating Poisson event systems. These examples are not chosen because of their importance or realism, but in order to explain the main concepts. For most examples, we show how to generate transition matrices manually. The method given in this chapter only work for small problem. More effective methods to generate transition matrices are given in Chapter 3. Moreover, the methods used for solving the equilibrium equations are not very efficient. More efficient methods are given in Chapter 9. Moreover, in Chapter 7, we describe an easily implementable method to find transient solutions.

2.1 The Binomial and the Poisson Process

One way to generate events is through the Bernoulli process, that is, events can only occur at times $0, 1, 2, \ldots$, and the probability that the event in question occurs at a given time is p. Here, p is independent of whatever happened in the past. By definition, the Bernoulli process is a sequence of independent Bernoulli trials. The binomial process $\{X(n), n > 0\}$ counts the number of successes in a Bernoulli process. $X(n)$ has a binomial distribution with parameters n and p. As discussed in Section 1.3.3, if one starts observing the process at any time t, the time to the next event of the same type has a geometric distributions, and the average time between two events is $1/p$. Since the Bernoulli trials of the binomial process are independent, no supplementary variable is necessary to record its past state.

By definition, a Poisson process is a counting process, that is, it counts the number of events from, say, 0 to t. It has one parameter μ, and by definition,

$$\frac{P\{1 \text{ event in } (t, t + h)\}}{h} = \mu + o(h).$$

Here, μ is independent of the past history. Consequently, the Poisson process can be considered a CTMC, with $q_{i,i+1} = \mu$, $q_{i,i} = -\mu$, and $q_{i,j} = 0$ otherwise. Provided the states are counted starting at 0, the transient equation (1.5) of Chapter 1 becomes

$$\pi_0'(t) = -\mu \pi_0(t) \tag{2.1}$$

$$\pi_n'(t) = \mu \pi_{n-1}(t) - \mu \pi_n(t). \tag{2.2}$$

If $\pi_0(0) = 1$, and $\pi_n(0) = 0$, $n > 0$, then we find the following solution:

$$\pi_n(t) = e^{-\mu t} \frac{(\mu t)^n}{n!}. \tag{2.3}$$

This is the Poisson distribution with parameter μt, hence the name Poisson process. To prove that (2.3) is indeed the solution of (2.1) and (2.2), one differentiates (2.3), and verifies that $\pi_0(0) = 1$.

In the Poisson process, the time to the first event is greater t if there is no renewal until t, that is,

$$P\{T > t\} = p(0, \mu t) = e^{-\mu t}.$$

Consequently, the distribution of the time to the first event is $f_T(t) = \mu e^{-\mu t}$, which is the exponential distribution.

Suppose now there is a Poisson process with a rate of μ. An observer arrives at time 0, and then he waits for τ minutes, but no event occurs. What is the probability that he still has to wait over an additional t minutes until the next event occurs? In other words, what is the probability that his waiting time T exceeds $t + \tau$, it has exceeded τ? To find this probability, note that if $\tau > 0$, $P\{T > t + \tau$ and $T > \tau\} = P\{T > t + \tau\}$. Consequently

$$P\{T > t + \tau \mid T > \tau\} = P\{T > t + -\}/P\{T > \tau\} = e^{-\mu(t+\tau)}/e^{-\mu\tau} = e^{-\mu t}.$$

In other words, the probability of having to wait an additional t minutes after having waited τ minutes is still $e^{-\mu t}$. Hence, how long the observer has already waited does not matter, the time to the next event is independent of the waiting time already spent: In Markov chains, the past is irrelevant, and waiting time already suffered is information of the past. It is therefore irrelevant whether the observer arrived just after the occurrence of an event or some time later. Specifically, if T_i is the time between event number i and $i + 1$, then all T_i have the exponential distribution with parameter μ.

2.2 Specification of Poisson Event Systems

In Poisson event systems, all events are generated by Poisson processes. Poisson processes are Markovian, that is, for the purpose of predicting future distributions, the past is irrelevant. Consequently, we do not need supplementary variables to reflect the past. The physical variables $X_1, X_2, \ldots, X_d$ are sufficient, and the state is given by the vector $X = [X_1, X_2, \ldots, X_d]$.

As indicated in Section 1.1, there are e different event types, denoted by Ev_1, Ev_2, ..., Ev_e. Each event type Ev_k has an event function $f_k(X)$, and an event condition $C_k(X)$. In addition, we have to specify at which rate the event will occur.

We distinguished between *deterministic* and *stochastic* event types (see Section 1.1 of Chapter 1). In Poisson event systems, one can eliminate all stochastic event types by splitting them into several different deterministic event types, each with its own rate and its own event function. For instance, if Ev_k has a rate of μ_k, and event function $f_{k,1}$ with probability p_1, and $f_{k,2}$ with probability p_2, then one can create two event types, say Ev_{k_1} with rate $p_1\mu_k$ and event function $f_{k,1}$, and $Ev_{k,2}$ with rate $p_2\mu_k$ and event function $f_{k,2}$. This turns out to be inconvenient when events are generated by processes that are not Poisson.

In this chapter, we will also discuss *immediate events*. These are events that occur without any delay once their event condition is satisfied. In principle, immediate events can always be combined with the events that trigger them. However, models are easier to understand when immediate events are used.

In conclusion, in a Poisson event system, provided there are no immediate events, we have

1. A number of variables that describe the present state of the system.
2. A number of event types, each with

   a. A Poisson process which is described by its rate. The rate may depend on the state of the system.
   d. A condition indicating in what states the event type can occur.
   c. An event function, or, for stochastic event types, several event functions, each with its own probability.

The set of the state variables forms the *state space*. If there is more than one state variable, the state space consists of a set of vectors, with the state variables representing the entries of the vectors. The state space will be denoted by S.

To demonstrate the concepts of state space and events, consider the following example. There are two waiting lines, and their lengths are denoted by X_1 and X_2. Consequently, the state of the system is given by the vector $X = [X_1, X_2]$. Anything that can affect the state is an event. One possible event, say Ev_2, would be a move from line 1 to line 2. For this event, we must provide a rate, an event function, and an event condition. The rate may be denoted by $\mu_2(X)$, the event function $f_2(X) = f_2(X_1, X_2) = [X_1 - 1, X_2 + 1]$, and the condition may be $C_2(X) = X_1 > 0, X_2 \leq N_2$, where N_2 is a given upper limit for line 2. The state $f_2(X)$ is called *target*, whereas X is the original state.

We represent the events, together with their event functions, event conditions, and rates by a table, which we call the *event table*. Table 2.1 provides the general form of an event table. The event table completely specifies the model in question, and programs can be written to find the transition matrix from event tables.

We will use the following:

Definition 2.1 If the event condition of event Ev_k is true in state X, we say that Ev_k is *enabled*. Otherwise event Ev_k is *disabled* in state X.

Table 2.1 Event table for Poisson event systems

Event type	Event function	Event condition	Rate
Ev_1	$f_1(X)$	$C_1(X)$	$\mu_1(X)$
Ev_2	$f_2(X)$	$C_2(X)$	$\mu_2(X)$
$\vdots$	$\vdots$	$\vdots$	$\vdots$
Ev_e	$f_e(X)$	$C_e(X)$	$\mu_e(X)$

It is often convenient to automatically disable all events that leave the state space. For instance, if X_i denotes the number in a queue, then X_i must not be negative. Hence, any event reducing X_i by 1 when it is already 0 would bring X_i to -1, which is outside the state space, and in this case, the event should be disabled. Still, it does not hurt to explicitly specify in the event table the conditions that are needed to prevent leaving the state space.

Since all event processes are assumed to be Poisson processes, it will be unnecessary to specify this in the examples below. We do not specify that arrivals are Poisson, or service times are exponential, because this is implied.

2.3 Basic Principles for Generating Transition Matrices

Once the state space S and the event table is known, we go to the next step, which is the creation of the transition matrix. Note that the transition matrix is completely defined by the state space and the event table. Once the transition matrix has been constructed, transient and equilibrium solutions can be calculated. To summarize, the process of modeling Poisson event systems consists of three steps:

1. The model formulation.
2. The creation of the transition matrix.
3. The calculation of transient and equilibrium solutions.

To create the transition matrix, one enumerates all states of the state space S. Each state $X \in S$ corresponds to a row of the transition matrix. Let i be this row. Each active event produces an entry in row i: The position of this entry is given by its event function, and the value of the entry by its rate. If there is the possibility that two different event functions $f_k(X)$ are identical for some X, we add the new rate to the rates already found. Once all events have been applied to all states, one still has to calculate the $q_{i,i} = -\sum_{j \in S} q_{i,j}$. The entire process is summarized by the algorithm below. For the conventions used in this algorithm, see Appendix A.

Algorithm 1. Generating a transition matrix

```
Q = [q(i,j)] = 0
for all states i ∈ S
    for k = 1 : e
        If C(k) true
            q(i,f(k,i)) = q(i,f(k,i)) + mu(k)
            q(i,i) = q(i,i) - mu(k)
        end if
    end for
end for
```

This algorithm hides an important issue, and this is how to enumerate all states. Algorithm implementing the enumeration of states is discussed in Chapter 3.

The algorithm above generates the transition matrix Q, and this matrix can be used to find the equilibrium probabilities. When the system is small, one can use equation (1.8), which allows one to apply matrix-inversion tools as they are available in MATLAB. Also, many spreadsheets can be used to invert a matrix. The function for matrix inversion is typically MINVERSE. This makes the solution of the equilibrium equations easy.

Another way to obtain results quickly is to download software systems that are built to determine equilibrium solutions for discrete event systems. These systems are typically using Petri net terminology, or stochastic activity terminology. In particular, one of the authors has used Möbius [79].

2.4 One-dimensional Discrete Event Systems

Some events have only one physical state variable, such as a single queue, or the stock of a single product. Such systems are discussed in this section. First, we provide a review of some important one-dimensional discrete event systems, and then we give some examples of one-dimensional discrete event systems. In these examples, our three-step approach is illustrated by very simple examples, but the reader should keep in mind that similar steps will be used in the more complex examples introduced in the next section.

2.4.1 Types of One-dimensional Discrete Event Systems

There are many variations that can be observed in systems with one queue. Such systems may have multiple servers. Arrivals may be individuals, but arrivals in groups are also possible. Service typically starts as soon as the server is idle, but it is also possible that service does not start until the number in a queue reaches a certain threshold. Typically, it is assumed that each customer is served individually, but service in groups or bulk have also been considered.

Discrete event models are also prevalent in inventory systems. There is only one state variable if one considers only a single item, which is ordered as soon as the stock falls below a certain level s. Typically, the order quantity is denoted by Q. The stock is reduced whenever there is a usage or a sale. Such systems are Poisson event systems if the usages form a Poisson process, and the lead time is exponential. The lead time is the time from the moment an order goes out, until the item is delivered.

In some ways, inventory theory is quite different from queueing theory, even though its models can be cast as discrete event systems. First, in many cases, the systems are assumed to be deterministic, at least approximately. Secondly, stock control relies much more than queueing theory on optimization, a topic not covered in this book. Thirdly, much emphasis is given to data collection. Of course, the two last issues are important, and maybe queueing theory should learn from inventory control theory in this respect. Since these topics are outside the scope of this book, the reader should consult specialized monographs on inventory control, such as [82].

For queueing and inventory models, we will always do the first two steps of our three-step model: the formulation of the system, as well as the generation of the transition matrix. Moreover, we will apply the classical method, and directly formulate and solve the equilibrium equations for these systems.

2.4.2 The $M/M/1/N$ Queue

The $M/M/1/N$ queue has one state variable, the line length X, and two event types, arrivals (Ev$_1$) and departures (Ev$_2$). The arrival process has a rate of λ, and the departure process a rate of μ. Arrivals are lost once the waiting line reaches N. The state space S now consists of the integers from 0 to N. The event function for arrivals is $f_1(X) = X + 1$. The loss of arrivals when $X = N$ can be expressed by the event condition $X < N$. The event function for departures is $f_2(X) = X - 1$, and since there are no departures when $X = 0$, its event condition is $X > 0$. The event table of this model is given in Table 2.2.

Table 2.2 The $M/M/1/N$ queue

Event type	Event function	Event condition	Rate
Arrival	$X + 1$	$X < N$	λ
Departure	$X - 1$	$X > 0$	μ

Let us now apply Algorithm 1 to the $M/M/1$. The states can easily be enumerated: a simple for-loop will do, with X ranging from 0 to N. For ranges, we will use the MATLAB$^{©}$ notation, that is, $0 : N$ means the range of integers for 0 to N. Here is the implementation of Algorithm 1.

Algorithm 2. Generating the transition matrix of the $M/M/1/N$ queue

Q = [q(i,j)] = 0
for i = 0 : N

> Arrival
> If i < N
> > j = i + 1
> > q(i,j) = lambda
> > q(i,i) = q(i,i) + lambda
> end if
> Departure
> If i > 0
> > j = i – 1
> > q(i,j) = mu
> > q(i,i) = q(i,i) + lambda
> end if
> q(i,i) = –q(i,i)

end for

The result is as follows:

$$
Q = \begin{bmatrix}
-\lambda & \lambda & 0 & \cdots\cdots & 0 \\
\mu & -(\lambda + \mu) & \lambda & \ddots\ \ddots & \vdots \\
0 & \mu & -(\lambda + \mu) & \ddots\ \ddots & \vdots \\
0 & \ddots & \ddots & \ddots\ \ddots & 0 \\
0 & \ddots & \ddots & \ddots\ \ddots & \lambda \\
0 & 0 & \cdots\cdots & 0\ \ \mu & -\mu
\end{bmatrix}. \tag{2.4}
$$

This transition matrix allows us to find transient solutions by algorithms discussed in Chapter 7. To find steady-state probabilities, we use the equilibrium equations, which are

$$
\pi_0 \lambda = \pi_1 \mu \tag{2.5}
$$

$$
\pi_i(t)(\lambda + \mu) = \pi_{i-1}(t)\lambda + \pi_{i+1}(t)\mu, \ 1 \le i < N \tag{2.6}
$$

$$
\pi_N(t)\mu = \pi_{N-1}(t)\lambda. \tag{2.7}
$$

Note that an unnormed solution of this system of equations can be obtained as follows: One sets $\pi_0 = 1$, and finds π_1 from (2.5). Once this is done, we use equation (2.6) with $i = 1$ to find π_2. Next, we use equation (2.6) with $i = 2$ to find π_3 and, in this way, one continues until all π_i are found. Actually, this method can be streamlined as follows: If equation (2.5) is added to (2.6) with $i = 1$, one finds

$$
0 = -\lambda\pi_1 + \mu\pi_2.
$$

Add this equation to (2.6) with $i = 2$ to find

$$0 = -\lambda\pi_2 + \mu\pi_3,$$

and in this way, one continues. In general, one has

$$0 = -\lambda\pi_i + \mu\pi_{i+1}, \ i \geq 0,$$

or, if $\rho = \frac{\lambda}{\mu}$

$$\pi_{i+1} = \rho\pi_i, \ i = 0, 1, \ldots, N. \tag{2.8}$$

This formula allows us to find a solution for the π_i in the form

$$\pi_i = \rho^i \pi_0.$$

The norming condition becomes now

$$\sum_{i=0}^{N} \pi_i = \sum_{i=0}^{N} \rho^i \pi_0 = \frac{1 - \rho^{N+1}}{1 - \rho} \pi_0 = 1$$

or

$$\pi_0 = \frac{1 - \rho}{1 - \rho^{N+1}}.$$

Hence

$$\pi_i = \frac{1 - \rho}{1 - \rho^{N+1}} \rho^i.$$

Note that if $\rho < 1$, we can let $N \to \infty$. In this case ρ^{N+1} converges to 0, and we get the following well-known formula for the $M/M/1$ queue:

$$\pi_i = (1 - \rho)\rho^i. \tag{2.9}$$

It follows that the number in the system has a geometric distribution, that is, it has the same distribution as the number of failures before the first success in a sequence of Bernoulli trials with $p = 1 - \rho$ and $1 - p = \rho$. Its expectation is therefore $\frac{\rho}{1-\rho}$, and its variance $\frac{\rho}{(1-\rho)^2}$.

2.4.3 Birth–Death Processes, with Extensions

In the birth–death process, there is a population, such as a group of people or a collection of things. If the population is of size $X = i$, births occur at rate λ_i, that is, with rate λ_i the population increases by 1. Similarly, deaths occur at rate μ_i, and they decrease the populations by 1. Accordingly, the event types are births and deaths, and their effects and rates are given in Table 2.3. Note that the rates depend on the state.

The $M/M/1$ queue and the $M/M/c$ queue are both birth–death processes. In the $M/M/c$ queue with $X = i$, λ_i is a constant λ, and μ_i is $i\mu$ when $i < c$ and $c\mu$ when $i \geq c$.

Table 2.3 The Birth–Death process

Event type	Event function	Event condition	Rate
Birth	$X + 1$		λ_X
Death	$X - 1$	$X > 0$	μ_X

Another example of a birth–death process is given by the finite-source queueing system. In this model, there are N elements in total, and each element not yet in line needs service at a rate of λ. Hence, the arrival rate is $\lambda_i = \lambda(N - i)$.

The equilibrium equations of the birth–death process are

$$0 = -\pi_0(t)\lambda_0 + \pi_1(t)\mu_1$$
$$0 = \pi_{i-1}(t)\lambda_{i-1} - \pi_i(t)(\lambda_i + \mu_i) + \pi_{i+1}(t)\mu_{i+1}, \ 1 \le i \le N$$
$$0 = \pi_{N-1}(t)\lambda_{N-1} - \pi_N(t)\mu_N.$$

Using the same methods that were explained in connection with the $M/M/1$ queue, one finds

$$0 = -\lambda_i\pi_i + \mu_{i+1}\pi_{i+1}, \ i = 0, 1, \ldots, N$$

or

$$\pi_{i+1} = \frac{\lambda_i}{\mu_{i+1}}\pi_i, \ 0 \le i < N.$$

This allows one to express all π_n as multiples of π_0, and once this is done, π_0 can be determined using the condition that the sum of all probabilities must be 1. Effectively, if $\pi_i = \tilde{\pi}_i\pi_0$, then one has

$$\pi_i = \frac{\tilde{\pi}_i}{\sum_{j=0}^{N} \tilde{\pi}_j}.$$

This determines all $\pi_i, 0 \le i \le N$.

2.4.4 A Simple Inventory Problem

Consider now an inventory problem as follows: Usages from an inventory occur at a rate of μ as long as there is still stock left. Also, once the stock reaches s, and no order is outstanding, O units are ordered. This order arrives after an exponential delay with an average delay time of $1/\lambda$. The event table is now given in Table 2.4. Note that the stock can never exceed $s + O$. If s and O are known, the transition matrix can be derived from the event table. For $s = 2$ and $O = 4$, the transition matrix is as follows:

$$Q = \begin{bmatrix} -\lambda & 0 & 0 & 0 & \lambda & 0 & 0 \\ \mu & -(\lambda+\mu) & 0 & 0 & 0 & \lambda & 0 \\ & \mu & -(\lambda+\mu) & 0 & 0 & 0 & \lambda \\ 0 & \mu & -\mu & 0 & 0 & 0 \\ 0 & 0 & \mu & -\mu & 0 & 0 \\ 0 & 0 & 0 & \mu & -\mu & 0 \\ 0 & 0 & 0 & 0 & \mu & -\mu \end{bmatrix}.$$

Note that a queue with arrivals in groups of 4 would lead to the same transition matrix.

Table 2.4 A simple inventory problem

Event type	Event function	Event condition	Rate
Usage	$X - 1$	$X > 0$	μ
Replenishment	$X + O$	$X \le s$	λ

If $\tilde{\pi}_i, i = 0, 1, \ldots, s + O$ express π_i as a multiple of π_0, then one can calculate the $\tilde{\pi}_i$ recursively, starting with $\tilde{\pi}_0 = 1$. This is true for any transition matrix with all entries below the subdiagonal equal to 0. Of course, if the usage is not restricted to one unit, such recursive solutions are no longer possible.

The equilibrium equations come in several flavors

$$0 = -\lambda\pi_0 + \mu\pi_1$$
$$0 = -(\lambda + \mu)\pi_i + \mu\pi_{i+1}, \ i < O, \ 0 < i < s$$
$$0 = \lambda\pi_{i-O} - (\lambda + \mu)\pi_i + \mu\pi_{i+1}, \ i \ge O, \ 0 < i < s$$
$$0 = \lambda - \mu\pi_i + \mu\pi_{i+1}, \ i < O, \ i \ge s$$
$$0 = \lambda\pi_{i-O} - \mu\pi_i + \mu\pi_{i+1}, \ i \ge O, \ i \ge s$$
$$0 = \lambda\pi_s - \mu\pi_{s+O}.$$

2.5 Multidimensional Poisson Event Systems

In this section, we discuss models with several state variables. We first give some overview of the types of systems that occur, and then discuss some models in detail. These examples show the flexibility of our approach, and they show what kind of transition matrices can be expected. Additional examples can be found in [7].

In this section, we will also discuss immediate events, that is, events that will be executed without delay once their event conditions are met. These events are important if a server serves different types of customers.

2.5.1 Types of Multidimensional Systems

Systems with multiple queues lead to multidimensional systems, with one state variable for each waiting line. We distinguish between a single service center with multiple queues, and multiple service centers with each center having its own queue. A service center may have one or multiple servers.

First, consider service centers with multiple queues. In either case, servers must choose which queue to serve next once the service of the present customer is completed. In some cases, some queues have priorities over others, which leads to the area of *priority queues* [52]. One distinguishes between *preemptive priorities* and *non-preemptive priorities*. In a preemptive priority, elements with higher priority can interrupt the service of lower priority elements, but in non-preemptive priorities, they cannot. After a preemption, service can either resume where it was interrupted, or start over. When starting over, a new service time is created. This new service time usually has the same as distribution as the one before the interruptions. In some systems, the priorities depend on the state. For instance, longer queues may get higher priorities.

The queue to be served next may also be determined by adhering to certain rules. In *polling systems*, the server typically moves from one queue to the next one, and once it has reached the last line, it moves to the first line [90]. There are several variations of this scheme. In a *one-limited* discipline, exactly one customer is served at each visit of the server. Service can also be *exhaustive* in the sense that the server serves all customers waiting in the line it is presently serving.

If the service center has several servers, not all servers may be qualified to serve all customers. For instance, in some telephone answering systems in Canada, there are unilingual servers that can serve only in English, and bilingual servers that can serve both in English and French [83]. Also, in repair facilities, repair persons may only be qualified to repair certain items, and not others.

Most setups encountered in systems with a single state variable, such as a single waiting line, can also be found when there are several state variables or several waiting lines. In particular, arrivals to a particular service center can come in bulk, and service can be done in bulk. Moreover, there may be single and multiple servers at the individual centers. There are also new features in systems with multiple waiting lines. For instance, there is the possibility of matching, such as matching sellers and buyers. In a production system, one may have a stock of part A, and a stock of part B, and the stock level of each contributes a state variable. Customers arrive, and whenever they buy part A, they also buy a certain number of part B.

In many queueing systems, one must know which type of customer is currently served, in particular, when the service rates of the customers of the different queues are different. To indicate this, a new state variable is required, a state variable that indicates the type of the customer.

Next, consider multiple service centers, each with its own waiting line. In this case, one needs to know to which service center the customer will move next. This choice may be fixed, such as in the case of a tandem queue. In this case, there are M

centers, which have to be visited in order, starting with the first center, then visiting the second center, and so on, until the entity has visited the Mth center. After that, the entity leaves the system. In other cases, the customer can choose the center or centers he/she will visit. This may depend on the state of the system. The prime example for this is the shorter queue problem, a system with two severs, each with its own waiting line. Any arriving customer chooses to join the shorter of the two queues [51] [98]. In more complex systems, customers prefer certain routes, such as routes with less traffic.

In many models, the choice which center to see next is random. In particular, a customer leaving center i will move to center j with probability $r_{i,j}$. In this case, the event "depart from center i" is an stochastic event, because the state after the event is random.

2.5.2 First Example: A Repair Problem

As our first example of a multidimensional system, consider the following model. A shop repairs a certain machine. In each repair, one unit of component 1 and 2 units of component 2 need to be replaced. The rates for these repairs are μ. The components are delivered from outside, and the times between deliveries are exponential, with a rate of λ_1 for component 1 and a rate of λ_2 for component 2. The components are kept in bins of size $N_i, i = 1, 2$, and once the bin is full, no components are delivered any more. To formulate this discrete event system, we need two state variables, X_1 and X_2, representing the number of units in the two bins. The event table is given by Table 2.5. In this table, one sees that when a "Replenish 1" occurs, then the number in bin 1 increases by 1 as long as its content is less than N_1, and this event type happens at rate λ_1. The event type "Replenish 2" can be explained in a similar fashion. Finally, if a repair occurs, then one item is taken from bin 1, and two from bin 2, but this can only happen if there is at least one unit in bin 1, and two in bin 2. Otherwise, the repair will not be performed.

Table 2.5 Repair shop

Event type	Event function	Event condition	Rate
Replenish 1	$X_1 + 1, X_2$	$X_1 < N_1$	λ_1
Replenish 2	$X_1, X_2 + 1$	$X_2 < N_2$	λ_2
Repair	$X_1 - 1, X_2 - 2$	$X_1 > 0, X_2 > 1$	μ

We can enumerate all states in lexicographic order, that is, all states with $X_1 = i$ precede states with $X = i + 1$, and within the set of states with $X_1 = i$, the states with $X_2 = j$ precede the states with $X_2 = j + 1$. Let us assume that $N_1 = 2$ and $N_2 = 4$. The states are then $[0, 0], [0, 1], [0, 2], [0, 3], [0, 4], [1, 0], [1, 1], \ldots, [2, 3], [2, 4]$.

Each of these states corresponds to a row and a column in the transition matrix. For state $[0,0]$, only "Replenish 1" and "Replenish 2" satisfy their event conditions, that is, only they must be executed. For "Replenish 1", the target state is $[1,0]$, and in the column corresponding to state $[1,0]$, λ_1 is to be entered. Similarly, "Replenish 2" has a target state $[0,1]$, and in the column corresponding to this state, λ_2 must be entered. The state $[1,0]$ is to be dealt with in a similar way. Its target states are $[2,0]$ and $[1,1]$, and in the corresponding columns, one has to enter λ_1, respectively, λ_2. This continues until one reaches state $[1,2]$, in which case all three events are applicable, with targets $[2,2]$, $[1,3]$, and $[0,0]$. This continues. The entire transition matrix is given in Table 2.6. This transition matrix was entered into an Excel$^{\text{©}}$ spreadsheet, converted into a system of equations as discussed in Section 1.2.2, and solved by the tools provided by Excel$^{\text{©}}$. We used $\lambda_1 = 1$, $\lambda_2 = 2$, $\mu = 1.5$. Table 2.7 gives the results. Also of interest is the probability that $X_1 < 1$ and $X_2 < 2$, because, in this case, no repair can be done due to a lack of parts. This probability is 0.44521.

Table 2.6 Transition matrix of the repair problem

	0,0	0,1	0,2	0,3	0,4	1,0	1,1	1,2	1,3	1,4	2,0	2,1	2,2	2,3	2,4
0,0	–	λ_2	0	0	0	λ_1	0	0	0	0	0	0	0	0	0
0,1	0	–	λ_2	0	0	0	λ_1	0	0	0	0	0	0	0	0
0,2	0	0	–	λ_2	0	0	0	λ_1	0	0	0	0	0	0	0
0,3	0	0	0	–	λ_2	0	0	0	λ_1	0	0	0	0	0	0
0,4	0	0	0	0	–	0	0	0	0	λ_1	0	0	0	0	0
1,0	0	0	0	0	0	–	λ_2	0	0	0	λ_1	0	0	0	0
1,1	0	0	0	0	0	0	–	λ_2	0	0	0	λ_1	0	0	0
1,2	μ	0	0	0	0	0	0	–	λ_2	0	0	0	λ_1	0	0
1,3	0	μ	0	0	0	0	0	0	–	λ_2	0	0	0	λ_1	0
1,4	0	0	μ	0	0	0	0	0	0	–	0	0	0	0	λ_1
2,0	0	0	0	0	0	0	0	0	0	0	–	λ_2	0	0	0
2,1	0	0	0	0	0	0	0	0	0	0	0	–	λ_2	0	0
2,2	0	0	0	0	0	μ	0	0	0	0	0	0	–	λ_2	0
2,3	0	0	0	0	0	0	μ	0	0	0	0	0	0	–	λ_2
2,4	0	0	0	0	0	0	0	μ	0	0	0	0	0	0	–

Note that if the states are in lexicographic order, the transition matrix is banded in the sense that $q_{i,j} = 0$ if $|j - i|$ exceeds a certain value. For the transition matrix of our problem, non-zero entries can only be found in the band given by $j - i \le 5$, and $i - j \le 7$.

The transition matrix of Table 2.6 is sparse in the sense that most entries are 0. For many purposes, it would be best to store only the non-zero elements. This is shown in Table 2.8.

Table 2.7 Results of the repair problem

$X_1 \setminus X_2$	0	1	2	3	4	Total
0	0.0396779	0.0413930	0.0973719	0.0778967	0.190171	0.4465106
1	0.0311107	0.028286	0.0793558	0.0298821	0.1395531	0.3081880
2	0.0259642	0.0229185	0.035769	0.028977	0.1316719	0.2453014
Total	0.0967528	0.0925978	0.2124970	0.1367562	0.4613962	1

Table 2.8 Transition matrix in row/column format

State	Target	Rate	State	Target	Rate	State	Target	Rate
0,0	1,0	λ_1	1,0	2,0	λ_1	2,0	2,1	λ_2
0,0	0,1	λ_2	1,0	1,1	λ_2	2,1	2,2	λ_2
0,1	1,1	λ_1	1,1	2,1	λ_1	2,2	2,3	λ_2
0,1	0,2	λ_2	1,1	1,2	λ_2	2,2	1,0	μ
0,2	1,2	λ_1	1,2	2,1	λ_1	2,3	2,4	λ_2
0,2	0,3	λ_2	1,2	1,3	λ_2	2,3	1,1	μ
0,3	1,3	λ_1	1,2	0,0	μ	2,4	1,2	μ
0,3	0,4	λ_2	1,3	2,3	λ_1			
0,4	1,4	λ_1	1,3	1,4	λ_2			
			1,3	0,1	μ			
			1,4	2,4	λ_1			
			1,4	0,2	μ			

2.5.3 Second Example: Modification of the Repair Problem

In our second example, we use a different representation of the transition matrix, and we also introduce the notion of a *state number*, a number that can be used to identify both the row and the column of the state within the transition matrix. This example also provides an unexpected result for the equilibrium distribution which we will explore. The example is a modification of the previous example.

As before, there are two bins, containing components of types 1 and 2, and a repair needs one unit of types 1 and 2 of type 2. However, as soon as the numbers in bin i, $i = 1, 2$, is below s_i, and there is no order outstanding, the amount of O_i is ordered. Once ordered, the orders arrive after an exponential time with the average time of $1/\lambda_i$, $i = 1, 2$. Since a repair always requires 2 units of type 2, it seems reasonable to consider for this type only order quantities that are divisible by 2, and the same holds for s_2. This convention allows us to use a package of 2 as the unit for this component. To formulate this model, we need two state variables: X_1 is the amount in bin 1 and X_2 is the number of packages of size 2 in bin 2 (Table 2.9).

Table 2.9 Repair shop, version 2

Event type	Event function	Event condition	Rate
Type 1 arrives	$X_1 + O_1, X_2$	$X_1 \leq s_1$	λ_1
Type 2 arrives	$X_1, X_2 + O_2$	$X_2 \leq s_2$	λ_2
Repair	$X_1 - 1, X_2 - 1$	$X_1 > 0, X_2 > 0$	μ

This event table indicates that if the number in bin i, $i = 1, 2$, is less than s_i, an order is on its way, and it arrives at a rate of λ_i, increasing X_i by O_i. The event "Repair" is similar to the one we mentioned in the previous example.

To obtain the transition matrix, we first write, for each state, how the events would change the state. With each state, we associate a state number, which is given in column 1. The state number corresponds to the row in the transition matrix. To find the column of an entry, one first coverts the state number to the state vector. Given the state vector, one can find all targets of each state, and find the state number of each target. This provides the position of the target entries in the transition matrix. The rate at this position is then given by the rate of the event table. To show this, let $s1 = s_2 = 1$ and $O_1 = O_2 = 2$. For these values, the position of all entries is given by the following table:

State Number	Vector	Arrivals Comp. 1	Arrivals Comp. 2	Repair	Arrivals Comp. 1	Arrivals Comp. 2	Repair
1	0 0	2 0	0 2		9	3	
2	0 1	2 1	0 3		10	4	
3	0 2	2 2			11		
4	0 3	2 3			12		
5	1 0	3 0	1 2		13	7	
6	1 1	3 1	1 3	0 0	14	8	1
7	1 2	3 2		0 1	15		2
8	1 3	3 3		0 2	16		3
9	2 0		2 2			11	
10	2 1		2 3	1 0		12	5
11	2 2			1 1			6
12	2 3			1 2			7
13	3 0		3 2			15	
14	3 1		3 3	2 0		16	9
15	3 2			2 1			10
16	3 3			2 2			11

For instance, row 1 of the transition matrix corresponds to the state $[0, 0]$, which has the state number 1. With rate λ_1, we have an arrival of component 1, bringing us to state $[1, 0]$, which is state 9. Consequently, the entry of row 1, column 9 is λ_1. The other entries can be found in a similar way.

We calculated the equilibrium probabilities with $\lambda_1 = \lambda_2 = 1.5$, and $\mu = 1$, using Excel$^{©}$. To our surprise, half of the equilibrium probabilities are 0! In fact, we obtain the following probabilities:

State	0 0	0 1	0 2	0 3	1 0	1 1	1 2	1 3
Probability	0.0340	0	0.0748	0	0	0.1020	0	0.0612
State	2 0	2 1	2 2	2 3	3 0	3 1	3 2	3 3
Probability	0.0748	0	0.4082	0	0	0.0612	0	0.1837

What happened? First, notice that the states of the non-zero probabilities all show an even difference between X_1 and X_2. A little reflection shows that if $X_1 - X_2$ is even in the original state, $X_1 - X_2$ is also even in all possible targets. For instance, an arrival of component 1 increases X_1 by 2, leaving X_2 unchanged. Similarly, an arrival of component 2 increases X_2 by 2, leaving X_1 unchanged. Finally, a repair decreases both X_1 and X_2 by 1, leaving the difference between X_1 and X_2 unchanged. Hence, if the difference is even, it stays even, and there is no way to ever reach a state with an odd difference between X_1 and X_2. Similarly, if one starts with a state where $X_1 - X_2$ is odd, the difference will remain odd.

The above is an example of a Markov chain with a decomposable transition matrix. We will discuss such systems in Chapter 8. In a way, we have two different Markov chains that share a single transition matrix. In fact, we could have first found the equilibrium probabilities of the Markov chain with an even difference, and then the one with an odd difference.

2.6 Immediate Events

In many discrete event systems, there are states that cannot last for any length of time. They are, in this sense, unstable. For instance, in many queueing systems, the server cannot be idle when there are customers waiting for service. Consequently, if, following an event, the server becomes idle, the server will immediately start serving the next customer if there is one. Also, after an arrival to an empty system, the server will immediately start service. Hence, "starting service" is an immediate event. It is similar to other events in that it has an event function and an event condition. It has no rate, or, if we assign a rate, the rate would have to be infinite.

When there are immediate events, we have to distinguish between two spaces: the space before any immediate event occurs, and the state space after all immediate events that can occur in the different states have been executed. The state space before immediate events occurs can have states satisfying the event conditions of the immediate events, and these states disappear once the immediate events are completed. Hence, there is a stable state space, the state space of the states where immediate events are disabled, and a pre-stable state space, a state space that includes unstable states.

We have not yet used immediate events in our queueing models. In fact, in Poisson event systems, most immediate events can be avoided. However, they are needed when the elements in service have to be distinguished from the elements in queue. Such a distinction is necessary as soon as the type of customer being in service affects the future. To make a record of this situation, an additional physical state variable, say X_k, is needed. X_k is redundant when the server is idle. This is usually indicated by setting X_k to 0. For instance, if one has two types of customers, with different service rates, then $X_k = 0$ while the server is idle, $X_1 = 1$ while customer type 1 is served, and $X_k = 2$ while customer type 2 is served.

The condition $X_k = 0$ splits the stable state space into two subspaces, with one subspace containing the state or states where states with $X_k = 0$ are stable, and the set of states where such states are unstable. The first set includes all states where the server is idle, whereas the second one includes the states where the server is busy. The existence of these two sets of states makes the enumeration of all states more difficult.

There are different ways to handle immediate events. One can create the pre-stable state space first, and then eliminate all unstable states. An alternative is to eliminate the immediate event by combining them with the events that trigger them. Technically, this can be done already when formulating the event table. If the transition matrix is generated on a computer, one can implement immediate events as subroutines, which are called by any event that triggers them.

Immediate events are extensively used in simulation modeling [3]. In particular, the classical model for simulating one-server queues contains three events: arrivals, start service, and depart. Obviously, service starts immediately as soon as the server is idle and there are customers requiring service. There are two situations where this can happen: either there is an arrival to the empty system, or there is a departure while there is still a queue. In either case, the event "start service" happens without any delay.

2.6.1 An Example Requiring Immediate Events

The following example explains how immediate events work. There is a computer that executes jobs for two customers, say customer D_1 and customer D_2. There is a buffer for the jobs of customer D_1 of size N_1, and one for customer D_2 of size N_2. Jobs of customer D_1 arrive at rate λ_1, and the ones of customers D_2 at rate λ_2. If the buffer is full, arrivals are lost. The computer can only execute one job at a time, and the expected execution times are $1/\mu_1$ for D_1 and $1/\mu_2$ for D_2. To reduce the number of jobs lost, the jobs of the customer with the smallest space left in their buffer are given preference. In case of a tie, jobs of customer 2 are done first. To express this mathematically, let X_1 be the number of jobs of customer D_1, excluding the one being executed, and let X_2 similarly be the number of jobs of customer D_2. Consequently, a job from customer D_1 is started if $N_1 - X_1 < N_2 - X_2$, or $X_2 - X_1 < N_2 - N_1$. Otherwise, a job of customer D_2 is started. To formulate this model, we need three state variables: X_1 and X_2 as they were defined earlier, and X_3, which records the customer whose job is in service: $X_3 = 1$ when the job comes from customer D_1, and

$X_3 = 2$ when it comes from customer D_2. When the system is empty, $X_3 = 0$. This leads to the model given in Table 2.10. Here, the events "Start job D_1" and "Start job D_2" are immediate events. "Start job D_1" will be enabled either after "D_1 arrives", after "End job D_1", or after "End job D_2", and it can be combined with these events.

Table 2.10 A dynamic priority model

Event type	Event function	Event condition	Rate
D_1 arrives	$X_1 + 1, X_2, X_3$	$X_1 < N_1$	λ_1
D_2 arrives	$X_1, X_2 + 1, X_3$	$X_2 < N_2$	λ_2
Start job D_1	$X_1 - 1, X_2, 1$	$X_1 > 0, X_3 = 0, N_1 - X_1 < N_2 - X_2$	
Start job D_2	$X_1, X_2 - 1, 2$	$X_2 > 0, X_3 = 0, N_2 - X_2 \leq N_1 - X_1$	
End job D_1	$X_1, X_2, 0$	$X_3 = 1$	μ_1
End job D_2	$X_1, X_2, 0$	$X_3 = 2$	μ_2

Table 2.11 provides the state-by-state event table for $N_1 = N_2 = 2$. From this table, the transition matrix is easily generated, and once this is done, transient and equilibrium probabilities can be calculated without difficulty. An algorithm for generating the transition matrix is given in Chapter 3.

Table 2.11 List of states and their target states

State	Arrival D_1	Arrival D_2	Dep. D_1	Dep D_2
0 0 0	0 0 1	0 0 2		
0 0 1	1 0 1	0 1 1	0 0 0	
0 0 2	1 0 2	0 1 2		0 0 0
0 1 1	1 1 1	0 2 1	0 0 2	
0 1 2	1 1 2	0 2 2		0 0 2
0 2 1	1 2 1		0 1 2	
0 2 2	1 2 2			0 1 2
1 0 1	2 0 1	1 1 1	0 0 1	
1 0 2	2 0 2	1 1 2		0 0 1
1 1 1	2 1 1	1 2 1	1 0 2	
1 1 2	2 1 2	1 2 2		1 0 2
1 2 1	2 2 1		1 1 2	
1 2 2	2 2 2			1 1 2
2 0 1		2 1 1	1 0 1	
2 0 2		2 1 2		1 0 1
2 1 1		2 2 1	1 0 1	
2 1 2		2 2 2		1 0 1
2 2 1			2 1 2	
2 2 2				2 1 2

2.6.2 A Second Example with Immediate Events: A Three-way Stop

In this section, we will introduce a more complex model, namely, a three-way stop. At such a stop, three roads meet, roads 1, 2, and 3, and arrivals at each road are Poisson, with rates λ_i for road i, $i = 1, 2, 3$. After coming to a stop, the cars cross the intersection in the same order as they have reached the head of the respective queue. The time to cross is exponential, with a rate μ, which is independent of the road they came from. The arrival rate of cars at road i is λ_i, $i = 1, 2, 3$.

As state variables, we use X_i, the number of cars lined up on road i, as well as their rank, R_i, indicating the rank of the road with the car that has the next turn. Thus, if $R_i = 1$, then the car on road i is the next to move, if $R_i = 2$, then the car on road i will cross right after the next car has cleared the intersection, and if $R_i = 3$, two cars must have cleared the intersection before the car on road i can move. If there are no cars waiting on road i, $R_i = 0$. We also use B to indicate if the intersection is busy, that is, $B = 0$ if the intersection is clear, and 1 otherwise.

We have two finite rate events, namely, "Arrival" and "Clear intersection", and one immediate event, "Enter intersection". No arrival occurs on road i if $X_i \geq N_i$. Consequently, $X_i \leq N_i$. The formulation of the event function is rather complicated, and we therefore give a short description of each event. We use X_i^+ and R_i^+ to describe the state after the event. Aside from the state variables, we also need the auxiliary variable "Num", which indicates the number of roads with cars waiting. Here is the complete description.

Arrival to line i, $i = 1, 2, 3$.
  Event condition: $X_i \leq N_i$.
  Event rate: λ_i
  Event function:
  $X_i^+ = X_i + 1$
  if $X_i = 0$, increase Num by 1, and set R_i^+ to Num
Enter Intersection.
  Event condition: Num > 0, $B = 0$.
  Immediate event
  Event function:
  Let k be the road with $R_k = 1$
  Set $X_k^+ = X_k - 1$, and $B = 1$ to indicate that intersection is occupied
  Reduce all R_i by 1 unless $R_i = 0$, $i = 1, 2, 3$
  If $X_k^+ > 0$ then $R_k^+ =$ Num, else $R_k^+ = 0$, Num= Num -1.
Clear Intersection.
  Event condition: $B = 1$.
  Event rate: μ
  Event function:
  $B = 0$

This is a very complicated state space which requires special methods to enumerate all states. However, once the states are enumerated, the creation of the transition matrix is relatively easy, and transient and equilibrium solutions can be obtained.

2.7 Event-based Formulations of the Equilibrium Equations

In most queueing papers, the equilibrium equations are stated explicitly. Since, in many cases, this is the best option, we discuss this approach next. We should mention, however, that the numerical evaluation of equilibrium probabilities does not require one to formulates the equilibrium equations. To find the numerical values of the equilibrium probabilities, the transition matrix is sufficient.

To formulate the equilibrium equations, we need the set of all states that allow events that turn the present state into state i, where $i = [i_1, i_2, \ldots, i_d]$. If $\mathcal{U}_k(i)$ is part of this set, we thus have

$$\mathcal{U}_k(i) = \{\ell \mid f_k(\ell) = i\}.$$

Often, $f_k(i) = i + a^{(k)}$, where i is a state vector, and $a^{(k)}$ is a vector of integers with the same dimension as i. For instance, consider a system with two waiting lines which at time t have length i_1 and i_2, respectively. A move from line 1 to line 2 results in line 1 decreasing by 1, and line 2 increasing by 1, that is, $f(i_1, i_2) = [i_1, i_2] + [-1, 1]$, consequently, $a^{(k)} = [-1, 1]$. If $j = i + a^{(k)}$, then $i = j - a^{(k)}$, and, as long as $i - a^{(k)}$ is part of the state space, the set $\mathcal{U}_k(i)$ only contains the element $i - a^{(k)}$. If $i - a^{(k)}$ is outside the state space, $\mathcal{U}_k(i) = \{\}$. In conclusion, $\mathcal{U}_k(i)$ is either empty or it contains a single state.

The general form of the equilibrium equation for state i is now, when $\mathbf{I}(A)$ equals 1 if A is true, and $\mathbf{I}(A) = 0$ otherwise:

$$\pi_i \sum_{k=1}^{e} \mathbf{I}(C_k(i))\mu_k(i) = \sum_{k=1}^{e} \sum_{\ell \in \mathcal{U}_k(i)} \pi_\ell \mu_k(\ell). \tag{2.10}$$

Here, the left side gives the rate of leaving state i, and the right side the rate of entering state i. To clarify this equation, note that in state i, the rate of leaving is equal to the sum of the rates of the enabled events, that is, of the events with $C_k(i)$ true. This sum must be multiplied by π_i to obtain the unconditional rate of leaving state i. For the rate of entering state i, we have to find the states from which state i is reached due to event Ev_k, and if ℓ is such a state, then we have to multiply π_ℓ with the rate of event Ev_k, and add all these rates.

Writing equilibrium equations is particularly useful if one or more state variables have no upper bound. For this reason, we now demonstrate how to find the equilibrium equations for a tandem queue with no upper bound for the first line. We define X_1 to be the length of line 1, and X_2 the length of line 2, with $X_2 \leq N_2$. The event table of this system is given in Table 2.12.

Table 2.12 Tandem queue with no limit on first line

Event type	Event function	Event condition	Rate
Arrival	$X_1 + 1, X_2$		λ
Move	$X_1 - 1, X_2 + 1$	$X_1 > 0, X_2 < N_2$	μ_1
Depart	$X_1, X_2 - 1$	$X_2 > 0$	μ_2

Here, we have three events, "Arrival", "Move", which is a move from line 1 to line 2, and "Depart". As one can see, arrivals join line 1, increasing in this way X_1. After that, they move from line 1 to line 2, provided $X_2 < N_2$, thereby decreasing line 1 and increasing line 2. Once served at station 2, they leave, and X_2 decreases by 1. For convenience, we use i for the present value of X_1, and j for the present value of X_2.

We start with states $[i, j]$ where all events are enabled, and for every event, there is a state such that Ev_k results in $[i, j]$. In this case, the rate into state $[i, j]$ becomes

$$\pi_{i,j}(\lambda + \mu_1 + \mu_2) = \pi_{i-1,j}\lambda + \pi_{i+1,j-1}\mu_1 + \pi_{i,j+1}\mu_2. \tag{2.11}$$

Here, the left side is the rate of leaving state $[i, j]$, either because of an arrival, a move from 1 to 2, or a departure. The right side indicates that state $[i, j]$ is reached if the system is in state $[i - 1, j]$ before, and an arrival occurs, or if the system is in state $[i + 1, j - 1]$, and a move from line 1 to line 2 occurs, or if the system is in state $[i, j + 1]$, and a departure occurs.

In state $[i, j]$, terms corresponding to disabled events must be dropped from the left side. Thus, if $i = 0$, $j > 0$, then μ_1 must be dropped, and if $i > 0$ and $j = 0$, then μ_2 must be dropped, and if i and j are both 0, then μ_1 and μ_2 must both be dropped. Finally, μ_1 must be dropped if $X_2 = N_2$ because of blocking. On the right of (2.11), one has to drop any term that contains the probability of a state outside the state space. When $i = 0$, the term $\pi_{i-1,j}\lambda$ must be dropped, when $j = 0$, the term $\pi_{i+1,j-1}\mu_1$ must be dropped, when $j = N_2$, the term $\pi_{i,j+1}\mu_2$ must be dropped. Keeping this in mind, we obtain the following equations in addition to (2.11):

$$\pi_{0,0}\lambda = \pi_{0,1}\mu_2 \tag{2.12}$$

$$\pi_{0,j}(\lambda + \mu_2) = \pi_{1,j-1}\mu_1 + \pi_{0,j+1}\mu_2, \ 0 < j < N_2 \tag{2.13}$$

$$\pi_{i,0}(\lambda + \mu_1) = \pi_{i-1,0}\lambda + \pi_{i,1}\mu_2,, \ i > 0 \tag{2.14}$$

$$\pi_{0,N_2}(\lambda + \mu_2) = \pi_{1,N_2-1}\mu_1 \tag{2.15}$$

$$\pi_{i,N_2}(\lambda + \mu_2) = \pi_{i-1,N_2}\lambda + \pi_{i+1,N_2-1}\mu_1, \ i > 0. \tag{2.16}$$

We will refer to equation (2.11) as *interior* equation. Equations (2.12)–(2.16) are the *boundary equations*.

As our second example to show how to formulate the equilibrium equations, consider the repair model described in Table 2.5. For this purpose, let $\pi_{i,j}$ be the probability that $X_1 = i$, $X_2 = j$. First, consider equation (2.10). If all events apply, both for the flow into the state and the flow out of the state, then we obtain the following equation:

$$\pi_{i,j}(\lambda_1 + \lambda_2 + \mu) = \pi_{i-1,j}\lambda_1 + \pi_{i,j-1}\lambda_2 + \pi_{i+1,j+2}\mu.$$

This is the interior equation. There are a great number of boundary equations. Instead of listing them all, we indicate under what conditions terms of the interior equation must be dropped.

On the left side, the outflow, the following changes to the interior equation must be made:

1. The term λ_1 must be dropped for $i = N_1$.
2. The term λ_2 must be dropped for $j = N_2$.
3. The term μ must be dropped for $i < 1$ or $j < 2$.

On the right side, we have to make sure that the state $f_k^{-1}(X)$ is inside the state space. This leads to the following changes:

1. The term $\pi_{i-1,j}\lambda_1$ must be dropped for $i = 0$.
2. The term $\pi_{i,j-1}\lambda_2$ must be dropped for $j = 0$.
3. The term $\pi_{i+1,j+2}\mu$ must be dropped for $i = N_1$ or $j \geq N_2 - 1$.

It follows that when writing the boundary equations, we have to consider a total of six conditions, three for the outflow and three for the inflow. Since each condition is either true or false, this provides $2^6 = 64$ cases. Of course, some conditions overlap, but writing down all boundary equations still requires an considerable effort.

2.8 Binomial Event Systems

In a binomial event system, all processes generating events are binomial processes, and the times between the events in the different event processes are therefore geometric. In a way, binomial event systems are very much like Poisson event systems. In fact, all examples of the Poisson event systems we presented have their counterpart in binomial event systems, except that probabilities are used instead of rates. Thus, for event Ev_k, one has, instead of the rate μ_k, the probability p_k. The $Geom/Geom/1/N$ queue, for instance, has the same state variable X, the same event functions, and the same event conditions as the $M/M/1/N$ queue, except that instead of λ and μ we have probabilities, say p_A for arrivals and p_D for departures. However, the generation of transition matrices for binomial event systems is more complicated than for Poisson event systems.

Since binomial event systems are discrete in time, more than one event can occur at any time t. For instance, in the $Geom/Geom/1/N$ queue, an arrival may occur at the same time as a departure. To understand the issues caused by simultaneous events, we present two examples, as well as a general discussion. Simultaneous events are relevant in any discrete-time system and because of this, the discussion that follows provides an introduction to the material of Chapter 4, which presents more complicated discrete event systems.

2.8.1 The $Geom/Geom/1/N$ Queue

The event table of the $Geom/Geom/1/N$ queue is as follows:

Event type	Event function	Event condition	Probability
Arrival	$\min(X + 1, N)$		p_A
Dept.	$X - 1$	$X > 0$	p_D

Here, p_A is the probability of an arrival, and p_D the probability of a departure. Also, instead of using the event condition $X < N$ for arrivals, we changed the arrival event function from $f_A(X) = X + 1$ to $f_A(X) = \min(X + 1, N)$. It is important to recognize the difference between the two approaches. If we say $f_A(X) = \min(X + 1, N)$, then the arrival has occurred, even when $X = N$, whereas in the case where we use the event condition $C_A(X) = (X < N)$, no arrival is allowed to occur. If arrivals are Poisson, the two methods lead to the same result because the distribution until the next arrival is independent from the fact that an arrival has happened. Of course, if inter-arrival times are not exponential, things are different.

To account for the fact that at time t, both an arrival and a departure can occur, let us complement the event table as follows:

Event type	Event function	Event condition	Probability
Arrival only	$\min(X + 1, N)$		$p_A(1 - p_D)$ if $X > 0$ else p_A
Dept. only	$X - 1$	$X > 0$	$p_D(1 - p_A)$
Arrival & Dept.	X	$X > 0$	$p_A p_D$

Note that the probability that an arrival, but no departure occurs is $p_A(1 - p_D)$, but only when $X > 0$. When $X = 0$, there is no departure, and therefore the probability of an arrival is p_A rather than $p_A(1 - p_D)$.

Next, we have to find the event function of the event consisting of both arrival and departure. Generally, the effect of several events occurring simultaneously is obtained by taking the composition of the event functions, but the composition operator is not commutative. If $f_A(X)$ is the event function of arrivals, and $f_D(X)$ is the event function of departures, then if the arrival function is executed first, the result is $f_D(f_A(X))$, whereas if departures are executed first, the result is $f_A(f_D(X))$. Let us see how this works out with $f_A(X) = \min(X + 1, N)$ and $f_D(X) = X - 1$. We have, if departures come first

$$f_A(f_D(X)) = f_A(X - 1) = \min(X, N) = X.$$

It follows that if departures are executed first, the combined effect is the identify function, that is, X does not change. If X^+ the state immediately after the events are completed, and if departures are executed first, then $X^+ = X$.

If arrivals are executed first, one has

$$f_D(f_A(X)) = f_D(\min(X + 1, N)) = \begin{cases} f_D(X + 1) = X & \text{if } X + 1 \le N \\ f_D(N) = N - 1 & \text{if } X + 1 > N. \end{cases}$$

Consequently, $f_D(f_A(N)) = N - 1$, whereas $f_A(f_D(N)) = N$, which is different. If follows that if arrivals are executed first, and $X = N$, then $X^+ = N$, but if departures are executed first, $X^+ = N - 1$. We conclude that for $X = N$, the order in which the events are executed makes a difference.

We now can list all the non-zero transition probabilities as follows:

$$p_{0,0} = 1 - p_A \tag{2.17}$$

$$p_{0,1} = p_A \tag{2.18}$$

$$p_{i,i-1} = p_D(1 - p_A), \ 0 < i < N \tag{2.19}$$

$$p_{i,i} = (1 - p_D)(1 - p_A), \ 0 < i < N \tag{2.20}$$

$$p_{i,i+1} = p_A(1 - p_D), \ 0 < i < N \tag{2.21}$$

$$p_{N,N-1} = \begin{cases} p_D & \text{arrival first} \\ p_D(1 - p_A) & \text{departure first} \end{cases} \tag{2.22}$$

$$p_{N,N} = \begin{cases} 1 - p_D & \text{arrival first} \\ 1 - p_D(\ - p_A) & \text{departure first.} \end{cases} \tag{2.23}$$

Let us now consider the case where the event function of arrivals is $f_X(X) = X + 1$, and there is the event condition $X < N$. The event condition prevents arrivals when $X = N$. Consequently, $p_{N,N-1} = p_D$, whereas $p_{N,N} = 1 - p_D$.

Note that the transition matrix of the $Geom/Geom/1$ queue is tridiagonal, like the birth–death process. This allows one to use the same solution procedures as in the birth–death process. One obtains from equations (2.17)–(2.23), with departures first:

$$\pi_1 = \frac{p_A}{p_D(1 - p_A)} \pi_0$$

$$\pi_{i+1} = \frac{p_A(1 - p_D)}{p_D(1 - p_A)} \pi_i, \ 1 \le i \le N.$$

If we set

$$\rho = \frac{p_A(1 - p_D)}{p_D(1 - p_A)},$$

then

$$\pi_1 = \frac{\rho}{1 - p_D} \pi_0, \quad \pi_i = \frac{-}{1 - p_D} \rho^i \pi_0, \ i = 1, 2, \ldots, N - 1.$$

π_0 can now be found by using the fact that $\sum_{i=0}^{N} \pi_i = 1$. For $N \to \infty$, one has

$$\sum_{i=1}^{\infty} \pi_i = \pi_0 \left(1 + \sum_{i=1}^{\infty} \frac{1}{1 - p_D} \rho^i \right) = \pi_0 \left(1 + \frac{\rho}{(1 - p_D)} \frac{1}{(1 - \rho)} \right).$$

When using the definition of ρ, this simplifies to

$$\sum_{i=0}^{\infty} \pi_i = \pi_0 \left(1 + \frac{p_A}{p_D - p_A} \right).$$

Consequently

$$\pi_0 = \left(1 + \frac{p_A}{p_D - p_A} \right)^{-1}.$$

With this, the problem is solved.

2.8.2 Compound Events and Their Probabilities

As shown above, the composition of event functions leads to *compound events*, which are defined as follows:

Definition 2.2 If several events occur at the same time, we combine all these events to form a new event, which we call *compound event*. Events that are not compound are called *atomic*. The atomic events forming a compound event are referred to as their *constituent events*.

Like atomic events, compound events have event functions, event conditions, and event probabilities. The event functions of compound events are the compositions of the event functions of the constituent events, and their event conditions are the conjunction of the individual event conditions. The probability of a compound event will be discussed below.

To solve binomial event systems, one first has to enumerate all compound events. Next, one has to find the probabilities of the compound events. To find these probabilities, one first has to identify the set of the atomic events that are enabled in the present state. Let $\mathcal{A}$ be this set. Also, let $\mathcal{B}$ be the set of atomic events that form the compound event in question. If Ev_h is the compound event corresponding to $\mathcal{B}$, its probability $p_h(X)$ becomes

$$p_h(X) = \prod_{k \in \mathcal{B}} p_k(x) \prod_{k \in \mathcal{A} - \mathcal{B}} (1 - p_k(X)). \tag{2.24}$$

Here, $\mathcal{A} - \mathcal{B}$ is the set of all active events that are not part of the compound event. For example, in the $Geom/Geom/1$ queue, we had two events, Arrival and Departure. In states where both events are active, $\mathcal{A} = \{\text{Arrival}, \text{Departure}\}$. The event "Arrival only" is a compound event, consisting of the single atomic event "Arrival", that is, $\mathcal{B} = \{\text{Arrival}\}$. Consequently,

$$\mathcal{A} - \mathcal{B} = \{\text{Arrival}, \text{Departure}\} - \{\text{Departure}\} = \{\text{Arrival}\}.$$

It follows

$$p_h = p_A(1 - p_D).$$

2.8.3 The Geometric Tandem Queue

Let us now look at the tandem queue in Table 2.12, except that we also have a bound on the size of the first line. The processes generating events are all binomial, with parameter p_A for the arrival process, p_M for the move process, and p_D for departures process.

Event type	Event function	Event condition	Probability
Arrival	$X_1 + 1, X_2$	$X_1 < N_1$	p_A
Move	$X_1 - 1, X_2 + 1$	$X_1 > 0, X_2 < N_2$	p_M
Dept.	$X_1, X_2 - 1$	$X_2 > 0$	p_D

Now we have three events, and we have to consider the case where only one event occurs, the case where two different events occur, and the case where all three events occur. For each of these compound events, we have to find their joint event functions. To reduce congestion, we will always execute departures first. All these events, and their combinations, are given in Table 2.13. To save space, the probabilities of these compound events are given in Table 2.14.

Table 2.13 Geometric tandem queue

Event type	Event function	Event condition
Arrival only	$\min(X_1 + 1, N_1), X_2$	
Move only	$X_1 - 1, X_2 + 1$	$X_1 > 0, X_2 < N_2$
Dept. only	$X_1, X_2 - 1$	$X_2 > 0$
Arrival & Move	$X_1, X_2 + 1$	$X_1 > 0, X_2 < N_2$
Arrival & Dept	$\min(X_1 + 1, N_1), X_2 - 1$	$X_2 > 0$
Move & Dept.	$X_1 - 1, X_2$	$X_1 > 0, X_2 < N_2, X_2 > 0$
All events	$X_1, X_2 - 1$	$X_1 > 0, X_2 > 0, X_2 < N_2$

We now find the probabilities for all these compound events. Clearly, if all events are enabled, then the probability of a set of events happening simultaneously is given by equation (2.24), with $\mathcal{A}$ representing the events that are enabled, and $\mathcal{B}$ the events that are disabled. The set of enabled events depends on the state, and so do the probabilities of the compound events. In particular, we have

Move	Depart		States
enabled	enabled		$X_1 > 0, 0 < X_2 < N_2$
enabled	disabled		$X_1 > 0, X_2 = 0$
disabled	enabled		$X_1 = 0$, or $X_2 = N_2$
disabled	disabled		$X_1 = 0, X_2 = 0$

Note that arrivals are always enabled, and their presence in the table above is implied. Each of the four combinations of enabled and disabled events has its own formula for obtaining the probabilities that the compound event in question occurs, as indicated in Table 2.14.

Table 2.14 Probabilities of geometric tandem queue

Event	Probabilities under the following conditions			
	$X_1 > 0,$ $0 < X_2 < N_2$	$X_1 > 0,$ $X_2 = 0$	$X_1 = 0$ or $X_2 = N_2$	$X_1 = 0,$ $X_2 = 0$
Arrival only	$p_A(1 - p_M)(1 - p_D)$	$p_A(1 - p_M)$	$p_A(1 - p_D)$	p_A
Move only	$(1 - p_A)p_M(1 - p_D)$	$(1 - p_A)p_M$		
Dept. only	$(1 - p_A)(1 - p_M)p_D$		$(1-p_A)(1-p_M)$	
Arrival & Move	$p_A p_M(1 - p_D)$	$p_A p_M$		
Arrival & Dept	$p_A(1 - p_M)p_D$		$p_A p_D$	
Move & Dept.	$(1 - p_A)p_M p_D$			
All events	$p_A p_M p_D$			

This information can be obtained from the event table given at the beginning of this section, provided departures are processed first. Once Tables 2.13 and 2.14 are completed, the transition matrix can be generated.

Problems

2.1 Consider a one-server queue where arrivals are in groups of 2, but each individual of the group is served on its own. The arrivals of the groups are Poisson with a rate of λ, and the service time is exponential with a rate of μ.

1. Formulate the event table.
2. Formulate the equilibrium equations.
3. Suppose $\lambda = 2$, $\mu = 5$, and no arrivals occur as soon as the number in the system reaches 6. Find the equilibrium probabilities.

2.2 There are N terminals served by a mainframe computer. Each terminal uses the mainframe at a rate of λ per hour for an average of $1/\mu$ hours. The times between usages are exponential, and so are the times a terminal uses the computer.

1. Formulate the event table.
2. Formulate the equilibrium equations
3. Calculate the equilibrium probabilities, with $N = 10$, $\lambda = 5$ and $\mu = 3$.

2.3 In a two-station tandem queue each station has two servers. Arrivals are Poisson with a rate of λ, and the service times of each server in station i are exponential with the parameters μ_i, $i = 1, 2$. The buffer sizes, including the elements being served, are N_i, $i = 1, 2$. Once the buffer size of station 2 is full, server 1 stops working.

1. Formulate the event table.
2. Formulate the transition matrix when $N_1 = 5$ and $N_2 = 3$.
3. Find the equilibrium probabilities by using either a spreadsheet or MATLAB. Use $\lambda = 0.8$ per hour, $\mu_1 = 1$ per hour, and $\mu_2 = 0.9$ per hour.
4. What is the equilibrium probability that an arrival occurs when the first buffer is full, and the arrival is lost? How many items leave the system per hour?

2.4 In a two-station tandem queue, both servers have the same service rate and the same buffer size. If you can change the rate of only one server, which service rate would you change? Give arguments.

2.5 Bring the event table into a computer-readable form, and write a program to generate the equilibrium equations. You may assume that the event functions are of the form $f_k(X) = X + a^{(k)}$, where $a^{(k)}$ is a vector of conforming dimension.

2.6 In a country there are two official languages, the majority language and the minority language. A telephone service has two types of server: servers who are either unilingual, that is, they only speak the majority language, and servers that bilingual, that is, they speak both the majority and the minority language. Calls from speaker of the majority language can be served by either server, but minority language speaker must be served by a bilingual server. Bilingual servers always serve the customer at the head of the minority language server as long there are customers in the minority language queue. When the majority language queue exceeds 3, and the minority language queue is empty, bilingual servers also accept majority language customers. The buffer size is N_1 unilingual customers, and N_2 bilingual customers.

1. Formulate the event table of this system. Use λ_i and μ_i, $i = 1, 2$ for the arrival rates of the two types of customers, and μ_i, $i = 1, 2$ for the two types of servers.
2. Formulate the equilibrium equations, assuming there is no upper bound on either waiting line.

Chapter 3
Generating the Transition Matrix

Clearly, the only practical way to generate the transition matrix of any discrete event systems with more than a hundred states is by using a computer program. This chapter therefore describes efficient algorithms for the matrix generation.

In order to create the transition matrix, we first have to enumerate all state vectors. Next, we have to give each state a state number, which can be used both as a row and a column number for the transition matrix. After that, the target states for each row must be determined, and these target states must be converted into state numbers in order to place the entries into the correct positions in the transition matrix.

To serve as row and column designations, the state numbers must be a one-to-one onto function between the state vectors and the set of integers between 1 and N, or between 0 and $N - 1$, where N is the number of states. Sometimes, it is also helpful to create numbers, we call them *codes*, which are one-to-one, but not onto, and which are helpful in finding the state numbers. In particular, we introduce what we call the *lexicographic code*.

In some cases, the lexicographic code can be used as state number, and this makes the generation of the transition matrix particularly easy and efficient. The resulting algorithms are then applied to find the transition matrices for a number of models. For models where this is not possible, we can still use the lexicographic code to find the state numbers corresponding to the different states as shown by a number of examples.

The outline of this chapter is as follows. We first define the lexicographic code, and we will then apply it to generate a number of transition matrices. In particular, we suggest to enumerate all lexicographic codes, and if this code can be used as a state number, we find the corresponding state vector, which then allows us to find all targets. These targets can then be encoded to find the column number. In this way, the transition matrix can be generated easily and efficiently. In other models, not all lexicographic codes correspond to states, but we can still enumerate all codes, and ignore the codes that do not correspond to any state. Aside from this enumeration method, we describe also other methods, including multiple loops and a method we call the *adder method*.

© The Author(s), under exclusive license to Springer Nature Switzerland AG 2022   57
W. Grassmann and J. Tavakoli, *Numerical Methods for Solving Discrete Event Systems*,
CMS/CAIMS Books in Mathematics 5, https://doi.org/10.1007/978-3-031-10082-6_3

We also introduce a basically different method to enumerate the states and generate the transition matrix: the *reachability method*. In this method, one starts with a single state, or a small set of states. The states that can be reached in one step from these states are then generated, which provides a new set of states. The states that can be reached from this new set are then enumerated and added to the states already reached before. This continues until all states are reached. Whenever a state is reached, the corresponding row of the transition matrix can be built.

3.1 The Lexicographic Code

We use the term *code* for a number that associates with each state vector a number that identifies it uniquely. The codes are not necessarily consecutive numbers. On the other hand, if there are N states, a *state number* is a number that associates with each state vector a number between 1 and N. This number must not be shared by any other state vector. In mathematical terms, a code is a one-to-one function, but a state number is one-to-one and onto. Instead of numbering the states from 1 to N, it is often convenient to number them from 0 to $N - 1$, or use any other sequence of N successive integers for numbering.

We now present a code, we call it *lexicographic code*, that can be used to arrange the states in lexicographic order. A *lexicographic order* is defined as follows:

Definition 3.1 The vectors $x^{(m)} = [x_1^{(m)}, x_2^{(m)}, \ldots, x_d^{(m)}]$ precedes vector $x^{(n)} = [x_1^{(n)}, x_2^{(n)}, \ldots, x_d^{(n)}]$ in lexicographic order if and only if there is a value k, $1 \leq k \leq d$, such that $x_j^{(m)} = x_j^{(n)}$, $j < k$, and $x_k^{(m)} < x_k^{(n)}$.

Suppose there are d state variables, X_i, $i = 1, 2, \ldots, d$, with X_i at least 0, and at most N_i for all i. In this case, we have the following theorem.

Theorem 3.1 *If*

$$c(X) = \sum_{i=1}^{d} u_i X_i, \tag{3.1}$$

with

$$u_d = 1, \ u_{i-1} = (N_i + 1)u_i, \ i = d, d - 1, \ldots, 1, \tag{3.2}$$

then $c(x^{(m)}) < c(x^{(n)})$ *if* $x^{(m)}$ *precedes* $x^{(n)}$ *in lexicographic order. Also,* $c(X)$ *is one-to-one onto from the set of vectors X to the integers from 0 to $u_0 - 1$, where X_i, $i = 1, 2, \ldots, d$ are integers between 0 and N_i, and u_0 is the total number of codes.*

Proof We first prove the following equation:

$$u_k = \left(\sum_{i=k+1}^{d} u_i N_i \right) + 1, \ k = 0, 1, \ldots, d. \tag{3.3}$$

This equation is true for $k = d$ since $u_d = 1$. Suppose (3.3) holds for k, and we now show that it holds for $k - 1$. Using (3.2) and then (3.3) we get

$$u_{k-1} = u_k(N_k + 1) = u_k N_k + u_k = u_k N_k + \sum_{i=k+1}^{d} u_i N_k + 1.$$

Consequently,

$$u_{k-1} = \sum_{i=k}^{d} u_i N_i,$$

which proves (3.3) for all $k \geq 0$.

To prove the theorem, we show that f $x_i^{(m)} = x_i^{(n)}$, $i < k$, and $x_k^{(n)} > x_k^{(m)}$, then $c(x^{(n)}) > c(x^{(m)})$, or $c(x^{(n)}) - c(x^{(m)}) > 0$. Since $x_i^{(n)} - x_i^{(m)} = 0$ when $i < k$, we have

$$c(x^{(n)}) - c(x^{(m)}) = u_k(x_k^{(n)} - x_k^{(m)}) + \sum_{i=k+1}^{d} u_i(x_i^{(n)} - x_i^{(m)}).$$

Since $x_k^{(n)} > x_k^{(m)}$, $x_k^{(n)} - x_k^{(m)}$ is at least 1. Consequently, $u_k(x_k^{(n)} - x_k^{(m)})$ is at least u_k. On the other hand, we find that for $i > k$, $x_i^{(n)} - x_i^{(m)} \geq -N_i$ and, by using (3.3), we get

$$\sum_{i=k+1}^{d} u_i(x_i^{(n)} - x_i^{(m)}) \geq - \sum_{i=k+1}^{d} u_i N_i = -u_k + 1.$$

Consequently,

$$c(x^{(n)}) - c(x^{(m)}) \geq u_k - u_k + 1 = 1.$$

This limit can be reached by setting, for $i > k$, $x_i^{(n)} = 0$ and $x_i^{(m)} = N_i$, that is, we have no gaps. This makes $c(X)$ one-to-one onto from the set of vectors X to the integers from 0 to $u_0 - 1$, where X_i, $i = 1, 2, \ldots, d$ are integers between 0 and N_i. $\square$

We now consider the case where the code c is given, and we wish to find the corresponding state vector $X = [X_1, X_2, \ldots, X_d]$. The following algorithm accomplishes this:

Algorithm 3. Decode Algorithm

```
decode(b)
rem = b
for k = 1 : d
    X(k) = rem/u(k), rounded down
    rem = rem - X(k)* u(k)
end for
```

In this algorithm, we use "rem" for the remainder, except initially, we set "rem" to b. The vector to be found is stored in the array "X".

To prove that the algorithm yields values X(i), which stand for X_i , let "rem(k)" be the remainder (rem) at the beginning of the for loop. We now prove by complete induction that rem(k) $= \sum_{i=k}^{d} X_i u_i$, $k = 1, 2, \ldots, d$.

1. To show that this is true for $k = 1$, note that rem(1) = b, and by assumption, b $= \sum_{i=1}^{d} X_i u_i$.
2. By the statement "rem = rem - X(k)* u(k)", we have, given rem = rem(k) $= \sum_{i=k}^{d} X_i u_i$:

$$\text{rem(k+1)} = \text{rem(k)} \ - X_k u_k = \sum_{i=k}^{d} X_i u_i - X_k u_k = \sum_{i=k+1}^{d} X_i u_i.$$

It follows that rem(k) $= \sum_{i=k}^{d} X_i u_i$ for all $k \leq d$.

Next, we prove that X(k) = rem/u(k), rounded down. Since rem(k) $= X_k u_k + \sum_{i=k+1}^{d} X_i u_i$, X_k comes out correctly provided that

$$0 \leq \sum_{i=k+1}^{d} u_i X_i < u_k,$$

and with u_k given by equation (3.3), this inequality obviously holds. This concludes the proof that the algorithm works correctly.

If all vectors $X = [X_i, 1 \leq i \leq d]$ that satisfy $0 \leq X_i \leq N_i$ are states, then one can use the lexicographic code as state number. In this case, we have, what we call a *Cartesian state space*. For non-Cartesian state spaces, the state number and the lexicographic code can be different. Still, we can use lexicographic codes to simplify the assignment of state numbers as will be shown later.

3.2 The Transition Matrix for Systems with Cartesian State Spaces

In systems with Cartesian state spaces, that is, in systems where all vectors of integers between 0 and N_i, $i = 1, 2, \ldots, d$, are states, the generation of a transition matrix is particularly easy and fast because in this case, the lexicographic code, as given by $c(X)$ as defined by equation (3.1), can also be used as state number. The idea of generating the transition matrix is now as follows: One enumerates all codes, and uses them as the row number for the transition matrix. One decodes the code to find X, and one applies all events that satisfy their event conditions. To clarify this, we convert this verbal description into an algorithm. This algorithm uses two functions. The function "encode(X)" calculates $c(X)$ for any X according to equation (3.1), and the function "decode(i)" (Algorithm 3) solves $c(X) = i$ for X. We also use the vector "Xplus" to represent the target state of an event. Here is the algorithm:

Algorithm 4.

```
for i = 0 : u(0) – 1
    X = decode(i)
    do event 1 if its condition holds:
        Xplus = f(1,X)
        j = encode(Xplus)
        q(i,j) = rate of event 1
    do event 2 if its condition holds:
        Xplus = f(2,X)
        j = encode(Xplus)
        q(i,j) = rate of event 2
        ⋮
    do event e if its condition holds:
        Xplus = f(e,X)
        j = encode(Xplus)
        q(i,j) = rate of event e
    q(i,i) = –sum of the rates of all events
end for
```

In many cases, there is no need to find and encode the vector Xplus. The reason is that typically, the event functions have the form $f_k(X) = X + a^{(k)}$. In this case,

$$c(f_k(X)) = c(X + c^{(k)}) = c(X) + c(a^{(k)}),$$

that is, the target of event k is $j = i + c(a^{(k)})$. In particular, if event k adds 1 to X_g, leaving the other state variables unchanged, then the lexicographic code of the target state is $j = i + u(g)$, and the corresponding entry in the transition matrix is the rate of event k in position i, i+u(g). The execution of event k can now be expressed as follows:

If event condition holds, $c(i, i+u(g))$ = rate of event k.

We will now present an example of how Algorithm 4 is applied. In a tandem queue, customers need service from three different servers, first from server 1, then from server 2, and finally from server 3. After that, customers leave the system. For those customers waiting for server i, $i = 1, 2, 3$, there is a finite waiting room N_i. Arrivals are lost if $X_1 = N_1$, where X_i is the number of customers in line i. Server 1 will start service only if the there are fewer than N_2 customers in line 2, and server 3 starts service only if there are fewer than N_3 customers in line 3. Arrivals have a rate of λ, and the service rate of server i is μ_i, $i = 1, 2, 3$. The system is summarized in Table 3.1.

Let us now apply this method to the tandem queue given by Table 3.1.

Table 3.1 A tandem queue

Event type	Event function	Event condition	Rate
Arrival	$X_1 + 1, X_2, X_3$	$X_1 < N_1$	λ
From line 1 to line 2	$X_1 - 1, X_2 + 1, X_3$	$X_1 > 0, X_2 < N_2$	μ_1
From line 2 to line 3	$X_1, X_2 - 1, X_3 + 1$	$X_2 > 0, X_3 < N_3$	μ_2
Departure	$X_1, X_2, X_3 - 1$	$X_3 > 0$	μ_3

$$u_0 = (N_1 + 1)(N_2 + 1)(N_3 + 1)$$
$$u_1 = (N_2 + 1)(N_3 + 1)$$
$$u_2 = N_3 + 1$$
$$u_3 = 1.$$

Also,

$$f_1(X_1, X_2, X_3) = X_1 + 1, X_2, X_3 \Rightarrow c(a^{(1)}) = c(1, 0, 0) = u_1$$
$$f_2(X_1, X_2, X_3) = X_1 - 1, X_2 + 1, X_3 \Rightarrow c(a^{(2)}) = c(-1, 1, 0) = -u_1 + u_2$$
$$f_3(X_1, X_2, X_3) = X_1, X_2 - 1, X_3 + 1 \Rightarrow c(a^{(3)}) = c(0, -1, 1) = -u_2 + u_3$$
$$f_4(X_1, X_2, X_3) = X_1, X_2, X_3 - 1 \Rightarrow c(a^{(4)}) = c(0, 0, -1) = -u_3.$$

The algorithm for generating the transition matrix is given below. For simplicity, we omit the determination of the diagonal entry.

Algorithm 5.

```
q(i, j)=0 for all i and j
for i = 0 : u(0) −1
    [X(1), X(2), X(3)] = decode(i)
    If X(1) < N(1)
        q(i, i+u(1)) = lambda
    end if
    If X(1)> 0 and X(2) < N(2)
        q(i, i−u(1)+u(2)) = mu(1)
    end if
    If X(2)> 0 and X(3) < N(3)
        q(i, i−u(2)+u(3) ) = mu(2)
    end if
    If X(3)> 0
        q(i, i−u(3)) = mu(3)
    end if
end for
```

Using this program, we can generate all q(i, j), i ≠ j extremely rapidly. Indeed, we believe that Algorithm 5 is faster than many of the more sophisticated methods, and, where applicable, it is preferable on account of its simplicity and efficiency. Once the transition matrix is created, the transient and equilibrium equations can be solved readily as shown in Section 7.5.

We should note that since the transition matrix is sparse, one may consider storing only the non-zero entries. In particular, one can use three arrays, one array for the row number, one array for the column number, and one array for the rates. If these array are called "rows", "cols", and "rates", then the nth non-zero entry would be "rows(n)", "cols(n)", and "rates(n)". Instead of "q(i, j) = rate of event k" one would then have the following statements, where "n" would have to set equal to 0 before any entry is inserted

n = n + 1
rows(n) = i
cols(n) = j
rates(n) = rate of event k.

We will use this form later for calculating transient solutions of the tandem queue above.

In Algorithms 5, some of the events are very similar, and we now show how they can be combined. Indeed, we now demonstrate how a tandem queue with an arbitrary number of servers can be modeled and its transition matrix can be generated. There are now d different servers, and customers are first served from server 1, then from server 2, and so on, until they leave the system once they have obtained service from the last server, which is server d. To formulate this system, let X_i, $i = 1, 2, \ldots, d$, be the number of elements in line i. For those customers waiting for server i, $i = 1, 2, \ldots, d$, there is a finite waiting room N_i. Except for server d, server i will start service only if there is room in the next line, that is, only if $X_{i+1} < N_{i-1}, i < d$. If $X_{i+1} = N_{i+1}$, the server does not start service. Arrivals have a rate of λ, and the service rate of server i is μ_i. Arrivals are lost if $X_1 = N_1$. Table 3.2 describes this model.

Table 3.2 A tandem queue

Event type	Event function	Event condition	Rate
Arrival	$X_1 + 1, X_2, \ldots, X_d$	$X_1 < N_1$	λ
From line i to $i + 1$	$X_1, X_2, \ldots, X_i - 1, X_{i+1} + 1, \ldots, X_d$	$X_i > 0, X_{i+1} < N_{i+1}$	μ_i
Departure	$X_1, X_2, \ldots, X_{d-1}, X_d - 1$	$X_d > 0$	μ_d

The algorithm to generate the transition matrix is now as follows:

Algorithm 6.

```
 input d
q(i, j)=0 for all i and j
for i = 0 : u(0) −1
      X = decode(i)
      If X(1) < N(1)
```

```
            q(i, i+u(1)) = lambda
            q(i, i) = q(i, i) – lambda
        end if
        for Line = 1 : d – 1
            If X(Line) > 0 and X(Line+1) < N(line+1)
                q(i, i – u(line) + u(line+1)) = mu(line)
                q(i, i) = q(i, i) – mu(line)
            end if
        end for
        If X(d)> 0
            q(i, i–u(d)) = mu(d)
            q(i, i) = q(i, i) – mu(d)
        end if
end for
```

Now, "d" can be input, that is, this algorithms work for tandem queues of any length.

3.3 The Lexicographic Code Used for Non-Cartesian State Spaces

Systems with immediate events typically have non-Cartesian state spaces, and the same is true if the state spaces is restricted by linear inequalities, such as $X_1 + X_2 + \ldots X_{d-1} \leq N$, where N is given. In such systems, one cannot use the lexicographic code as state number because some codes do not correspond to any state. Still, one can use the lexicographic code to make the generation of the transition matrix more efficient. In fact, we suggest the following procedure, where u(0) is the number of codes.

Algorithm 7.

```
for code = 0 : u(0) – 1
    X = decode(code)
    if X is a state
        find i, the state number of X
        for event = 1 : e
            if event condition holds
                Xplus = f(event,X)
                find j, the state number of Xplus
                q(i, j) = rate of event
            end if
        end for
    end if
end for
```

There are two open questions with this algorithm sketch:

1. How to find i, the state number of X.
2. How to find j, the state number of the target Xplus.

We propose creating two arrays, say "stNums", an array that stores the state number for each lexicographic code, and "Codes", an array that stores the lexicographic codes, one for each state number. Stated differently, stNums(c) is the state number corresponding to the code c, and Codes(s) is the code corresponding to state number s. If the code c does not correspond to any state, we set stNums(c) = −1. The following algorithm can be used to generate these two arrays.

Algorithm 8.

```
s=0
for code = 0 : u(0) − 1
    X = decode(code)
    If X is a state
        s = s + 1
        Codes(s) = code
        stNums(code) = s
    else stNums(code) = −1
    end if
end for
N = s
```

The number N provides the number of states. Note that the array "codes" starts with 1, but the array "stNums" starts with 0. Using these two arrays, it is easy to implement Algorithm 7. We show that by using the following example: There are two one-server queues that share a common buffer of size N, but they are independent otherwise. If the common space is full, arrivals are lost. We define X_1 to be the length of line 1, and X_2 to be the length of line 2. The event table of this system is given in Table 3.3.

Table 3.3 Parallel queue problem

Event type	Event function	Event condition	Rate
Arrival 1	$X_1 + 1, X_2$	$X_1 + X_2 < N$	λ_1
Arrival 2	$X_1, X_2 + 1$	$X_1 + X_2 < N$	λ_2
Dept. 1	$X_1 - 1, X_2$	$X_1 > 0$	μ_1
Dept. 2	$X_1, X_2 - 1$	$X_2 > 0$	μ_2

Since $X_1 \leq N$ and $X_2 \leq N$, we find that $u_2 = 1$, $u_1 = N + 1$, $u_0 = (N + 1)^2$. Algorithm 7 becomes

Algorithm 9.

```
s=0
for code = 0 : u(0) − 1
    X = decode(code)
    If X(1) + X(2) <= N
```

```
            s = s + 1
            Codes(s) = code
            stNums(code) = s
        else stNums(code) = −1
        end if
end for
```

Also, Algorithm 7, when applied to our example, looks as follows:

Algorithm 10.

```
for code = 0 : u(0) − 1
    X = decode(code)
    if X(1)+X(2) <= N
        i = stNums(code)
        if X(1) + X(2) < N
            Xplus = [X(1)+1, X(2)]
            jcode = encode(Xplus)
            j = stNums(jcode)
            q(i, j) = lambda(1)
            Xplus = [X(1), X(2)+1]
            jcode = encode(XPlus)
            j = stNums(jcode)
            q(i, j) = lambda(2)
        end if
        if X(1) > 0
            Xplus = [X(1) − 1, X(2)]
            jcode = encode(XPlus)
            j = stNums(jcode)
            q(i, j) = mu(1)
        end if
        if X(2) > 0
            Xplus = [X(1), X(2) − 1]
            jcode = encode(XPlus)
            j = stNums(jcode)
            q(i, j) = mu(2)
        end if
    end if
end for
```

With this, the program for generating the transition matrix for this example is complete.

As our second example, we consider a closed queueing network. There are a total of M entities in the network, and d stations. Entities move from station 1 to station 2, from station 2 to station 3, and so on, until they reach station d. Once they finish at station d, they return to station 1. Only one entity can be served at each station, and the service rate is μ_i for station i. As before, the server is blocked when

$X_{i+1} \geq N_{i+1}$. The event table for this system is given in Table 3.4. Such a network can be found in production, where the products are placed on palettes as they go through the production process. The number of palettes is typically kept constant. It is also applicable to model rotating parts in the airline industry. Therefore, parts stay in the aircraft until actual or expected failure, and after that, they go through some fixing stations until they are stored for further use. Interestingly without blocking, this system has an equilibrium solution that is relatively easy to obtain, as will be discussed in Section 11.3.3 of Chapter 11. There is obviously no blocking if all buffers are of size M.

Table 3.4 A closed tandem queue

Event type	Event function	Event condition	Rate
From line i to $i+1$	$X_1, X_2, \ldots, X_i - 1,$ $X_{i+1} + 1, \ldots, X_d$	$X_i > 0, X_{i+1} <$ N_{i+1}	μ_i
From line d to 1	$X_1 + 1, X_2, \ldots, X_{d-1}, X_d - 1$	$X_d > 0, X_1 < N_1$	μ_d

To generate the transition matrix, we first have to find the arrays "Codes" and "stNums". Since this is done practically the same way as before, this step is omitted. It is important to note, however, that these two arrays depend on the number in the system. This will be discussed in the next section, where we also provide a different algorithm for finding "Codes" and "stNums". Also note that there is a slight difference between our algorithms. Whereas before, the loop variable was the code, whereas now, it is the state number. This requires finding the code from the state number, which results in additional effort. On the other hand, the number of times the loop is repeated is reduced.

Algorithm 11. Transition Matrix of a Closed Network

```
input d
q(i,j) = 0 for all i and j
for i = 1 : N
    code = Codes(i)
    X = decode(code)
    q = 0
    for Line 1 : d
        NextLine = Line + 1
        if NextLine = d + 1 then NextLine = 1
        Xplus = X
        Xplus(Line) = X(Line) – 1
        Xplus(NextLine) = X(NextLine) + 1
        newcode = encode(Xplus)
        j = stNums(newcode)
        q(i, j) = mu(Line)
        q(i, i) = q(i, i) – mu(Line)
    end for
```

end for

This provides a general pattern for generating transition matrices. Once the matrices are generated, transient and equilibrium solutions can be found.

3.4 Dividing the State Space into Subspaces

Sometimes, it is advantageous to divide the state space into subspaces, and deal with each subspace separately. This device can be used to reduce the size of the array stNums. Moreover, in some cases, it is possible to calculate the state number of the target space by adding or subtracting a constant from the state number of the original state, just like in models with Cartesian subspaces. Thus, if there are m subspaces, the states of subspace 1 can be numbered from 1 to v_1, the states of subspace 2 from $v_1 + 1$ to v_2, and so on, ending with the state numbers of subspace m which has the state numbers from $v_{m-1} + 1$ to v_m. Care must be taken if an event creates a target of a different subspace. To demonstrate the entire matrix generation when using subspaces, consider the example given in Table 2.10 of Chapter 2, which we repeat here for convenience.

Table 3.5 A dynamic priority model

Event type	Event function	Event condition	Rate
D_1 arrives	$X_1 + 1, X_2, X_3$	$X_1 < N_1$	λ_1
D_2 arrives	$X_1, X_2 + 1, X_3$	$X_2 < N_2$	λ_2
Start job D_1	$X_1 - 1, X_2, 1$	$X_1 > 0, X_3 = 0, N_1 - X_1 < N_2 - X_2$	
Start job D_2	$X_1, X_2 - 1, 2$	$X_2 > 0, X_3 = 0, N_2 - X_2 \le N_1 - X_1$	
End job D_1	$X_1, X_2, 0$	$X_3 = 1$	μ_1
End job D_2	$X_1, X_2, 0$	$X_3 = 2$	μ_2

In this example, we have jobs from two sources, and two queues X_1 and X_2, which are limited to N_1, respectively, N_2. If $N_1 - X_1 < N_2 - X_2$, then line 1 is given priority, otherwise, priority is given to line 2. The queues do not include the element being served. We also need a state variable, say X_3, to indicate the state of the server. X_3 is 1 if the element being served comes from line 1, it is 2 if the element being served comes from line 2, and $X_3 = 0$ if the server is idle.

In this case, we can divide the state space into two subspaces, depending whether or not $X_3 = 0$. As can be seen from the event table, states with $X_3 = 0$ are unstable unless the system is empty. Consequently, the state space with $X_3 = 0$ consists of one state only, the state $[0, 0, 0]$. We thus set $v_1 = 1$, and start numbering the other states starting with $v_1 + 1 = 2$. The subspace with $X_3 \ne 0$, we call it the busy subspace, is then Cartesian, and the earlier procedures can be used with minor modifications. In the busy subspace, X_i, $i = 1, 2$, range from 0 to N_i, whereas X_3 is either 1 or 2. The fact that $X_3 \ge 1$ requires an adaption of our encoding mechanism: When using the function "encode", we use encode (X(1), X(2), X(3)–1). Similarly, given a code c, we can use $[X(1), X(2), X(3)] = $ decode(c), but subsequently, we must use the statement $X(3) = X(3)+1$.

We now find the transition rates, starting from the idle subspace, that is, the subspace consisting of state $[0, 0, 0]$ only, which has the state number 1, and constitutes row 1 of the transition matrix. In this state, only two events are possible: arrivals of D_1 and arrivals of D_2. Arrivals of D_1 immediately triggers start job D_1, leading to state $[0, 0, 1]$. Similarly, arrivals of D_2 immediately lead to state $[0, 0, 2]$.

Now let us look at the other states, which we encode as normal. In fact, $u_3 = 1$, $u_2 = 2$, $u_1 = 2(N_2 + 1)$, $u_0 = 2(N_2 + 1)(N_1 + 1)$. Using these values, encoding can now proceed in the normal way. In particular, the state $[0, 0, 1]$ has code 0, and state $[0, 0, 2]$ has code 1. However, since state 1 is the empty state, we add 2 to its code to find its state number. Hence, state $[0, 0, 1]$ is state number 2, and state $[0, 0, 2]$ is state number 3. For the remaining states, one can number the states from 3 to $N = 2(N_1 + 1)(N_2 + 1) + 1$. This enumeration forms the basis of the algorithm below. For simplicity, we omit the diagonal elements.

Algorithm 12. Creation of Transition Matrix of Table 3.5

```
q(i,j) = 0 for i,j = 1 : N
q(1,2) = lambda(1)
q(1,3) = lambda(2)
for c = 0 : u(0) - 1
    [X(1), X(2), X(3)] = decode(c)
    X(3) = X(3) - 1
    i = c + 2
    Arrivals
    If X(1) < N(1) then q(i, i+u(1)) = lambda(1)
    If X(2) < N(2) then q(i, i+u(2)) = lambda(2)
    Departures
    If X(1) = 0 and X(2) = 0 and X(3) > 0 then q(i, 1) = mu(X(3))
    else
        if X(2) = 0 or N(1) - X(1) < N(2) - X(2)
            j = i - u(1)
            if X(3) = 2 then j = j - 1
            q(i, j) = mu(X(3))
        end if
        if X(1) = 0 or N(2) - X(2) ≤ N(1) - X(1)
            j = i - u(2)
            if X(3) = 1 then j = j + 1
            q(i, j) = mu(X(3))
        end if
    end if
end for
```

Once the transition matrix has been generated, transient and equilibrium probabilities can be calculated without difficulty. This completes our discussion of dividing the state space.

3.5 Alternative Enumeration Methods

When creating the transition matrix, we somehow have to enumerate all states. In fact, the matrix generation methods we discussed so far used enumeration methods by enumerating all codes for the state. This section provides alternatives. The easiest way to enumerate all state vectors is to use d nested for loops, one for each state variable. This works fine if d is a constant, but not when d changes. Still, multiple loops are easily implemented, as the next example shows.

We have a system with three state variables, X_1, X_2, and X_3. The state space is restricted by the condition that $X_1 + X_2 + X_3 = M$, where M is given. The following algorithm enumerates all states and creates the arrays "Codes" and "stNums":

Algorithm 13.

```
stNums(i) = -1 for all i
n=0
for X(1)= 0 : M
    for X(2) = 0 : M- X(1)
        n = n+1
        code = encode([X(1), X(2)])
        stNums(code) = n
        Codes(n) = code
    end for
end for
```

Note that when we know X_1 and X_2, X_3 must be $M - X_1 - X_2$, that is, no loop for X_3 is required. Technically, X_3 is not a state variable.

Nested loops can also be used for more than two state variables, but they are inconvenient when the number of state variables is unknown. There are, however, algorithms that can handle arbitrary values of d. Before presenting one of them, let us first review the way we count. Essentially, counting means adding repeatedly 1 to a number. Thus, for counting from say, 08 to 11, we start with 08, add 1 to obtain 09. When adding 1 to 09, we use a *carry*, that is, we increase the first digit by 1, and we set the second digit to 0 to obtain 10. Adding 1 to 10 merely increase the last digit to obtain 11. The same trick can be used for counting vectors. Specifically, when we have vectors $[X_1, X_2, \ldots, X_d]$, with $0 \le X_i \le N_i$ for all i, then we first increase X_d, starting with $X_d = 0$, and adding 1 to X_d until we reach $X_d = N_d$. When adding one at this point we use a carry, that is, we increase X_{d-1} by 1, while setting $X_d = 0$. Next, we continue, adding 1 to X_d, until the next carry is needed. Of course, as this continues, X_{d-1} will eventually reach N_{d-1}, at which time we have a carry to

X_{d-2}. This continues until $X(1)$ reaches its upper bound $N(1)$, and all vectors are enumerated. The following algorithm implements this idea. The variable "carry" is true when a carry occurs. Also, "digit" is the subscript of the state variable to be changed.

Algorithm 14. The Adder Enumeration Method

```
X(i) = 0 for i = 1 : d
done = false
n = 1
do
    | Vector obtained |
    digit = d
    carry = true
    while carry
        if X(digit)< N(digit)
            X(digit)= X(digit) + 1
            carry = False
        else
            if digit >1
                carry = true
                X(digit) = 0
                digit = digit – 1
            else
                carry = false
                done = true
            end if
        end if
    end while
    n = n + 1
until done
```

In this algorithm, $X(\text{digit})$ is increased either because digit = d, or because of a carry. When a carry occurs, the variable digit is decreased. We will call this method the *adder* method, because it always adds one. The algorithm also finds the number of states. Indeed, once the algorithm is done, the variable n exceeds the number of states by 1.

As in the nested loops implementation for generating vectors, one can make the upper bound $N(i)$ of $X(i)$ a function of $X(k)$, $k < i$. In Algorithm 14, one merely has to recalculate N(digit) before using it. For instance, if $X(1) + X(2) + X(3) \leq M$, then $N(1) \leq M$, $N(2) \leq M - X(1)$ and $N(3) \leq M - X(1) - X(2)$. These values can readily be calculated inside Algorithm 14.

More complex restrictions can be implemented in a similar fashion, just the same way as one can implement them by using nested "for" loops. Indeed, writing an algorithm by using for loops first is helpful when one needs to adapt the adder method to deal with more complex linear restrictions. Also note that one can initialize

X(digit) with a lower bound L(digit) instead of 0. In this case, the statement "X(digit) = 0" becomes "X(digit) = L(digit)". L(digit) may depend on X(k), k < digit.

3.6 The Reachability Method

In this section, we will introduce a method that numbers the states in the order they are reached from some initial state or from some initial set of states. In the process of numbering the states, we also generate the transition matrix. We call resulting enumeration and matrix generating method *reachability method*. To implement the reachability method, we divide the set of all states that have been reached so far from some initial state into three subsets:

1. The "done set" is the set of states where all target states have been generated.
2. The "do now" state is the state whose target states are being generated now.
3. The "waiting set" is set of states that have been reached, but no target states are yet generated.

The states in the done set, the *done states*, are states that have been given state numbers and all their targets have been identified. These target states also have been given state numbers, which makes it possible to enter the corresponding rates into the transition matrix.

The "do now" state is the state that is currently processed. This involves finding all targets, and determining their state numbers. If a target has previously been reached, it has been given a state number which can now be used. If the state has not been reached, a new state number is created, and the state is added to the waiting set. The states in the waiting set have been reached, and they therefore have a state number, but no targets have yet been generated.

To number all states, and to create the transition matrix, we proceed as follows: Initially, the "done set" is empty, and there is no "do now" state. However, we create a waiting set, which typically consists of a single state, the starting state, and its state number is set to 1. If there are several possible starting states, they all must be given state numbers. Once the initial waiting set has been created in this fashion, the first state of the waiting set is removed from the waiting set and is made the "do now" state: All the targets of the "do now" state are generated, and a check is made to find whether or not they are part of the waiting set, in which case they have a state number. Otherwise, the target is given the next available state number, and it is added to the waiting set. Once this is done, the "do now" state is added to the "done" set. This completes the first iteration.

The *n*th iteration, $n > 1$ progresses along the same lines: The first state of the waiting set is removed from the waiting set and it is made the "do now" state and processed: All targets are generated, and it is checked whether or not they are part of the done set or of the waiting set. In either case, they have a state number. States that do not belong to any of these two sets have not been reached yet. They must therefore be given a state number and added to the waiting set. In either case, we have a state number for the target state, which allows us to enter the rate of the event in question into the transition matrix. The algorithm ends as soon as the waiting set is empty.

The entire algorithm is given below. In the algorithm, we use again the lexicographic code. As in our earlier algorithms, we use two arrays, called "Codes" and "stNums". Initially, all elements of the array "stNums" are set to -1 to indicate that no state corresponds to the code in question. Also, "Codes(n)" will eventually contain the code of the state with the number n. Of course, the code can only be determined once the state is reached.

In the algorithm, the done set, the waiting set, and the "do now" state are represented as follows. The "do now" state is given the state number "doNow". The done set includes all states with numbers below doNow, and the waiting set is formed by the states with state numbers above doNow. The process is initialized with state X0. After that, doNow is assigned values from 1 to N, where N is the number of states. As doNow creates its row in the transition matrix, new states are created as described above. The algorithm below illustrates this process. Note that e : -1 : 1 means the integers e, e-1, e-2, ..., 1, as described in Appendix A.

Algorithm 15. Reachability Algorithm to Find Transition Matrix

```
stNums(i) = −1 for all i
Provide initial state X0
c = encode(X0)
Codes(1) = c
stNums(c) = 1
top = 0
for doNow = 1 : N
    codeNow = Codes(doNow)
    stVector = decode(codeNow)
    for event k = e : -1 : 1
        if event condition holds for stVector
            TargetV = f(k, stVector)
            TargetC = encode(TargetV)
            if stNums(TargetC) = −1 (that is, state not yet reached)
                top = top + 1
                stNums(TargetC) = top
                Codes(top) = TargeC
                j = top
            else
                j = stNums(TargetC)
            end if
            q(doNow, j) = mu(k)
        end if
    end for
```

All done?

     if doNow $\geq$ top then break
end for

Note that we processed the events in reverse order. The reason moving from e to 1 is that we want to do departures before arrivals, and when formulating a system, one typically starts with arrivals. By processing departures first, the states closest to zero are numbered before the ones further away from 0. This reduces the band width, which reduces the time complexity of all direct methods to find equilibrium solutions.

Problems

3.1 Consider the queueing system described in Table 2.12, but introduce a bound of $N_1 = 4$ for X_1. Enumerate all states, first in lexicographical order, then by the reachability method.

3.2 Trace the adder method in the following cases:

1. $d = 3$, $X_i \leq 3$, for $i = 1, 2, 3$.
2. $X_1 + X_2 + X_3 + X_4 \leq 4$.

3.3 One can also enumerate all vectors $[X_1, X_2, \ldots, X_d]$ by recursion as follows: Suppose you have an algorithm for enumerating all vectors $[X_1, X_2, \ldots, X_k]$, where $0 \leq X_i \leq N_i$, $i = 1, 2, \ldots, k$. Create an algorithm that enumerates all vectors $[X_1, X_2, \ldots, X_{k+1}]$, where $0 \leq X_i \leq N_i$, $i = 1, 2, \ldots, k + 1$. Use this idea to create a recursive algorithm to enumerate all vectors. Hint: start with $k = 1$.

3.4 A company has N_1 machines of type A and N_2 machines of type B. These machines require checkups at a rate of λ_A and λ_B per day, respectively. The machines are managed by three workers. Worker 1 can only handle machines of type A, worker 2 only machines of type B, and worker 3 hand handle either type. The expected time required per check is exponential, with rate μ_A for machines of type A, and rate μ_B for machines of type B. Write a program to generate the transition matrix.

3.5 Modify Algorithm 14 to find and print all vectors satisfying $0 \leq X(i)$, $i = 1, 2, \ldots, d$, with $X(i)$, $i = 1, 2, \ldots, d$ integer and $\sum_{i=1}^{d} X(i) \leq M$.

3.6 A queueing system has the following setup: Arrivals to the system occur at a rate of λ. All arrival will first be checked at rate μ_0. After that, they move with probability p_1 to repair station 1 and with probability p_2 to repair station 2 to be worked on. The time for repair is μ_i for repair station i, $i = 1, 2$. After this, the parts leave the system. Write a program to generate the transition matrix.

Chapter 4
Systems with Events Created by Renewal Processes

A frequent assumption in discrete event systems is that the times between events of a certain type, such as the times between arrivals, are independent random variables. Technically, this means that the process generating the event type in question forms a renewal processes, with the events being the renewals. We call a discrete event system a *renewal event system* if the times between events are renewal processes. Here, we include renewal processes that can be suspended, but that can be restarted at a later time. For instance, departures from a waiting line can form a renewal process, but this process ends as soon waiting line is empty, to be resumed as soon as there is an arrival. Generally, whether the renewal process generating an event is interrupted and resumed later is determined by the event conditions.

In this chapter, we will start with a description of the renewal process. Of course, renewal event systems are affected by several event types, all with their own renewal process, and we show how they all act together to change the state of the system. Renewal processes are not Markovian per se, but they can be made Markovian by adding *supplementary state variables*. If the times between events are continuous random variables, then so are the supplementary state variables, and this makes the analysis rather complicated. To make our task easier, we discretize the distributions, in which case we obtain Markov chains discrete in state and time. Once the Markov chain is discrete in time, several events may happen simultaneously, and as shown in connection with the binomial event systems in Section 2.8, this complicates the generation of the transition matrix.

Renewal systems have an enormous number of states, and this restricts their applicability for modeling. When only equilibrium solutions are required, the number of states can be substantially reduced by only considering the times when the physical state changes. Readers only interested in equilibrium solutions of renewal event systems should therefore consult Section 10.2 for ways to reduce the computational complexity for solving the equilibrium equations in question.

The outline of this chapter is as follows. We will first discuss the renewal process, with emphasis on the time since the immediately preceding renewal, called the *age* and the times until the immediately following renewal, called the *remaining lifetime*. These times are important because either of one can be used as a supplementary vari-

W. Grassmann and J. Tavakoli, *Numerical Methods for Solving Discrete Event Systems*, CMS/CAIMS Books in Mathematics 5, https://doi.org/10.1007/978-3-031-10082-6_4

able. This is followed by the description of renewal event processes. This description must be converted into a transition matrix. As in Poisson event systems, we assume that any event changes the physical state in a predictable way. In contrast to Poisson event systems, we now have to consider the state space of the supplementary state variables, the *supplementary state space*, and the events can affect this supplementary state space in different ways. We discuss how this affects the enumeration of the states, and the generation of the transition matrix.

4.1 The Renewal Process

Renewal processes are processes where the times between events of a given type are independent random variables. Mathematically, they are defined as follows: Let T_n, $n \geq 1$ be a sequence of independent, identically distributed (iid) random variables, and let

$$S_n = \sum_{m=1}^{n} T_m.$$

Let $N(t)$ be the integer satisfying

$$S_{N(t)} \leq t < S_{N(t)+1}.$$

This relation defines $N(t)$ as the number of renewals at or before t. The process $\{N(t), \ t > 0\}$ is called a renewal process. Since the T_n are iid, we often omit the subscript n, and just write T.

An example of a renewal process is the following: there is a lamp with a new light bulb, and as soon as it burns out, it is replaced by a new one, and if this second one burns out, it is again replaced, and this continues. In this case, T_n is lifetime of the nth bulb, $n = 1, 2, \ldots$, and $N(t)$ is the number of bulbs that had to be replaced from time 0 to t, or, to use the terms of renewal theory, $N(t)$ is the number of *renewals*.

If the inter-arrival times are independent random variables, then the arrivals form, by definition, a renewal process. The same holds for any other event of type Ev_k. Specifically, $N(t)$ will be the number of occurrences of Ev_k up to and including t.

Of importance is the time since the immediately preceding event, that is, $t - S_{N(t)}$. This time is often called *backward renewal time*. In the case of the light bulb, the backward renewal time would be the age of the bulb presently in use. We will also use the word age in the context of the renewal processes of events, that is, the age of Ev_k at time t is the time that has elapsed since the occurrence of Ev_k immediately preceding t.

For each t, $S_{N(t)+1} - t$ is the time until the next renewal. In the case of light bulbs, this would be the remaining lifetime, a term that we will use in the context of the process generating events. Another word for the remaining lifetime is *forward recurrence time*. In renewal theory [15], the main interest is $N(t)$ the number of renewal until t, a topic also discussed in this section.

Both the age and the remaining lifetime describe Markov chains, that is, the distribution of the age at the time $t + h$ is completely determined by the age at time t, and a similar statement is true for the remaining lifetime. If T is a continuous random variable, then so are both the age and the remaining lifetime. This means that their Markov chains have a continuous state space. Of course, a continuous distribution $f_T(t)$ can be approximated by a discrete distribution: We divide the interval from 0 to t into small subintervals of length Δ, and associate the probability p_n with $\Delta f_T(n\Delta)$. If the discrete intervals are numbered starting with one, we essentially deal with random variables that can only assume the values 1, 2, ... Hence, let p_i be the probability that $T = i$. We will also assume that T is bounded, that is, there is a value $\bar{T}$ such that $T \leq \bar{T}$. The p_i fully determine the distributions of the age and the remaining lifetimes.

4.1.1 Remaining Lifetimes

We denote the remaining lifetime by $Y(t)$. If there is a renewal at t, $Y(t)$ changes from 0 to T, where T is a random variable representing the inter-renewal time. It turns out to be convenient to set $Y(t) = 0$ at this point, with $Y(t^+) = T$, and $Y(t + 1) = T - 1$. Here, t^+ denotes the time right after t.

Given the $p_i, i = 1, 2, \ldots, \bar{T}$, and using the fact that $Y(t+1) = T-1$ when $Y(t) = 0$, the transient equations for $\pi_i(t) = P\{Y(t) = i\}$ become

$$\pi_j(t + 1) = \pi_0(t)p_{j+1} + \pi_{j+1}(t), \ j < \bar{T} - 1$$

$$\pi_{\bar{T}-1}(t + 1) = \pi_0(t)p_{\bar{T}}.$$

In words: the remaining life time at $t + 1$ is j, if there is a renewal at time t, and the inter-renewal time T is $j + 1$, or if the remaining life time is $j + 1$ one time unit before. However, $Y(t + 1)$ can assume the value $\bar{T} - 1$ only if there was a renewal at t.

In equilibrium, the equations above become

$$\pi_j = \pi_0 p_{j+1} + \pi_{j+1}, \ 0 \leq j < \bar{T} - 1 \tag{4.1}$$

$$\pi_{\bar{T}-1} = \pi_0 p_{\bar{T}}. \tag{4.2}$$

We now find the equilibrium probabilities for these equations by solving for π_j. To do this, we take the sum of the equilibrium equations, starting at i. This yields

$$\sum_{j=i}^{\bar{T}-1} \pi_j = \sum_{j=i}^{\bar{T}-1} \pi_0 p_{j+1} + \sum_{j=i}^{\bar{T}-2} \pi_{j+1}.$$

Clearly, $\sum_{j=i}^{\bar{T}-1} \pi_j - \sum_{j=i}^{\bar{T}-2} \pi_{j+1} = \pi_i$. Consequently

$$\pi_i = \pi_0 \sum_{j=i}^{\bar{T}-1} p_{j+1}, \ i \geq 0. \tag{4.3}$$

Since $\sum_{j=1}^{\bar{T}} p_j = 1$, this equation also holds for $i = 0$. π_0 can now be determined by requiring that $\sum_{i=0}^{\bar{T}-1} \pi_i = 1$, that is,

$$\sum_{i=0}^{\bar{T}-1} \pi_i = 1 = \pi_0 \sum_{i=0}^{\bar{T}-1} \sum_{j=i+1}^{\bar{T}} p_j. \tag{4.4}$$

Now, as shown in Section 1.3.1,

$$\sum_{i=0}^{\bar{T}-1} \sum_{j=i+1}^{\bar{T}} p_j = \sum_{i=0}^{\bar{T}} i\pi_i = \mathrm{E}(T).$$

Equation (4.4) thus reduces to $\pi_0 \mathrm{E}(T) = 1$, implying $\pi_0 = 1/\mathrm{E}(T)$. In words, the probability of a renewal at t is the reciprocal of the expected time between renewals. Also, from equation (4.3), we conclude

$$\pi_i = \sum_{j=i+1}^{\bar{T}} p_j / \mathrm{E}(T). \tag{4.5}$$

With this, we have found the distribution of the remaining life time in equilibrium.

4.1.2 The Age Process

Like the remaining life process, the age process $A(t)$ forms a Markov chain. When there is a renewal at time t, then $A(t)$ is assigned the time since the previous renewal, and $A(t^+)$ is set to 0. For the age process, we do not need to limit the size of T, that is, $\bar{T}$ can be infinite.

We now formulate the transient equations of the Markov chain $A(t)$, with $\pi_i(t) = \mathrm{P}\{A(t) = i\}$. To do this, we need to know the probability of a renewal given the time since the previous renewal is i. Let r_i be this probability. We have, if T is the time between renewals:

$$r_i = \mathrm{P}\{T = i \mid T \geq i\} = \frac{\mathrm{P}\{T = i \text{ and } T \geq i\}}{\mathrm{P}\{T \geq i\}} = \frac{p_i}{\sum_{j=i}^{\infty} p_j}. \tag{4.6}$$

If $A(t) = i$, then there is either a renewal, and this happens with probability r_i, or there is none, and this happens with probability $1 - r_i$. If there is a renewal, then $A(t^+) = 0$, and $A(t + 1) = 1$, and if there is none, then $A(t + 1) = A(t) + 1$. The transient equations are therefore:

$$\pi_1(t + 1) = \sum_{i=1}^{\infty} \pi_i(t)r_i$$

$$\pi_i(t + 1) = \pi_{i-1}(t)(1 - r_{i-1}), \ i > 1.$$

This leads to the equilibrium equations

$$\pi_1 = \sum_{i=1}^{\infty} \pi_i r_i$$

$$\pi_i = \pi_{i-1}(1 - r_{i-1}), \ i > 1. \tag{4.7}$$

From (4.7) we conclude

$$\pi_2 = \pi_1(1 - r_1)$$
$$\pi_3 = \pi_2(1 - r_2) = \pi_{\ } (1 - r_1)(1 - r_2)$$
$$\pi_4 = \pi_3(1 - r_3) = \pi_{\ } (1 - r_1)(1 - r_2)(1 - r_3).$$

Continuing this way, one finds

$$\pi_i = \pi_1(1 - r_1)(1 - r_2)\ldots(1 - r_{i-1}), \ i = 1, 2, \ldots .$$

Now

$$1 - r_i = 1 - \frac{p_i}{\sum_{j=i}^{\infty} p_j} = \frac{\sum_{j=i+1}^{\infty} p_j}{\sum_{j=i}^{\infty} p_j}.$$

Using this relation, one finds, since $\sum_{j=1}^{\infty} p_j = 1$:

$$\pi_i = \frac{\sum_{j=2}^{\infty} p_j}{\sum_{j=1}^{\infty} \pi_j} \frac{\sum_{j=3}^{\infty} p_j}{\sum_{j=2}^{\infty} p_j} \cdots \frac{\sum_{j=i}^{\infty} p_j}{\sum_{j=i-1}^{\infty} p_j} \pi_1 = \pi_1 \sum_{j=i}^{\infty} p_j, \ i \geq 1.$$

To find π_1, we use $\sum_{i=1}^{\infty} \pi_i = 1$:

$$\sum_{i=1}^{\infty} \pi_i = \pi_1 \sum_{i=1}^{\infty} \sum_{j=i}^{\infty} p_j = \pi_1 E(T).$$

This implies $\pi_1 = 1/E(T)$ and

$$\pi_i = \sum_{j=i}^{\infty} p_j / E(T), \ i \geq 1. \tag{4.8}$$

Note that the age is 1 if and only if there was a renewal one time unit before, that is, $\pi_1 = \frac{1}{E(T)}$ is the probability of a renewal. Also, in a statistical equilibrium, the distribution of the age is related to the distribution of the lifetime, as one sees by comparing (4.5) with (4.8). One has

$$P\{A = i + 1\} = P\{Y = i\}.$$

4.1.3 The Number of Renewals

In the renewal process, the only physical variable is $N(t)$, the number of renewals. In addition to that, we use either the age or the remaining lifetime as supplementary variable. Consider first the remaining lifetime $Y(t)$ as supplementary variable, that is, we have to find the transient solution for the discrete-time process $(N(t), Y(t))$. Note that if $Y(t) > 0$, then there is no renewal at t, and $N(t + 1) = N(t)$, while $Y(t_1) = Y(t) - 1$. If $Y(t) = 0$, there is a renewal at time t, and then $N(t+1) = N(t)+1$, and $Y(t + 1) = i$ with probability p_{i+1}. It follows that the transient equations for $P\{N(t) = n, Y(t) = i\}$ become

$$P\{N(t + 1) = n, \ Y(t + 1) = i\} = P\{N(t) = n, \ Y(t) = i + 1\}$$
$$+ P\{N(t) = n - 1, \ Y(t) = 0\}p_{i+1}.$$

This equation allows one to find all $P\{N(t) = n, \ Y(t) = i\}$ for all n, i, and t. The distribution of $N(t)$ is then

$$P\{N(t) = n\} = \sum_i P\{N(t) = n, \ Y(t) = i\}.$$

Now we use $A(t)$, the age at time t, as the supplementary variable. If $N(t) = n$, and $A(t) = i$, then with probability $1 - r_i$, there is no renewal at t, implying $N(t + 1) = n$, $A(t + 1) = i + 1$. On the other hand, with probability r_i, there is a renewal at t, and then $N(t + 1) = N(t) + 1$ and $A(t + 1) = 1$. Consequently, the transient equations become

$$P\{N(t + 1) = n, \ A(t + 1) = j\} = P\{N(t) = n, \ A(t) = j - 1\}(1 - r_{j-1}), \ j > 1$$

$$P\{N(t + 1) = n, \ A(t + 1) = 1\} = \sum_{i=1}^{\infty} P\{N(t) = n - 1, \ A(t) = i\}r_i.$$

These equations allow one calculate, for all t, $P\{N(t) = n, \ A(t) = i\}$ for all n and i. The distribution of $N(t)$ can then be found by summing over all i.

4.2 Renewal Event Systems

In renewal event systems, there is a separate process for each type of event, that is, if there are e event types, there are e processes. These processes are essentially renewal processes, with the time between events of type Ev_k, $k = 1, 2, \ldots e$ being independent, identically distributed random variables $T^{(k)}$. Moreover, the processes generating the different event types are mutually independent.

The process generating Ev_k events differs from a standard renewal process in one respect: It can be interrupted after an event, to be restarted at a later time, and for this reason, we call such processes *interrupted renewal processes*. For instance, the departure process from a waiting line is interrupted as soon as all customers in line have been served and the server is idle.

In this section, we show how to formulate renewal event systems, and we discuss the dynamics of such systems, that is, we show how they develop over time. This requires that we know the state of the processes generating events, and to record this information, we use *supplementary state variables* Y_k, one for each event type Ev_k. By using these supplementary state variables, the process is made Markovian. The question then arises as to how to generate the transition matrix of the resulting Markov chain, a topic that is discussed in Section 4.3.

4.2.1 Description of Renewal Event Systems

Like in any discrete event system, the physical state is given by a set of physical state variables $X_1, X_2, \ldots, X_d$. In addition to that we need supplementary state variables, one for each event type, say Y_k for event type Ev_k, $k = 1, 2, \ldots, e$. The events change the physical state variables, that is, in state X, event Ev_k changes the state from X to $X^+ = f_k(X)$, where X^+ is the state immediately after event Ev_k has happened. For each event type Ev_k, we have a separate renewal process, with inter-renewal times $T^{(k)}$. We assume that $1 \leq T^{(k)} \leq \bar{T}_k$, where $\bar{T}_k$ is fixed for each k. The probabilities $P\{T^{(k)} = i\}$ are denoted by $p_i^{(k)}$, $i = 1, 2, \ldots, \bar{T}_k$, $k = 1, 2, \ldots, e$.

As before, we use event conditions $C_k(X)$, and event Ev_k cannot happen when $C_k(X)$ is not satisfied. Instead of saying "$C_k(X)$ is satisfied", we also say "$C_k(X)$ holds", or "$C_k(X)$ is true", and when $C_k(x)$ is not satisfied, we say "$C_k(X)$ is false" or "$C_k(X)$ is fails". Event conditions are implemented as follows. After each renewal of Ev_k, we check whether or not $C_k(X)$ is true. If $C_k(X)$ is false, then the Ev_k process is interrupted, and no events of type Ev_k are generated any more until the moment $C_k(X)$ becomes true again. When, at time t $C_k(X(t))$ changes from false to true, a new inter-event time $T^{(k)}$ is generated, and the Ev_k process will produce an event at time $t + T^{(k)}$, where $T^{(k)}$ has the same distribution as the other inter-renewal times. An event whose process is interrupted is said to be *disabled*, otherwise, the event is *enabled*.

As an example of the disabling and enabling processes, consider the departure process of a $GI/G/1$ queue. Clearly, no departures can occur when the length of the line X is zero. Consequently, the departure process continues until the moment when $X = 0$, and it becomes disabled at this point. The departure process is suspended, until the line is greater 0, at which time the process is restarted. Note that since only events can change the physical state X, only events can suspend or restart event processes.

Mostly, event processes are interrupted because there is no work. In particular, a server in a queueing system interrupts the departure process as soon as she becomes idle. In this way, one can say the event "depart" interrupts itself. It is also possible that other events interrupt the event process of Ev_k. For instance, if the server is a machine, then the event "server breaks down" would interrupt the normal departure process. A similar situation arises if a high priority customer interrupts the service time of a low priority customer. We do not discuss such interruptions, even though such interruptions are important and merit a detailed analysis.

From the discussion above, it follows that in order to formulate renewal event systems, we need, for each event Ev_k

1. An event function $f_k(X)$.
2. An event condition $C_k(X)$.
3. The distribution of the inter-renewal times.

The event tables of renewal event systems are thus very similar to the ones of Poisson event systems, except that instead of a rate, we have a distribution.

In order to explain the ideas above, we use some examples. The first example is the $GI/G/1/N$ queue as described in Table 4.1. This is a system with two events, arrivals (Ev_1) and departures (Ev_2) and a finite buffer of N. In this system, there is one physical state variable, the number in the system, and two supplementary variables, Y_1 for the arrival process, and Y_2 for the service process. These supplementary state variables can either be ages or remaining lifetime. The event table of this system looks very much like the one of the $M/M/1/N$ queue, except that the rates are replaced by the inter-event distributions.

Table 4.1 The $GI/G/1/N$ queue

Event type	Event function	Event condition	Distribution
Arrival (Ev_1)	$\min(X + 1, N)$		$p_i^{(1)}, i = 1, 2, \ldots, \bar{T}_1$
Departure (Ev_2)	$X - 1$	$X > 0$	$p_i^{(2)}, i = 1, 2, \ldots, \bar{T}_2$

Note that in contrast to the $M/M/1/N$ queue, there is no event condition for arrivals. Instead, the arrival event function is changed from $f_1(X) = X + 1$ to $f_1(X) = \min(X + 1, N)$. Using the function $\min(X + 1, N)$ means that arrivals are lost. However, when we use $f_1(X) = X + 1$ with the event condition $C_1(X) = (X < N)$, then the arrival process is delayed. To demonstrate the difference, assume that at time t, the number in the system is N, and an arrival takes place. If there is no event condition, but $f_1(X) = \min(X + 1, N)$, then the arrival is lost, and the next arrival will occur at $t + T^{(1)}$, where $T^{(1)}$ has the distributions given by $p_1^{(1)}, p_2^{(1)}, \ldots, p_{\bar{T}_k}^{(1)}$. If, on the other hand. $f_1(X) = X + 1$, and the event condition is $X < N$, then the arrival process interrupted at time t, and restarted at time $t + Y_2$, where Y_2 is the remaining service time. The next arrival then occurs at $t + Y_2 + T^{(1)}$ rather than at time $t + T^{(1)}$.

As the second example, consider the queue with dynamic priorities as described in Table 4.2, whose Poisson version is given in Table 3.5. Note that whereas in its Poisson version, we needed three physical variables, the renewal event system needs only two physical state variables. The reason is that in order to predict the future, one does not need to know the type of the job presently in service, as long as one knows the time of departure. Consequently, in contrast to the Poisson event system, we do not need the state variable X_3, indicating the job type in service. If one wishes, one can, however, use X_3 as an auxiliary variable. Not using X_3 as a state variable also removes the need for the immediate events, and we no longer have to record whether D_1 or D_2 has completed its service. To simplify the notation, we use the variable C_1, which is 1 as long as type D_1 jobs have priority, and $C_1 = 0$ when type D_2 jobs have priority. Remember that as long as there is less room left in the buffer for D_1 than in the one for D_2, then D_1 jobs are given priority. Clearly, if $C_1 = 1$, then after a departure, X_1 decreases by 1, while X_2 remains unchanged, and the service time distribution is given by $p_i^{(3)}$. On the other hand, when $C_1 = 0$, X_1 remains unchanged after a departure, while X_2 decreases by 1, and the service time distribution is given by $p_i^{(4)}$.

The careful reader may observe that service time distribution may change as the state of the system changes, and calling the resulting process a renewal process may be inadmissible. However, this does not affect the further treatment of our dynamic priority model.

Table 4.2 A dynamic priority model

Event type	Event function	Event condition	Distribution
D_1 arrives	$\min(X_1 + 1, N), X_2$		$p_i^{(1)}, i = 0, 1, \ldots, \bar{T}_1$
D_2 arrives	$X_1, \min(X_2 + 1, N)$		$p_i^{(2)}, i = 0, 1, \ldots, \bar{T}_2$
Departure	$X_1 - C_1, X_2 - (1 - C_1)$	$X_1 > 0$ or $X_2 > 0$	$C_1 = 1: p_i^{(3)}, i = 0, 1, \ldots, \bar{T}_3$
			$C_1 = 0: p_i^{(4)}, i = 0, 1, \ldots, \bar{T}_4$

Next, consider the tandem queue, which is given in Table 4.3. To save space, we have omitted the distributions. Again, there is no condition for arrivals. However, there is one for the event "From line i to $i + 1$". In this case, the service process is disabled, and it will be enabled again as soon as the event condition is met.

Table 4.3 A tandem queue

Event type	Event function	Event condition
Arrival	$X_1 + 1, X_2, \ldots, X_d$	
From line i to $i + 1$	$X_1, X_2, \ldots, X_i - 1, X_{i+1} + 1, \ldots, X_d$	$X_i > 0, X_{i+1} < N_{i-1}$
Departure	$X_1, X_2, \ldots, X_{d-1}, X_d - 1$	$X_d > 0$

4.2.2 The Dynamics of Renewal Event Systems

In addition to the physical state variables $X_1, X_2, \ldots, X_d$, we now need supplementary state variables $Y_1, Y_2, \ldots, Y_e$, one for each event type. Let $X(t) = [X_1(t), X_2(t), \ldots X_d(t)]$ and $Y(t) = [Y_1(t), Y_2(t), \ldots, Y_e(t)]$. To capture the dynamics of the system, we have to determine, for each $(X(t), Y(t))$, the distributions of $X(t+1)$ and $Y(t+1)$.

We always assume that $X(t)$ can change its value only when t is integer. Obviously, when t is integer, and there is an event at time t, then we have to make a choice as to which value we assign to $X(t)$. Here, we always use the value that $X(t)$ has before the change. Technically, we assume that our sample functions are *left continuous*, that is, we assign the value to the left of t to $X(t)$. The new value of $X(t)$ will be denoted by $X(t^+)$, where t^+ is the time right after t, the point where $X(t)$ changes. Hence, if the change is due to event Ev_k, then $X(t^+) = f_k(X(t))$. For simplicity, we use X^+ for $X(t^+)$, and X for $X(t)$, that is, $X^+ = f_k(X)$. Moreover, $Y(t)$ is the value of Y just before t and $Y(t^+)$ the value just after t. We also use Y for $Y(t)$ and Y^+ for $Y(t^+)$.

Any change of $X(t)$ affects the set of enabled events, and this has to be taken into account as well. All this will now be explored, first with ages used as supplementary variables, then the remaining lifetimes. In each case, we first discuss the ranges of the supplementary state variables and the size of the supplementary state space, which is followed by the discussion of the possible targets, and by the effects of disabling and enabling of events. Hence, if $\mathcal{A}$ is the set of enabled events, we have to discuss how $\mathcal{A}$ is changed by events.

4.2.2.1 The Age as Supplementary State Variable

First, consider the range of Y_k. When $Y_k(t) = i$, then $Y_k(t^+) = 0$ with probability $r_i^{(k)}$, and $Y_k(T^+)$ remains i with probability $1 - r_i^{(k)}$. If there is a renewal at t and Ev_k remains enabled, $Y_k(t+1) = 1$. It follows that as long as Ev_k is enabled, then $Y_k > 0$. The range of Y_k is therefore equal to the range of lifetimes, that is, $1 \le Y_k \le \bar{T}^{(k)}$. Since this holds for all $k \in \mathcal{A}$, the supplementary state space has a size of $\prod_{k \in \mathcal{A}} \bar{T}^{(k)}$.

Consider now the system while it is in state (X, Y) with the set of enabled states given by $\mathcal{A}$. To find the possible targets for this state, we need all subsets of $\mathcal{A}$. If $\mathcal{B}$ is such a subset, then the probability that all events in $\mathcal{B}$ will occur is given by

$$\prod_{k \in \mathcal{B}} r_i^{(k)} \prod_{k \in \mathcal{A} - \mathcal{B}} (1 - r_i^{(k)}).$$

To illustrate this formula, consider the tandem queue described in Table 4.3 with $d = 2$, that is, there are only two stations. We call the event "From 1 to 2" now "Move". Thus, the events are "Arrival", "Move", and "Departure", abbreviated by Arr, Move, and Dep, respectively. We select a specific state and calculate the probabilities for all possible subsets $\mathcal{B}$ for this state. Y_1, the age of the arrival process is 2; Y_2, the age of the Move process is 3; and Y_3, the age of the depart process is 2. In this case, all events are active, that is, $\mathcal{A} = \{\text{Arr, Move, Dep}\}$. This set has eight subsets $\mathcal{B}$, including the empty set. These sets, with their corresponding probabilities, are as follows:

Values for $\mathcal{B}$	Probability
{}	$(1 - r_2^{(1)})(1 - r_3^{(2)})(1 - r_2^{(3)})$
{Dep}	$(1 - r_2^{(1)})(1 - r_3^{(2)})r_2^{(3)}$
{Move}	$(1 - r_2^{(1)})r_3^{(2)}(1 - r_2^{(3)})$
{ Move, Dep}	$(1 - r_2^{(1)})r_3^{(2)}r_2^{(3)}$
{Arr}	$r_2^{(1)}(1 - r_3^{(2)})(1 - r_2^{(3)})$
{Arr, Dep}	$r_2^{(1)}(1 - r_3^{(2)})r_2^{(3)}$
{Arr, Move}	$r_2^{(1)}r_3^{(2)}(1 - r_2^{(3)})$
{Arr, Move, Dep}	$r_2^{(1)}r_3^{(2)}r_2^{(3)}$

Of course, if $X_1 = 0$, then event "Move" is disabled, and $\mathcal{A} = \{ \text{Arr, Dep}\}$, which has only four subsets: the empty set, { Dep}, { Arr} and { Arr, Dep}.

We now have to examine the target state when the events of the set $\mathcal{B}$ all occur. If $\mathcal{B}$ includes several events, then they all must be executed. In this case, the resulting target state depends on the order in which the events are executed. Here, we give priority to events that bring us closer to the empty state. In particular, we do "Dep" before "Move" and "Move" before "Arr". The joint effect of all events of $\mathcal{B}$ can be programmed as follows:

$X^+ = X$
for each k in $\mathcal{B}$
$\quad X^+ = f_k(X^+)$
end for

To show how this works, consider again the two-station tandem queue, and suppose $X_1 = 2$, $X_2 = 1$. Let $\mathcal{B} = \{\text{Arr, Move, Dep}\}$, and we first execute the event Dep, which results in state $X_1 = 2$, $X_2 = 0$, and the events Arr and Move are left to do. In the second iteration, we execute "Move" using $X^+ = [2, 0]$, which results in a new vector $X^+ = [1, 1]$. This is the starting state for the event "Arr", which results in state $X^+ = [2, 1]$. After this, we are done.

The change of X potentially causes $C_k(X(t))$ to change for some k. Generally, if $k \in \mathcal{B}$, then $Y_k(t + 1) = 1$ if $C_k(X^+)$ is true, and $Y_k(t + 1) = 0$ if $C_k(X(t^+))$ is false. If $k \notin \mathcal{B}$, then $Y_k(t + 1) = Y_k(t) + 1$. With this, we have determined all target states, together with their probabilities.

4.2.2.2 The Remaining Lifetime as Supplementary State Variable

If the remaining lifetime is used as supplementary variable, then the range for Y_k is given by $0 \le Y_k < \bar{T}^{(k)}$. It is true that $Y_k(t^+)$ can have a range from 1 to $\bar{T}^{(k)}$, but since

$Y_k(t + 1) = Y(t^+) - 1, 0 \le Y_k < \bar{T}^{(k)}$. It follows that the number of supplementary states is $\prod_{k \in \mathcal{A}} \bar{T}^{(k)}$, which is the same number one has when using ages. Since Y_k can reach 0 for event types that are enabled, we cannot use $Y_k = 0$ to indicate that Ev_k is disabled. Instead, we set $Y_k = -1$ when Ev_k is disabled.

When ages are used as supplementary variables, there are many possible sets $\mathcal{B}$. Now, however, $\mathcal{B}$ is unique, and it is given by the set of all k where $Y_k(t) = 0$, that is,

$$\mathcal{B} = \{k \mid Y_k(t) = 0\}.$$

To find X^+, one can proceed as in the age case. Once we have X^+, the physical state at time $t + 1$ is $X(t + 1) = X(t^+)$. Moreover, $X(t^+)$ may change some of the $C_n(X(t))$, $n = 1, 2, \ldots, e$. This affects the event processes that have to be renewed, that is, for some processes, $Y_n(t^+)$ has to be assigned new lifetimes $T^{(i)}$. If this occurs for Ev_n, we say Ev_n has to be *scheduled*. The scheduling of Ev_n only takes place when $C_n(X(t^+))$ is true and $Y_n(t)$ is zero or -1. If $\mathcal{H}$ is the set of events that must be scheduled, we thus have

$$\mathcal{H} = \{n \mid Y_n \le 0 \text{ and } C_n(X^+) \text{ is true}\}. \tag{4.9}$$

Since the lifetimes are random variables, the $Y_k(t^+)$ can assume many different values, and we have to find probabilities for each possible outcome. Since the probability that $T^{(n)} = i_n$ is $p_{i_n}^{(n)}$, the probability that for all $n \in \mathcal{H}$, $T^{(n)} = Y_n(t^+) = i_n$ becomes

$$P\{\forall_{n \in \mathcal{H}}(Y_n^+ = i_n)\} = \prod_{n \in \mathcal{H}} p_{i_n}^{(n)}. \tag{4.10}$$

To explain this formula, we use the tandem queue example and pick out some states for closer analysis. In state $X_1 = 1$, $X_2 = 1$, $Y_1 > 0$, $Y_2 = 0$, $Y_3 = 0$, two events occur: "Move" and "Dep", that is, $\mathcal{B} = \{Move, Dep\}$. However, after "Move", $X_1 = 0$, disabling the move process, and $\mathcal{H} = \{Dep\}$. It follows that there are $\bar{T}_3$ possible states. They are

$$X_1 = 0, \ X_2 = 1, \ Y_1^+ = Y_1, \ Y_2^+ = 0, \ Y_3^+ = i, \ i = 1, 2, \ldots, \bar{T}_3.$$

For given i, the probabilities of these states are equal to $p_i^{(3)}$.

Suppose now $X_1 = 2$, and then, "Move" is no longer disabled at t^+, and $\mathcal{H} = \{Move, Dep\}$. We thus have to find remaining lifetimes for both "Move" and "Dep", which yields a total of $\bar{T}_2\bar{T}_3$ alternatives. They are given by

$$X_1 = 0, \ X_2 = 1, \ Y_1^+ = Y_1, \ Y_2^+ = i, \ Y_3^+ = j, \ i = 1, 2, \ldots, \bar{T}_2, \ j = 1, 2, \ldots, \bar{T}_3.$$

The probabilities of these alternatives are $p_i^{(2)} p_j^{(3)}$. This example illustrates equation (4.10).

We now have found $X(t^+)$ and $Y(t^+)$ for the example, together with their probabilities. The values of $X(t + 1)$ and $Y(t + 1)$ can now be found as $X(t + 1) = X(t^+)$ and $Y_k(t + 1) = Y_k(t^+ - 1)$ for all k with $Y_k \ne -1$, and $Y_k(t + 1) = -1$ for $Y_k(t^+) = -1$.

4.3 Generating the Transition Matrix

As in Poisson event systems, we generate the transition matrix by enumerating all states, and for each state, we determine the different targets, together with their probabilities. The probabilities are then entered into the transition matrix at the place given by the target state. To simplify this process, we create two procedures: The procedure "Vec2Num" accepts a state vector, such as (X, Y), and returns the corresponding state number. On the other hand, the procedure "Num2Vec" accepts a state number and returns the corresponding state vector (X, Y). To construct these procedures, one can, for instance, make use of the lexicographic code and create the arrays "Codes" and "stNums", where "Codes(i)" is the lexicographic code of state number i, and "stNums(c)" is the state number of the lexicographic code c. To calculate the values of these arrays, one can use Algorithms 13 or 14. Once these arrays are found, "Vec2Num" becomes

function Vec2Num(X,Y)
c = encode(X,Y)
return stNums(c).

Similarly, "Num2Vec" can be implemented as

function Num2Vec(Num)
c = Codes(Num)
(X,Y) = decode(c)
return (X,Y).

There are, of course, other ways to implement these two functions. We also assume that we can find the number of states, which we denote by N. Indeed, Algorithms 13 and 14 also find N.

Using "N", "Vec2Num", and "Num2Vec", the structure of the procedure to find the transition matrix is as follows:

Algorithm 16.

for i = 1 : N
    (X,Y) = Num2Vec(i)
    Find the set of targets $\mathcal{H}$
    for each target in $\mathcal{H}$
        find Xplus, Ytplus1 and p of target
        j = Vec2Num(Xplus, Ytplus1)
        trMat(i, j) = p
    end for
end for

Of course, $Xplus = X(t^+)$, $Yplus = Y(t^+)$, and $Ytplus1 = Y(t + 1)$. Also, p is the probability of the target, and "trMat' is the transition matrix. It remains to flesh out the details of this algorithm. We do this first for the case where all Y_k are ages, then for the case where all Y_k are remaining lifetimes. Finally, we look at the case where some Y_k are ages, others are remaining lifetimes. First, however, we discuss problems involving the enumeration of all states.

4.3.1 The Enumeration of States in Renewal Event Systems

For algorithms that determine the equilibrium probabilities, the order in which the states are enumerated can make a significant difference. As a general rule, the aim is that the states close to the empty system should have lower state numbers than the states further away from the empty state. This, however, would necessitate some method to measure distances, which may not be available. Of course, in a queue, with X being the line length, a higher X should be considered further away from the empty state than a state with a low value of X. However, if there are different state variables, there is no clear cut method to measure distances. For instance, one could define the distance between two states i and j as the minimum number of events needed such that i is converted to j. However, with this measure, the distance from i to j may differ from the distance from j to i. Also, this definition ignores the probabilities of the different events. In discrete-time Markov chains, the ones we have here, distance could also be defined as the minimum time needed to go from i to j, but this definition leads to the same difficulties that we mentioned before. In conclusion, though the notion of a difference lacks mathematical rigor, it can be used as a guideline.

For the physical state variables, a higher value of X_i tends to mean that one is further away from the empty system, and we used this fact implicitly in our enumerations. However, this is not necessarily a good rule for supplementary variables. Clearly, long remaining times until the next departure indicate a greater distance from the empty state than a short remaining inter-departure times. However, this does not seem right for inter-arrival times, quite on the contrary. When the next arrival is far in the future, then we are closer to the empty state than when the next arrival is imminent. Hence, inter-arrival times must be treated differently than inter-departure times. If a lexicographic code is used, one should take this into account. Specifically, if Ev_1 is an arrival event and $\bar{T}^{(1)}$ is the maximum inter-arrival times, then $\bar{T}^{(t)} - Y_k$ rather than Y_k should be used when generating codes according to equation (3.1) of Chapter 3, which is the equation used for generating codes.

Event conditions are another source of difficulties when enumerating all states in renewal event systems. If Y_k represents an age, and Ev_k is enabled, then $1 \leq Y_k \leq \bar{T}_k$. When using the lexicographic code, we thus have to take into account that the lower bound is 1 and not 0 as for the physical state variables. Also, only supplementary state variables of enabled events should be used when calculating the code. If remaining lifetimes are used, then the lower bound of Y_k is zero, as is the case for physical state variables. Again, Y_k should only be used to calculate the lexicographic code if Ev_k is enabled.

4.3.2 Ages Used as Supplementary State Variables

If Y is an array of ages, then we first have to identify $\mathcal{A}$, the set of all enabled events, that is, all events with $Y_k > 0$. Next, we have to enumerate all subsets $\mathcal{B}$ of $\mathcal{A}$, each of which will generate a target. For each $\mathcal{B}$, we have to find its probability p. Also, we have to apply all event functions $f_k(X)$, $k \in \mathcal{B}$ to X in order to find X^+, and set all Y_k^+ to 0, $k \in \mathcal{B}$. Once this is done, we calculate $Y_k(t + 1)$, $k = 1, 2, \ldots, e$. If $C_k(X^+)$ is false, $Y_k(t + 1) = 0$. Otherwise, if $Y_k(t^+) \neq 0$, then $Y_k(t + 1) = Y_k(t) + 1$, and if $Y_k(t^+) = 0$, then $Y_k(t + 1) = 1$. This method is summarized by the following algorithm:

Algorithm 17.

$\mathcal{A} = \{k \mid Y_k > 0\}$
for all $\mathcal{B} \subseteq \mathcal{A}$
$\quad p = \prod_{n \in \mathcal{B}} r_{Y_n}^{(n)} \prod_{n \in (\mathcal{A} - \mathcal{B})} (1 - r_{Y_n}^{(n)})$
$\quad$ Xplus = X
$\quad$ for all k $\in \mathcal{B}$
$\quad\quad$ Xplus = f(k, Xplus)
$\quad\quad$ Yplus(k) = 0
$\quad$ end for
$\quad$ for k = 1 : e
$\quad\quad$ if C(k, Xplus)
$\quad\quad\quad$ if Yplus(k) = 0 then Ytplus1(k) = 1
$\quad\quad\quad$ else Ytplus1(k) = Y(k) + 1
$\quad\quad$ else Ytplus1 = 0
$\quad$ end for
end for

This algorithm does not indicate how to enumerate all subsets of $\mathcal{A}$, and how to enumerate all elements of $\mathcal{B}$. To remedy this, we suggest the following procedure. First, we represent set $\mathcal{A}$ by an array "Aset", where each entry in the set is the number of an event type. For instance, if there are four events Ev_i, $i = 1, 2, 3, 4$, and Ev_3 is disabled, then Aset = [1,2,4]. To find all subsets $\mathcal{B}$ of $\mathcal{A}$, we represent $\mathcal{B}$ by an array of the same size as the array "Aset", with Bset(i) = 1 if the event Aset(i) is in $\mathcal{B}$, and Bset(i) = 0 otherwise. To find all possible sets $\mathcal{B}$, we associate each array Bset with a binary number "altB", with Bset(i) being equal to the ith digit of altB. We now let "altB" range from 0 to $2^{n-1} - 1$, where n is the number of enabled events. To find Bset, we use decode(altB). These are the main ideas, and the details now follow. First, we find the array "Aset".

n=0
for k = 1 : e
$\quad$ If Y(k) = 0
$\quad\quad$ n = n + 1
$\quad\quad$ Aset(n) = k

```
        end if
end for
```

Once Aset, X and Y is found, the following algorithm finds p, Xplus and Ytplus1 for all $\mathcal{B} \subseteq \mathcal{A}$. Note that n is the size of Aset.

Algorithm 18.

```
for altB = 0 : 2^n -1
    Yplus = Y
    Xplus = X
    Bset = decode(altB)
    p = 1
    for indEv = 1 : n
        if Bset(indEv) = 1
            Event = Bset(indEv)
            Xplus = f(Event,Xplus)
            Yvalue = Y(Event)
            Yplus(Event) = 0
            p = p * r(Event, Yvalue)
        else
            p = p * (1 – r(Event, Yvalue))
        end if
    end for
    for k = 1 : e
        if C(k, Xplus) then Ytplus1(k) = Yplus + 1 else Ytplus1 = 0
    end for
    i = Vec2Num(X,Y)
    j = Vec2Num(Xplus, Ytplus1)
    trMat(i,j) = p
end for
```

We now implement this algorithm for the tandem queue with two stations. As before, "Arr" is Ev_1, "Move" is Ev_2, and "Dep" is Ev_3. We have

Algorithm 19.

```
for i = 1 : N
    (X,Y) = Num2Vec(i)
    n=0
    for k = 1 : e
        If Y(k) > 0
            n = n + 1
            Aset(n) = k
        end if
    end for
    for altB = 0 : 2^n – 1
        Yplus = Y
```

```
            Xplus = X
            Bset = decode(altB)
            p = 1
            for indEv = 1 : n
                if Bset(indEv) = 1
                    Event = Aset(indEv)
                    if Event = 3 then Xplus = Xplus + [0, −1]
                    if Event = 2 then Xplus = Xplus + [−1, 1]
                    if Event = 1 then Xplus(1) = min(N(1), Xplus(1) + 1)
                    Yvalue = Y(Event)
                    Yplus(Event) = 0
                    p = p * r(Event, Yvalue)
                else
                    p = p * (1 − r(Event, Yvalue))
                end if
            end for
            if X(1) > 0 and X(2) < N(2)
                Ytplus1(2) = Yplus(2) +1 else Ytplus1(2) = 0
            end if
            if X(2) > 0 then Ytplus1(3) = Yplus(3) + 1 else Ytplus1(3) = 0
            j = Vec2Num(Xplus, Ytplus1)
            trMat(i, j) = p
        end for
end for
```

The use of ages is particularly advantageous for distributions that have exponential tails. If $T^{(k)}$ is such a distribution, then from a certain point onward, say from $i > i_0$, $r_i^{(k)}$ is independent of the age i, that is, for $i_0 < i < j$, $r_i^{(k)} = r_j^{(k)}$. We can thus use $Y_k = i_0 + 1$ to indicate that the age is greater i_0. For an application of this idea, see [42].

4.3.3 Remaining Lifetimes used as Supplementary State Variables

We start this section with an algorithm that shows how to generate the transition matrix when using remaining lifetimes. The algorithm is followed by an explanation of each step. To facilitate this, we provide line numbers. In contrast to the earlier algorithms, we extensively use mathematical notation.

Algorithm 20.

1. for all $(X, Y) \in \mathcal{S}$
2. $i = \text{stNums}(X, Y)$
3. $X^+ = X$
4.   for $k = e{:}{-}1{:}1$

5.   if $Y_k = 0$ then $X^+ = f_k(X^+)$
6.   end for
7.   n=0
8.   for k = 1:e
9.   if $C_k(X^+), Y_k \leq 0$
10.   n = n + 1
11.   inB(n) = k
12.   end if
13.   end for
14.   m = n
15. $Y^+ = Y$
16.   for all $z = [z_n, \ 1 \leq n \leq m, \ 1 \leq z_n \leq \bar{T}_{inB(n)}]$
17.   p = 1
18.   for n = 1 : m
19.   >   k = inB(n)
20. $Y_k^+ = z_n$
21. $p = pp_{z_n}^{(k)}$
22.   end for
23.   for $n = 1 : e$
24. $Y_n^* = Y_n^+ - 1$
25.   if not $C_k(X^+)$ then $Y_n^* = -1$
26.   end for
27.   j = stNums(X^+, Y^*)
28.   p(i,j) = p
29.   end for
30. end for

The "for" loop in line 1 enumerates all states. Consequently, inside this loop, we have states $(X, Y) = ([X_1, X_2, \ldots, X_d], [Y_1, Y_2, \ldots, Y_d])$. In line 2, we find the state number for the present state, which requires the array "stNums" as explained earlier. We assume that this array exists. Lines 3–6 find the new value of X, denoted by X^+. Note that X changes only when $Y_k = 0$.

The next few lines find Y^+, the value of Y right after all events have taken place. First, let $\mathcal{B}$ be the set of all events that have to be scheduled. Only events that satisfy the event condition are included in $\mathcal{B}$, and of those only the ones that occur ($Y_k = 0$), or that they were disabled before ($Y_k = -1$). The array "inB" lists all events that are part of $\mathcal{B}$. Here, "m" is the number of elements in $\mathcal{B}$. For instance, if events Ev$_1$ and Ev$_3$ are in $\mathcal{B}$, the array "inB" has a dimension of m = 2 with inB(1) = 1 and inB(2) = 3.

The "for" loop in line 16 generates all possible values of a vector z, which contains the lifetimes of the events that are part of $\mathcal{B}$ and therefore must be scheduled. For instance, if inB = [1, 3], meaning events Ev$_1$ and Ev$_3$ must be assigned new values, then one possible vector z is [3, 2], that is, $Y_1^+ = 3$ and $Y_3^+ = 2$. This is accomplished in line 20. Line 21 finds the probability of this vector, which is $p = p_3^{(1)} P_2^{(3)}$. Note that p is initialized in line 17. Lines 23–26 find $Y_k^* = Y_k(t + 1)$. The remainder of this program then creates the transition matrix.

To enumerate all vectors $z = [z_1, z_2, \ldots, z_m]$, one needs an enumeration procedure, such as the code/decode method used in Algorithm 4 for enumeration of all members of a Cartesian product, or the adder method as given by Algorithm 14. With this, our discussion of how to find the transition matrix when using remaining lifetimes is complete.

4.3.4 Using both Age and Remaining Life as Supplementary State Variables

It is possible that some supplementary state variables are ages, and others are remaining lifetimes, but typically, the complications this causes make this method unattractive. There may be cases where it is advantageous to mix age and remaining lifetimes, but this should never be the first choice. Still, this method has been used in literature [2] in the discrete $GI/G/1$ queue, where there are only two events, namely, arrivals and departures. However, whenever there are more than two events, the algorithms for generating the transition matrix are rather involved and are beyond the scope of this book.

Problems

4.1 There are two independent renewal processes, say process 1 and process 2. Let $N_i(t)$, $i = 1, 2$ be the number of renewals of process i, $i = 1, 2$. Show that the process $N_1(t) + N_2(t)$ is typically not a renewal process.

4.2 In a renewal process, the time between renewals consists of two phases: a constant phase of d units and a geometric phase with distribution $p_i = (1 - p)p^i$, $i = 0, 1, \ldots$, where p is a parameter. Give a formula for the equilibrium distributions of the age and the remaining lifetime.

4.3 In a queueing system, arrivals come from two streams, with the inter-arrival time distributions given by $p_i^{(k)}$, $k = 1, 2$, $i = 1, 2, \ldots, h$. There is one server, with a service time distribution of $p_i^{(3)}$, $i = 1, 2, \ldots, g$. The maximum number of elements in stream 1 is 5, and in stream 2, it is 3.

1. Define the state space.
2. Write a program to generate the transition matrix.

4.4 Consider a $GI/G/1/4$ queue with the following inter-arrival time distribution, where p_i represents the probability that the time between two arrivals is i.

$$p_1 = 0.2, \ p_2 = 0.2, \ p_3 = 0.3, \ p_4 = 0.2, \ p_5 = 0.1.$$

The service time is always 3. Write a program to generate the transition matrix in lexicographical order, but define your state variables such that states needing fewer steps to reach the empty state come first.

4.5 Write a program to generate the transition matrix for a $GI/G/2/N$ queue. Here, the input consists of the inter-arrival time distribution, the service time distribution, and N. Both inter-arrival and service time distributions must be bounded, where the bounds are input.

Chapter 5
Systems with Events Created by Phase-type Processes

In the renewal processes discussed so far, we have used explicit distributions, given by p_i, the probability that times between successive events is i. In *phase-type systems*, on the other hand, the renewal processes have so-called phase-type inter-renewal times, that is, the inter-renewal time has phase-type or PH distributions. This overcomes a big disadvantage of explicit distributions: Unless the range of values they can assume is small, such distributions greatly increase the state space, making it impossible to deal with systems that contain many events. Phase-type (PH) distributions, on the other hand, can range over many values, without the need to store a probability for each value the distribution can assume. Phase-type distributions are either continuous or discrete. We will mainly use continuous phase-type distributions, which have the added advantage that unless an event is triggered by some other event, no two events can occur at the same time t. This greatly reduces the density of the transition matrices.

PH distributions consist of several segments. In continuous PH distributions, the segments last for an exponentially distributed time. In discrete-time PH distributions, the length of the segments is either constant, or it follows a geometric distribution. In continuous PH distributions, the times between phases are always exponential. The first PH distribution, the so-called Erlang distributions, was created by Erlang [19]. Later, Morse [70] suggested another PH distribution, the hyper-exponential distributions. The general theory PH distributions is due to Neuts [71].

PH distributions are particularly advantageous if one only has the mean and the variance of the inter-event times. Consequently, we will discuss how to set the parameters of the distributions considered such that a particular mean and variance can be obtained. Continuous PH distributions cannot exactly represent distributions of random variables with upper bounds or distributions with discontinuities. Approximations are available, though at the cost of an increased number of phases, and the ensuing additional execution times.

In this chapter, we will concentrate on relatively simple phase-type distributions, namely, phase-type distributions based on *sums* and/or *mixtures*. This provides enough flexibility to cover most needs. After the discussion of the PH distributions,

© The Author(s), under exclusive license to Springer Nature Switzerland AG 2022
W. Grassmann and J. Tavakoli, *Numerical Methods for Solving Discrete Event Systems*,
CMS/CAIMS Books in Mathematics 5, https://doi.org/10.1007/978-3-031-10082-6_5

we will use them to formulate PH event systems, that is, discrete event systems with PH distributions.

5.1 Phase-type (PH) Distributions

Technically, a discrete PH distribution is characterized by a substochastic matrix $\mathbf{T} = [t_{i,j}]$, $i, j = 1, 2, \ldots, N$ and an initial vector $\alpha = [\alpha_i, i = 1, 2, \ldots, N]$. Similarly, a continuous phase-type distribution is characterized by an infinitesimal subgenerator $\mathbf{T} = [t_{i,j}]$, $i, j = 1, 2, \ldots, N$ and an initial vector $\alpha = [\alpha_i, i = 1, 2, \ldots, N]$. It is often convenient to use the transition matrix $\mathbf{T}^+$ instead of $\mathbf{T}$, where $\mathbf{T}^+$ is obtained by adding an absorbing phase, say phase 0 to the phases $1, 2, \ldots, N$. This means that $\mathbf{T}^+$ is obtained by adding a row at the top of $\mathbf{T}$ and a column to the left. For continuous PH distributions, this means that all entries of the top row are 0. The same is true for discrete PH distribution, except that $t_{0,0} = 1$. In discrete PH distribution, $t_{i,0} = 1 - \sum_{j=1}^{N} t_{i,j}$, whereas in continuous PH distribution, $t_{i,0} = -\sum_{j=1}^{N} t_{i,j}$. We also use the vector α^+, which is obtained from α by adding α_0 to the left. The phase-type distribution denoted by $\text{PH}(\alpha, \mathbf{T})$ is then the distribution of the time until absorption of the Markov chain with the initial state α^+ and transition matrix $\mathbf{T}^+$. We assume that in the matrix $\mathbf{T}^+$, there is a path from each phase to the absorbing phase.

Note that for the renewal processes considered here, we exclude the case that the inter-renewal time is 0. This means that α_0 must always be 0. Therefore, we no longer need α_0 in the sense defined above, which allows us to use α_0 for other purposes.

Instead of discussing the PH distributions in all their generality, we concentrate on two methods to create phase-type distributions, namely, by using sums and/or by using mixtures. We mainly consider sums and mixtures of exponential random variables. Sums are used to reduce variances, whereas mixtures are typically used to increase variances. Remember that the expectation of the exponential distribution with parameter λ is $1/\lambda$, as is its standard deviation, see Section 1.3.3. The coefficient of variation (CV) is therefore 1.

There are closed-form formulas for the expectation and variance of sums of exponential distributions. For matrices $\mathbf{T}$ that cannot be interpreted as sums or mixtures of random variables, expressions for the means and variances require matrix inversions, a topic that will be covered in Section 9.2, a section that deals with absorbing Markov chains.

The same distribution can have several PH representations, that is, more than one matrix $\mathbf{T}$ and α lead to the same distribution. The best way to prove this is by using eigenvalues, which will be discussed in Section 8.2.8.

5.1.1 Phase-type Distributions based on Sums, and the Erlang Distribution

If Z is the sum of n independent random variables $X_i, i = 1, 2, \ldots, n$, then we have according to Section 1.3.2:

$$E(Z) = E\left(\sum_{i=1}^{n} X_i\right) = \sum_{i=1}^{n} E(X_i) \tag{5.1}$$

$$\text{Var}(Z) = \text{Var}\left(\sum_{i=1}^{n} X_i\right) = \sum_{i=1}^{n} \text{Var}(X_i). \tag{5.2}$$

If the X_i all have the same distribution, it follows that

$$E(Z) = nE(X) \tag{5.3}$$

$$\text{Var}(Z) = n\text{Var}(X) \tag{5.4}$$

$$\text{Std}(Z) = \sqrt{n}\,\text{Std}(X)$$

$$CV = \frac{1}{\sqrt{n}} \frac{\text{Std}(X)}{E(X)}. \tag{5.5}$$

Also, according to the central limit theorem, the distribution approaches the normal distribution as n increases.

The Erlang distribution is the sum of k independent exponential random variables, which all have the same rate λ. We will call k the shape parameter, and λ the rate parameter. An Erlang distribution with a shape parameter k will be called an Erlang-k distribution, abbreviated E_k distribution. If X follows an E_k distribution, we have

$$f_X(x) = F_X'(x) = \lambda e^{-\lambda x} \frac{(\lambda x)^{k-1}}{(k-1)!}, \quad x \geq 0 \tag{5.6}$$

$$E(X) = \frac{k}{\lambda} \tag{5.7}$$

$$\text{Var}(X) = \frac{k}{\lambda^2}. \tag{5.8}$$

Equations (5.7) and (5.8) follow directly from (5.3) and (5.4). To prove (5.6), note that the renewal process generated by the exponential distribution is the Poisson process. The probability $f_X(x)dx$ is now found as the probability that one has exactly $k - 1$ Poisson events until x and the kth event occurs between x and $x + dx$. Alternatively, one can find the complementary cumulative Erlang distribution $F_X^C(x)$ first and take the derivatives of $F_X(x) = 1 - F_X^c(x)$. Observe that $X > x$ if there are fewer than k Poisson events until t. Consequently

$$F_X^c(x) = \sum_{n=0}^{k-1} p(n; \lambda x).$$

If $\mu_X = E(X)$, and $\sigma_X = \text{Std}(X)$, one finds

$$\lambda = \mu_X/k, \ k = \frac{\mu_X^2}{\sigma_X^2}.$$

Strictly speaking, this works only if k turns out to be integer. However, as shown in Section 5.1.2, this problem can be overcome by using mixtures.

Note that the CV is $\frac{1}{\sqrt{k}}$. This means that the variation of the Erlang distribution is smaller than the one of the exponential distributions. Indeed, it is known that the least variable PH distribution with k phases is the Erlang-k distribution [1]. The use of the Erlang distribution goes back to Erlang [19] who in the years from 1919 to 1920 published a number of seminal results on telephone traffic.

To represent the Erlang distribution as a PH distribution, let

$$\alpha = [0, \ldots, 0, 1]$$

$$\mathbf{T} = \begin{bmatrix} -\lambda & 0 & \ldots & 0 & 0 \\ \lambda & -\lambda & \ddots & \ddots & \ddots \\ \vdots & \ddots & \ddots & \ddots & \vdots \\ 0 & \ddots & \ddots & -\lambda & 0 \\ 0 & \ddots & \ddots & \lambda & -\lambda \end{bmatrix}.$$

Note that instead of starting with k, and going down, we could have made k the absorbing state, and, starting from 0, going up to k. If this had been done, an upper triangular rather than a lower triangular matrix would have resulted.

The Erlang distribution can be generalized as follows. Instead of having sums of identical exponential distributions, one can also use sums of exponential distributions with different parameters $\lambda_i, i = 1, 2, \ldots, k$. The mean and variance then become

$$E(X) = \sum_{i=1}^{k} \frac{1}{\lambda_i}, \ \text{Var}(X) = \sum_{i=1}^{k} \frac{1}{\lambda_i^2}.$$

In the matrix $\mathbf{T}$, one only has to change λ to λ_i. The resulting distribution is called *generalized Erlang distribution*.

5.1.2 Phase-type Distributions Based on Mixtures, and the Hyper-exponential Distribution

Formally, a continuous distribution $f_X(x)$ is a *mixture* of the distributions $f_i(x)$, $i = 1, 2, \ldots, k$ if its density satisfies

$$f_X(x) = \sum_{i=1}^{k} p_i f_i(x).$$

Here, p_i is the probability that distribution $f_i(x)$ is chosen.

If $E(X \mid i)$ is the expectation corresponding to $f_i(x)$, and $\text{Var}(X \mid i)$ is its variance, then the expectation and the variance of their mixture are as follows:

$$E(X) = \sum_{i=1}^{k} p_i E(X \mid i) \tag{5.9}$$

$$\text{Var}(X) = \sum_{i=1}^{k} p_i \text{Var}(X \mid i) + \sum_{i=1}^{k} p_i (E(X \mid i) - E(X))^2. \tag{5.10}$$

In words, the expectation of X is the weighted sum of expectations, and the variance of X is the weighted sum of variances, plus the variance of the expectations. If the expectations of the individual distributions vary greatly, the variance of expectations would be very high and dominate the overall variance.

The proof of (5.9) for continuous PH distributions follows from the fact that if $f_X(x) = \sum_i p_i f_i(x)$, then $x f_X(x) = \sum_i x p_i f_i(x)$, hence

$$E(X) = \int_{-\infty}^{\infty} x f_X(x) dx = \int_{-\infty}^{\infty} x \sum_{i=1}^{k} p_i f_i(x) dx$$

$$= \sum_{i=1}^{k} p_i \int_{-\infty}^{\infty} x f_i(x) dx = \sum_{i=1}^{k} p_i E(X \mid i).$$

This proves equation (5.9).

We now prove (5.10). Note that equation (5.9) remains valid if x is replaced by x^2 or any function of x. Consequently, $E(X^2) = \sum_i p_i E(X^2 \mid i)$. If $\mu = E(X)$, $\mu_i = E(X \mid i)$, we thus have

$$\begin{aligned}
\text{Var}(X) &= \text{E}(X^2) - \mu^2 \\
&= \sum_i p_i \text{E}(X^2 \mid i) - \mu^2 \\
&= \sum_i p_i (\text{Var}(X \mid i) + \mu_i^2) - \mu^2 \\
&= \sum_i p_i \text{Var}(X \mid i) + \left(\sum_i p_i \mu_i^2 - \mu^2 \right).
\end{aligned}$$

The expression in parenthesis is equal to $\sum_i p_i (\mu_i - \mu)^2$, which completes the proof of (5.10). The proof is valid for both discrete and continuous PH distributions.

The hyper-exponential distribution, or H_2 distribution, is a mixture of two exponential distributions with rates λ_1 and λ_2, chosen with probability p and $1 - p$, respectively

$$f_X(x) = p \frac{1}{\lambda_1} e^{-\lambda_1 x} + (1 - p) \frac{1}{\lambda_2} e^{-\lambda_2 x}.$$

Equations (5.9) and (5.10) now yield

$$\text{E}(X) = \frac{p}{\lambda_1} + \frac{1 - p}{\lambda_2} \tag{5.11}$$

$$\text{Var}(X) = \frac{p}{\lambda_1^2} + \frac{1 - p}{\lambda_2^2} + p \left(\frac{1}{\lambda_1} - \text{E}(X) \right)^2 + (1 - p) \left(\frac{1}{\lambda_2} - \text{E}(X) \right)^2. \tag{5.12}$$

If $\text{E}(X)$ in (5.12) is replaced by (5.11), one finds after some calculation:

$$\text{Var}(X) = \frac{p}{\lambda_1^2} + \frac{1 - p}{\lambda_2^2} + p(1 - p) \left(\frac{1}{\lambda_1} - \frac{1}{\lambda_2} \right)^2.$$

The matrices α and $\mathbf{T}$ for the H_2 distribution are

$$\alpha = [p, 1 - p]$$
$$\mathbf{T} = \begin{bmatrix} -\lambda_1 & 0 \\ 0 & -\lambda_2 \end{bmatrix}.$$

The H_2 distribution has three parameters: p, λ_1, and λ_2. However, we only need two parameters to match the expectation and the variance of a given data set. Hence, we can satisfy an additional requirement. Morse [70] sets his values of λ_1 and λ_2 such that

$$\lambda_1 = \frac{2p}{\text{E}(X)}, \quad \lambda_2 = \frac{2(1 - p)}{\text{E}(X)}. \tag{5.13}$$

With these values, equation (5.11) is satisfied. If $\mu_X = \text{E}(X)$, the variance becomes

$$\begin{aligned}
\text{Var}(X) &= \frac{\mu_X^2}{4p} + \frac{\mu_X^2}{4(1-p)} + p(1-p)\left(\frac{\mu_X}{2p} - \frac{\mu_X}{2(1-p)}\right)^2 \\
&= \frac{\mu_X^2}{4}\left(\frac{1}{p(1-p)} + \frac{(1-2p)^2}{p(1-p)}\right) \\
&= \frac{\mu_X^2}{4p(1-p)}(1 + (1-2p)^2) \\
&= \frac{\mu_X^2}{2p(1-p)}(1 - 2p(1-p)).
\end{aligned}$$

Hence, if SCV is CV^2, the squared coefficient of variation, one has

$$\text{SCV} = \frac{\text{Var}(X)}{\text{E}^2(X)} = \frac{1 - 2p(1-p)}{2p(1-p)} = \frac{1}{2p(1-p)} - 1. \tag{5.14}$$

We now show that $\text{Var}(X) \geq \mu$, that is, SCV is greater 1. To prove this, note that $p(1-p)$ has its maximum at $p = \frac{1}{2}$, and when $p = \frac{1}{2}$, then $2p(1-p) = \frac{1}{2}$ and SCV = 1. As $2p(1-p)$ decreases, then $\frac{1}{2p(1-p)}$ increases, and so does the SCV.

If $\text{E}(X)$ and $\text{Var}(X)$ are given, one can solve (5.14) for p to find

$$p = \frac{1}{2}\left(1 \pm \sqrt{1 - \frac{2}{\text{SCV} + 1}}\right).$$

Once p is found, λ_1 and λ_2 are given by (5.13), and we have all parameters.

The hyper-exponential distribution has the disadvantage that the distribution has its highest density at 0. This can be avoided by taking a mixture of two Erlang distributions with the same shape parameter, but with different rate parameters. For instance, one could take the mixture of two Erlang distributions with the shape parameter $k = 2$. However, the second distribution has a rate that is larger than the rate of the first distribution. The probability of obtaining the first distribution is p, which gives a probability of $1 - p$ that the second distribution is selected.

Formally, any distribution with vectors α that contains more than one non-zero entry is a mixture, with $p_i = \alpha_i$, and the $f_i(x)$ corresponding to the distributions of the times to absorption obtained by starting in phase i. Not all mixtures of PH distributions can be expressed by this type of a mixture. To find the mixture of two arbitrary phase-type distribution, say the distributions given by $\alpha_1 T_1$ and $\alpha_1 T_2$, one uses

$$\alpha = [p, 1-p]$$
$$T = \begin{bmatrix} T_1 & 0 \\ 0 & T_2 \end{bmatrix}. \tag{5.15}$$

Note that these types of mixtures lead to transition matrices T of dimension $k_1 + k_2$, where k_1 and k_2 are the number of phases of the constituent distributions.

Mixtures are also useful in overcoming the restriction that the shape parameter k of the E_k distributions must be integer. If $k = \text{CV}^{-2}$ is not integer, but CV^{-2} is an integer plus a fraction f, it is frequently suggested to use the E_k distribution with probability $p = 1 - f$, and the E_{k+1} distribution with probability f. This is easily accomplished by changing the vector α of the E_{k+1} distribution as follows: instead of $\alpha_{k+1} = 1$, use $\alpha_{k+1} = f$ and $\alpha_k = 1 - f$. This provides the correct expectation. The variance, however, is no longer $\frac{k}{\lambda^2}$ as in the E_k distribution.

5.1.3 Coxian Distributions

A continuous PH distribution of dimension N with a bi-diagonal transition matrix $\mathbf{T}$ is called *Coxian*. This means that the phases can be arranged such that

$$
\mathbf{T} = \begin{bmatrix}
t_{1,1} & 0 & 0 & \cdots \\
t_{2,1} & t_{2,2} & \ddots & \ddots \\
0 & t_{3,2} & t_{3,3} & \ddots \\
\vdots & & \ddots & \ddots & \ddots
\end{bmatrix}.
$$

Hence, the only non-zero entries are $t_{i,i-1}$ and $t_{i,i}$, where $t_{i,i} < 0$, $i = 1, 2, \ldots$, and $0 \leq t_{i,i-1} \leq -t_{i,i}$, $i = 2, 3, \ldots$. A distribution will be called *strong Coxian* if $t_{i,i-1} = -t_{i,i}$ for all $i > 1$, that is, if absorption can only occur in phase 1. The strong Coxian distributions can be interpreted as the mixture of sums of exponential random variables. When absorption can also occur in states other than 1, we call the distribution *weak Coxian distribution*.

Most PH distributions discussed so far are Coxian. The Erlang distribution is strong Coxian, but the hyper-exponential distribution is weak Coxian. There are explicit formulas for strong Coxian distributions. If X has a strong Coxian distribution with $t_{i,i} = -\lambda_i$, $i \geq 1$, and $t_{i,i-1} = \lambda_i$, one has

$$
E(X) = \sum_{i=1}^{N} \alpha_i \sum_{j=1}^{i} \frac{1}{\lambda_i}
$$

$$
\text{Var}(X) = \sum_{i=1}^{N} \alpha_i \sum_{j=1}^{i} \frac{1}{\lambda_j^2} + \sum_{i=1}^{N} \alpha_i \left(\sum_{j=1}^{i} \frac{1}{\lambda_j} - E(X) \right)^2.
$$

To demonstrate this, note that if one starts in phase k, which happens with probability α_k, then X is really the sum of k exponentially distributed random variables, each having the parameter λ_i, $i = 1, 2, \ldots, k$. Consequently, one has the mixture of distributions, each being a sum of exponential random variables. Consequently, $E(X)$ and $\text{Var}(X)$ can be obtained from equations (5.9) and (5.10).

Due to the great flexibility of Coxian distributions, one may wonder if one needs any other PH distributions. According to [80] and [81], any distribution can be approximated arbitrarily close by what are effectively weak Coxian distributions. However, the number of phases needed to obtain a close fit may be high. The question becomes then how to represent a distribution with as few phases as possible.

5.1.4 Renewal Processes of Phase type

A renewal process where the times between renewals are independent PH random variables will be called *PH renewal process*. As before, $t_{i,j}$ is the probability of going from phase i to phase j, and $t_{i,0}$ is the probability that the inter-renewal time ends, which corresponds to absorption. Once the absorbing state is entered, a new inter-renewal time starts immediately, that is, the time spent in the absorbing state is zero. Consequently, if we are in phase i at time t, there are two ways to go to phase j at time $t + 1$: With probability $t_{i,j}$, there is a transition from i to j without a renewal, and a probability of $t_{i,0}\alpha_j$ of ending the inter-renewal time, and choosing a new inter-renal time with the initial phase j. The i, j entry of the transition matrix P of the phase the renewal process is therefore $t_{i,j} + t_{i,0}\alpha_j$. Expressed in matrix notation, P is therefore $\mathbf{T} + \alpha t_0$, where $t_0 = [t_{i,0}]$. When used as part of a discrete event system, going from i to j with a renewal also triggers an event, and the physical states change.

For renewal processes with continuous-time PH distributions, the situation is similar. There is a rate of going from i to j without a renewal in between, and a rate of $t_{i,0}$ that the inter-renewal time is over, and the next inter-renewal time starts in phase j. Consequently, the rate of the phase process as part of the renewal process has the transition matrix $Q = T + \alpha t_0$, as before. Again, when used as part of a discrete event system, going from i to j with a renewal triggers an event.

An advantage of the strong Coxian distribution is that $t_{i,0}$ is 0 unless $i = 1$. For this reason, $\alpha_j t_{i,0}$ is different from 0 only if $i = 1$. It follows that in the transition matrix $\mathbf{T} + \alpha t_0$, only row 1 changes. Similarly, if $\alpha_j = 0$ unless $j = k$, only row k of the transition matrix changes.

If an event is disabled, then the PH renewal process is temporarily stopped, and the phase remains in the absorbing state 0 until the event is enabled again. How this works in detail will be discussed later.

5.1.5 Discrete Distributions as PH Distributions

Any discrete distribution of integers ranging from 1 to M can be interpreted as a phase-type distribution. This implies that the explicit distributions used in Chapter 4 are phase-type distributions. Indeed, there are two different ways phases can be defined: based on the remaining life times or based on ages. These two alternatives will now be discussed. We also show that any discrete PH distribution with an upper bound can be represented either by using the age or the remaining lifetime as phases. Moreover, we prove that there is no continuous PH distribution with an upper bound. Distributions with finite upper bounds are encountered frequently. In particular, service times often have well-defined durations, possibly depending on the job to be done. These cannot be represented by continuous PH distributions.

5.1.5.1 Remaining Life Used as Phase

One can interpret the remaining lifetimes as a phase of a PH renewal process. Note that if Y is the remaining lifetime, then $Y = 0$ is not an absorbing phase. To be consistent with our earlier discussion, let $Y + 1$ rather than Y be the phase, with 0 replacing -1, and the non-absorbing phases running from 1 to M. Using these conventions, the matrix $\mathbf{T}$ becomes

$$\mathbf{T} = \begin{bmatrix} 0 & 0 & \ldots\ldots & 0 & 0 \\ 1 & 0 & \ddots \ddots & \vdots & \vdots \\ 0 & 1 & \ddots \ddots & \vdots & \vdots \\ \vdots & & \ddots \ddots \ddots \ddots & & \vdots \\ 0 & 0 & \ddots \ddots & 1 & 0 \end{bmatrix}. \tag{5.16}$$

If $t_{i,0}$ is the probability that in phase i, the inter-renewal time ends, then $t_{1,0} = 1$, and $t_{i,0} = 0, i = 1, 2, \ldots, M$. Moreover, we have, if p_i represents the probability that the lifetime is i:

$$\alpha = [p_1, p_2, \ldots, p_M]. \tag{5.17}$$

Consequently, the phase process has the following transition matrix $P = \alpha t_0 + \mathbf{T}$

$$P = \begin{bmatrix} p_1 & p_2 & \cdots\cdots & p_{M-1} & p_M \\ 1 & 0 & \ddots \ddots & \vdots & \vdots \\ 0 & 1 & \ddots \ddots & \vdots & \vdots \\ \vdots & \vdots & \ddots \ddots \ddots & & \vdots \\ 0 & 0 & \ddots \ddots & 1 & 0 \end{bmatrix}.$$

This shows that when using the remaining lifetime, we already used a phase-type distribution.

5.1.5.2 Age as Phase

If age is used as the phase, then one can use 0 as the absorbing state, because according to our convention, the age of non-absorbing states starts from 1 rather than from 0. Moreover, $\alpha_1 = 1$, and $\alpha_i = 0$, $i > 1$, that is,

$$\alpha = [1, 0, 0, \ldots]. \tag{5.18}$$

Also, the probability $t_{i,i+1}$ can be set to $1 - r_i$, where r_i is the probability of not surviving age i. We now have

$$T = \begin{bmatrix} 0 & 1 - r_1 & 0 & \ldots 0 & 0 \\ 0 & 0 & 1 - r_2 & \ddots & \vdots & \vdots \\ 0 & 0 & & \ddots & \ddots & \vdots & \vdots \\ \vdots & \vdots & \ddots & & \ddots & \vdots & \vdots \end{bmatrix}. \tag{5.19}$$

Now, $t_{i,0} = r_i$, and with $\alpha_1 = 1$, $\alpha_i = 0$, $i \neq 1$, the transition matrix of the phase process becomes

$$P = \begin{bmatrix} r_1 & 1 - r_1 & 0 & \ldots 0 & 0 \\ r_2 & 0 & 1 - r_2 & \ddots & \vdots & \vdots \\ r_3 & 0 & & \ddots & \ddots & \vdots & \vdots \\ \vdots & & \ddots & \ddots & & \ddots & \vdots & \vdots \end{bmatrix}.$$

This settles the case of the age used as phase.

5.1.5.3 PH Distributions with Finite Support

If X is a random variable that has an upper bound M such that $P\{X > M\} = 0$, then it cannot be a continuous PH distribution, because the time in each phase has an exponential distribution. For random variables with upper bound, one therefore has to use discrete PH distributions, but not all discrete-time PH distributions have an upper bound. To have an upper bound, the transition matrix T must have no loops, that is, if there is a possibility to reach j from i, it must be impossible to reach i from j. Also excluded are self-loops, that is, $t_{,i}$ must be zero for all i. If there are loops, then it is possible to repeat the same loop forever, and it can take an infinite amount of time until absorption, which means that there is no upper bound.

If a PH distribution has finite support, then one can calculate its explicit distribution p_i, $i = 1, 2, \ldots, M$ by calculating the transient solution of the DTMC defined by T. Once this is done, then one can use either the age or the remaining lifetime as a phase. This would bring us back to the methods discussed in Chapter 4.

Note that there are different PH representations for the same distribution. For example, every distribution with finite support can either be represented by the remaining lifetime or by the age. We will show in Section 8.2.9 that this is normal. Indeed, if there are two PH representations of the same distribution, any mixture of these representations is also a representation of this distribution. Consequently, the different representations of a particular distribution form a continuum.

5.2 The Markovian Arrival Process (MAP)

Aside from renewal processes, there are many other processes that can be used to generate events. In particular, we can use Markov chains to generate events, and this brings us to the *Markovian Arrival Process* or MAP. The name MAP is somewhat misleading because any event can be generated by a Markovian process, not only arrivals. The main idea is that events can be generated using a Markov chain with a finite number of states. There are two ways of generating events with a Markov chain:

1. If the Markov chain is in phase i, it generates events at a rate of λ_i, where λ_i does not depend on the past history.
2. If the Markov chain moves from i to j, it creates an event with probability $w_{i,j}$.

We will call the Markov chain that generates events the *Event Generating Chain* or EGC. For simplicity, we only deal with EGCs that are continuous.

Consider first the case that events occur as the EGC changes its phase: Whenever there is a change from i to j in the EGC, there is a probability $w_{i,j}$ that an event is triggered at this point. To express this, let $D = [d_{i,j}]$ be the transition matrix of the EGC. When $i \neq j$, the rate of going from i to j without creating events is $d_{i,j}(1 - w_{i,j})$, and the rate of going from i to j while creating an event is $d_{i,j}w_{i,j}$. When $i = j$, events are created at a rate of λ_i. We now define the matrices $D^{(0)} = [d_{i,j}^{(0)}]$ and $D_{i,j}^{(1)} = [d_{i,j}^{(1)}]$ with

$$d_{i,j}^{(0)} = d_{i,j}(1 - w_{i,j}), \ d_{i,j}^{(1)} = d_{i,j}w_{i,j}, \ i \neq j$$
$$d_{i,i}^{(0)} = d_{i,i} - \lambda_i, \qquad d_{i,i}^{(1)} = \lambda_i.$$

In this way
$$D = D^{(0)} + D^{(1)}.$$

If events occur at a rate of λ_i while the EGC is in phase i and not at state changes, then the resulting process is called *state-dependent Poisson process* . For state-dependent Poisson processes, we have

$$D^{(1)} = \begin{bmatrix} \lambda_1 & 0 & \dots 0 & 0 \\ 0 & \lambda_1 & \ddots \vdots & \vdots \\ \vdots & \ddots & \ddots \ddots & \vdots \\ \vdots & \ddots & \ddots \ddots & \vdots \\ 0 & 0 & \dots \dots 0 & \lambda_M \end{bmatrix}.$$

In this case, the entries of $D^{(0)}$ are $d_{i,j}^{(0)} = d_{i,j}$, $i \neq j$, and $d_{i,i}^{(0)} = d_{i,i} - \lambda_i$.

PH renewal processes can also be interpreted as MAPs. There, the rate of going from i to j without generating an event is $t_{i,j}$. An event occurs only when the absorbing state is entered, but instead cf an absorption, we renew the process. In PH renewal processes, the rate (probability) of going from i to j, with an event happening in between, is $t_{i,0}\alpha_j$. Consequently

$$D^{(0)} = \mathbf{T}, \ D^{(1)} = [t_{i,0}\alpha_j] = \alpha t_0,$$

where the vector $t_0 = [t_{i,0}]$.

A MAP allows one to create dependent inter-event times. To demonstrate this, consider the state-dependent Poisson process with 2 states, 1 and 2, where λ_2 is much smaller than λ_1. If the state changes are rare, then if the events follow each other rapidly, we are likely to be in state 1, and unless there is a state change, the time to the next event is likely to be short as well. Hence, the length of the past inter-event time influences the next inter-event time

The construct of the MAP can be extended to create dependencies between events of different types. If there are two events, Ev_1 and Ev_2, then we can use the same $D^{(0)}$ for both events, but make $D^{(1)}$ the matrix for the process of Ev_1, and create a different matrix $D^{(2)}$ for the process Ev_2. There is also the possibility of having batch arrivals (BMAP). In this case, $D^{(i)}$ represents the process that i events of the same type occur at the same time. This shows that MAPs provide great flexibility.

5.3 PH Event Systems

In this section, we describe how to formulate PH event systems, and how to generate their transition matrices. In contrast to renewal event systems, it is easy to use continuous-time PH systems effectively. This eliminated the complications due to the fact that several unrelated events can occur simultaneously. On the other hand, we lose a certain amount of predictability: Whereas in renewal event systems, the supplementary state variables indicate which event will happen, if any, this is no longer true for PH event systems. For this reason, the importance of distinguishing between supplementary and physical state variables is diminished. It thus seems natural to consider changes of supplementary variables also as events, and this is what we will do. Hence, besides *physical events*, the events that change the physical state, we now have *phase events*, events that only change phases. It still makes sense to use event tables containing only physical events as a first step, but in order to create the transition matrix, we have to include phase events as well. The resulting event tables will be called *joint event tables*.

When numbering the states, the state number should reflect the distance from the origin, that is, states further away from the origin should have a higher state number than the ones closer to the origin. Of course, the question as to what is further away, and what is closer is often difficult to determine. However, when using Erlang distributions for arrivals, it means that the state variable should be the number of phases done, whereas for service times, it should be the number of phases remaining.

In renewal event systems, we could typically avoid immediate events, because when a state triggered an event, we could do the event, and then check for all other events if their status changed from enabled to disabled, or from disabled to enabled. In this way, the supplementary state is updated immediately after each physical state change. These immediate actions affecting the supplementary state must now be modeled as immediate events.

The outline of this section is as follows. We first discuss the different types of immediate events, and then we formulate some discrete event systems. Finally, we provide some algorithms to generate the transition matrices of PH event systems.

5.3.1 Immediate Events in PH Event Systems

As stated in Section 2.6, immediate events are those that are executed immediately as soon as their event conditions are satisfied. For PH event systems, we will add to the event description any immediate events that could be triggered. When using programs to generate the transition matrix of a PH event system, we suggest implementing immediate events as procedures. Like any other event, immediate events have event functions and event conditions. They may also trigger other immediate events.

In some states, several immediate events may be enabled, and the question is which of the events must be executed first. In some cases, all immediate events are given a priority, and if several immediate events are triggered, only the one with the highest priority is executed. This changes the state, and this change may disable competing immediate events. In other cases, the events are given probabilities, and which event will be executed is determined randomly according to these probabilities.

To demonstrate the different types of immediate events, consider the arrival to an empty queuing system with two servers, each having a PH service time. In this case, two potential immediate events are enabled, namely, "start service server 1" or "start service server 2". One possibility in this case is to give one of the two servers priority. If the event "start service server 1" is given priority, then it is always server 1 who becomes busy when an arrival to the empty system occurs. Server 2 only starts serving if server 1 is busy. Alternatively, the arrival can choose a server at random. If the arrival chooses server 1 with probability p, and server 2 with probability $1 - p$ while both servers are idle, then the arrival event will trigger "start service server 1" with probability p, and "start service server 2" with probability $1 - p$.

We mentioned that immediate events can trigger other immediate events. This can happen when blocking occurs. Therefore, the start of a service may reduce one line, and this may unblock another server. When an immediate event enables several other immediate events, the situation may become quite complicated, especially when these events are chosen with certain probabilities. We will not consider these complications.

5.3.2 Two Examples

As a first example of a PH event system, consider the $H_2/E_k/2/N$ queue, which is a system with hyper-exponential arrivals, 2 servers, each having an Erlang-k service time, and a buffer size of N. In this case, we use the following state variables:

1. X_1 is the number in the queue, excluding the elements served. We assume $0 \leq X_1 \leq N$.
2. X_2 is the arrival phase, which is either 1 or 2. When in phase 1, the arrival rate is λ_1, and when in phase 2, it is λ_2.
3. X_3 is the phase of server 1, and we assume that the service time of server 1 has K_1 phases, that is, $0 \leq X_3 \leq K_1$.
4. X_4 is the phase of server 2. This server has K_2 phases, that is, $0 \leq X_4 \leq K_2$.

When an arrival occurs, the initial phase of the next inter-arrival time is 1 with probability p, and 2 with probability $1 - p$. Aside from arrivals, there are two finite rate events, "Next phase 1", meaning server 1 goes into the next lower phase, that is, X_3 decreases by 1, and "Next phase 2", which reduces X_4 by 1. If $X_3 = 0$, the immediate event "Start 1" is triggered, which reduces the queue by 1. The event of "Next phase 2" when $X_4 = 0$ has a similar effect. The rates of the events "Next phase i" are $\mu_i K_i, i = 1, 2$. Table 5.1 provides the event table of this system. For simplicity, we omit the word "phase" in "next phase i". Note that we split the event "Arrival" into four subevents, depending on the present state and on the target state. Also note that immediate events listed as part of the event description are only executed if their event conditions hold. Table 5.1 provides all information needed to generate the transition matrix, as we will show in the next section.

Table 5.1 The $H_2/E_k/2$ queue

Event	Event function	To call	Condition	Rate
Arr. 1 1	$X_1 + 1, 1, X_3, X_4$	Start 1, Start 2	$X_1 \leq N, X_2 = 1$	$\lambda_1 p$
Arr. 1 2	$X_1 + 1, 2, X_3, X_4$	Start 1, Start 2	$X_1 \leq N, X_2 = 2$	$\lambda_1(1 - p)$
Arr. 2 1	$X_1 + 1, 1, X_3, X_4$	Start 1, Start 2	$X_1 \leq N, X_2 = 1$	$\lambda_1 p$
Arr. 2 2	$X_1 + 1, 2, X_3, X_4$	Start 1, Start 2	$X_1 \leq N, X_2 = 2$	$\lambda_1(1 - p)$
Next 1	$X_1, X_2, X_3 - 1, X_4$	Start 1	$X_3 > 0$	$K_1\mu_1$
Next 2	$X_1, X_2, X_3, X_4 - 1$	Start 2	$X_4 > 0$	$K_2\mu_2$
Start 1	$X_1 - 1, X_2, K_1, X_4$		$X_3 = 0, X_1 > 0$	
Start 2	$X_1 - 1, X_2, X_3, K_2$		$X_4 = 0, X_1 > 0$	

The second example is the tandem queue with Poisson arrivals and Erlang service times. This model is similar to the one of Table 4.3, except that arrivals are lost as soon as the first queue (excluding the element in service) reaches N_1, and that the service times are Erlang. For the present model, we use X_1, X_3, and X_5 to represent the different queue lengths, and X_2, X_4, and X_6 to represent the phase the service is in. The joint event table is given in Table 5.2. Note that here, the immediate event "Start 2" may trigger "Start 1". This happens when server 1 is blocked. Similarly, "Start 3" may trigger "Start 2", which, in turn, may trigger "Start 1". We thus have a sequence of events that trigger one another. The implementation of this in computer languages poses no problem: If immediate events are programmed as procedures, then these procedures may, in turn, call other procedures.

Table 5.2 Tandem queue

Event	Event function	To call	Condition	Rate
Arrival	$X_1 + 1, 1, X_3, X_4, X_5, X_6$	Start 1	$X_1 \le N_1$	λ
Next 1	$X_1, X_2 - 1, X_3, X_4, X_5, X_6$	Start 1	$X_2 > 0$	$K_1 \mu_1$
Next 2	$X_1, X_2, X_3, X_4 - 1, X_5, X_6$	Start 2	$X_4 > 0$	$K_2 \mu_2$
Next 3	$X_1, X_2, X_3, X_4, X_5, X_6 - 1$	Start 3	$X_6 > 0$	$K_3 \mu_3$
Start 1	$X_1 - 1, K_1, X_3 + 1, X_4, X_5, X_6$		$X_2 = 0, X_1 > 0, X_3 < N_2$	
Start 2	$X_1, X_2, X_3 - 1, K_2, X_5 + 1, X_6$	Start 1	$X_4 = 0, X_3 > 0, X_5 < N_3$	
Start 3	$X_1, X_2, X_3, X_4, X_5 - 1, K_3$	Start 2	$X_6 = 0, X_5 > 0$	

5.4 Generating the Transition Matrix with Immediate Events

This section describes programs designed to create the transition matrix. They can be used as a pattern to write a program for any PH event system. This section may be skipped. unless you plan to write a program for generating transition matrices.

If there are immediate events, then there are unstable states, that is, states that are immediately left. Unstable states do not belong to the state space of the Markov chain, and this causes irregularities. As a consequence, the lexicographic code can no longer be used as state number. However, we can still use it to find the correct state. To do this, we prepare two arrays, one array, called "stNums", which associates a state number to each lexicographic code, and another array, called "Codes", which associated a code with each state number. For instance, if state (X_1, X_2) has the lexicographic code c, and the state number is n, then Codes(c) = n, and stNums(n) = c. If the state for a given code value c does not exist, or if the state corresponding to c is unstable, then the state number is set to -1. Given the state, one easily finds its lexicographic code c, and therefore its state number stNums(c). We also need the functions "decode" which, given a lexicographic code, finds the state vector, and "encode" which, given a state vector, finds its lexicographic code. We thus need to calculate the u_i, $i = 0, 1, \ldots, d$, where d is now the total number of state variables. Here, u_0 is the number of codes, and this number is given by the variable "numCodes". The algorithm for constructing the arrays "stNums" and "Codes" is as follows:

Algorithm 21.

```
n = 0
for c = 0 : numCodes − 1
    X = decode(c)
    if X is a stable state
        n = n + 1
        Codes(n) = c
        stNums(c) = n
    else
        stNums(c) = −1
    end if
end for
```

For languages where the lowest array subscript is 1, stNums(c) must be replaced by stNums(c+1).

Using the arrays "Codes" and "stNums", one can now generate the transition matrix as follows:

Algorithm 22.

```
for c = 0 : numCodes − 1
    if stNums(c) ≠ −1
        i = stNums(c)
        for k = 1 : e
            if C(k,X)
                Xplus = f(k,X)
                c = encode(Xplus)
                j = stNums(c)
                q(i, j) = rate(k, X)
            end if
        end for
    end if
end for
```

To demonstrate our method, we now fill in the details of the algorithms above to find the transition matrix of the $H_2/E_k/2/N$ queue. First note that numCodes $= (N+1)2(N_3+1)(N_4+1)$. To simplify the enumeration of the codes, we redefine X_2 to be 0 when the arrival process is in phase 1 of the inter-arrival time, and 1 otherwise. First, we create the arrays "stNums" and "Codes".

Algorithm 23.

```
n = 0
for c = 0 : numCodes − 1
```

```
        X = decode(c)
        if X(1) = 0 or X(1) > 0 and X(3) > 0 and X(4) > 0
            n = n + 1
            Codes(n) = c
            stnums(c) = n
        else
            stNums(c) = -1
        endif
endfor
```

Next, we calculate $Q = [q_{i,j}]$ as follows:

Algorithm 24.

```
for c = 0 : numcodes - 1
    if stNums(c) ≠ -1
        X = decode(c)
        i = stNums(c)
        if X(1) < N then
            Xplus = X
            Xplus(1) = X(1) + 1
            Xplus(2) = 1
            Xplus = start(Xplus,1)
            if X(3) > 0
                Xplus = start(Xplus,2)
            end if
            j = encode(Xplus)
            j = stNum(j)
            q(i, j) = lambda(X(2)+1)*p
            Xplus = X
            Xplus(1) = X(1) + 1
            Xplus(2) = 2
            Xplus = start1(Xplus)
            if X(3) > 0
                Xplus = start(Xplus,2)
            end if
            j = encode(Xplus)
            j = stNum(j)
            q(i, j) = lambda(X(2)+1)*(1-p)
        end if
        Xplus = X
        if X(3) > 0
            Xplus(3) = X(3) - 1
            Xplus = start1(Xplus)
            j = encode(Xplus)
            j = stNums(j)
```

```
        q(i, j) = mu(1)* K(1)
    end if
    Xplus = X
    if X(4) > 0
        Xplus(4) = X(4) – 1
        Xplus = start2(Xplus)
        j = encode(Xplus)
        j = stNums(j)
        q(i, j) = mu(2)* K(2)
    end if
  end if
end for

function start(Xplus,ev)
if Xplus(2+ev) = 0 and X(1)>0
    Xplus(1) = X(1) – 1
    Xplus(2+ev) = K(ev)
    return Xplus
end if
end function
```

This completes our discussion of the generation of the transition matrix of a PH event system.

Problems

5.1 Find the formula for the variance of the mixture of the E_k and the E_{k+1} distribution by using equation (5.10). Both distributions have the same rate parameter λ. Use p for the probability of selecting the E_k distribution, and $1 - p$ for selecting the E_{k+1} distribution.

5.2 Generate the transition matrix of the $H_2/H_2/1/3$ queue by hand. Next, modify Algorithm 24 to generate the transition matrix.

5.3 Generate the transition matrix of an $M/E_2/2/5$ queue.

5.4 Formulate the system given in Table 2.9, except the lead times are Erlang-2. The lead times are the times from the moment the item is ordered until it arrives.

5.5 Find the number of states of a three station tandem queue as follows. The limits of the lines are $N_1 = 10$, $N_2 = 5$, and $N_3 = 5$. Arrivals are Poisson, and all service times are E_3.

5.6 Often, only the non-zero entries of the transition matrix are stored. Change Algorithm 24 such that for each entry, only the row number, the column number, and the rate are stored.

Chapter 6
Execution Times, Space Requirements, and Accuracy of Algorithms

The quality of an algorithm can be judged according to the following three criteria: how long it takes to obtain an answer, how much memory is required, and how accurate are the results. All these criteria will be covered in this chapter. The time needed for finding an answer is covered under the heading of *time complexity*, and the requirement for memory needed to execute the algorithm is covered under the heading of *space complexity*. To estimate the time complexity of an algorithm, we count the floating point operations, also called flops, and the space complexity of an algorithm is estimated by counting the number of single precision words needed for its execution.

In view of the fact that even a simple laptop can do millions of floating points operations per second, and that it can store millions of variables, the time or space complexity may seem irrelevant. Since a laptop costs around a thousand dollars, and since it can store millions of data and perform millions flops per second, the cost for computing is completely negligible when compared to the cost of human work. However, this argument misses an important point: As the number of variables is increased, the number of flops increases exponentially, and the solution of what many consider a small model might require years or even centuries of computer time. One could still argue that running a program on a laptop for a year costs relatively little when compared to the yearly wage of the user. This is certainly true, but most people hate to wait even a few minutes to obtain an answer. Here, the cost of waiting must be considered as well.

The low cost of computing has important consequences. For programs or parts of programs that need only a few millions of flops to obtain an answer, the first consideration is the programming effort. Programs should be as clearly structured and as simple as possible, and techniques that decrease the flop count while simultaneously reducing program clarity should be avoided. Also, if the program is expected to take less than a second, it does not matter when the execution time doubles or even triples. The aim of the complexity analysis should thus be to identify any parts of programs that take hours rather than seconds to execute and concentrate efforts on these parts. A complexity analysis can help in this respect.

© The Author(s), under exclusive license to Springer Nature Switzerland AG 2022   115
W. Grassmann and J. Tavakoli, *Numerical Methods for Solving Discrete Event Systems*,
CMS/CAIMS Books in Mathematics 5, https://doi.org/10.1007/978-3-031-10082-6_6

Aside from the time needed, it is important that the results obtained from a computer program are reasonably accurate. Aside from programming errors, there are three reasons why the results obtained may have errors

1. The input data may be inaccurate.
2. Since computers store numbers with a finite number of digits, there may be rounding errors.
3. In iterative algorithms, one typically stops the iteration as soon as the answer obtained is close enough to the actual value. Also, truncation of infinite sums or integrals with upper bounds of infinity leads to inaccuracies. Errors due to limiting the number of iterations, limiting the number of terms, and making the range of integrals finite will all be covered under the heading of truncation errors.

It is good practice to identify the main source of these errors. For instance, if the input data are correct with a precision of $\pm 1\%$, then it makes no sense that the program is correct with an error of 1 out of a million. If one looks at the input data for queueing problems, such as arrival rates, relative errors are typically above 1%, and it appears that results presented with 5 to 6 digits of precision are overkill.

The problem with rounding errors is that they can be magnified greatly as the calculation progresses. In particular, subtractions tend to increase errors committed in earlier stages of the calculation. Fortunately, when dealing with probabilities, which are always non-negative, subtractions can often be avoided. In fact, many of the algorithms we will present in later chapters contain no subtractions at all, and this reduces the effect of rounding errors significantly. As far as truncation errors are concerned, they can typically be reduced by increasing the number of terms for sums, respectively, the number of iterations in iterative algorithms. Here, the traditional techniques developed in numerical mathematics can be used.

In this chapter, we will discuss space and time complexity, how they are measured, and how they can be reduced. We also will discuss the errors due to rounding and truncation.

6.1 Asymptotic Expressions

To estimate the time and space complexity, we do not need very accurate measures. In fact, it does not matter if we get results in half a second, a second, or even two seconds. What we need is approximate expressions, and in this regard, we discuss the *big O* notation, the *big Theta* (Θ) notion. We also discuss the *little o* notation, though it will be used in a somewhat different context.

If n is a non-negative integer, such as the dimension of the transition matrix, a function $f(n)$ is said to be "big O $g(n)$", written

$$f(n) = O(g(n))$$

if for $n \geq n_0$, $f(n) \leq Mg(n)$, where n_0 is fixed and M is a positive constant that can be freely chosen. Popular choices for $g(n)$ are powers of n like n^m. For instance, if

$g(n) = n$, the big O notation becomes $O(n)$, and if $g(n) = n^2$, the big O notation becomes $O(n^2)$. Also important is $O(1)$, which indicates that the upper bound for the function is a constant. Also note that since we are talking of upper bounds, the function $f(n)$ may grow at a slower rate than $g(n)$. For instance, if $f(n) = cn$, then $f(n) = O(n^2)$. To verify this, set $M = 1$, and then $cn < n^2$ for $n > c$.

It is important to note that the sign $=$ as used here is not really equality, but it has more the meaning of "is a" such as "Fido is a dog". In other words, the right side of the equation is cruder than the left side.

As an application of the big O notation, consider the sum of squares from 1 to n, that is

$$\sum_{i=1}^{n} i^2.$$

We will need this expression a number of times to evaluate the time complexity. It is known that

$$\sum_{i=1}^{n} i^2 = \frac{n(n+1)(2n+1)}{6}. \tag{6.1}$$

This can be proven by complete induction. It is true for $n = 1$, and a little calculation shows that if it is true for n, it is also true for $n + 1$. We would now like an approximations for the sum of squares. For this purpose, we first write equation (6.1) as follows

$$\sum_{i=1}^{n} i^2 = \frac{n^3}{3} + \frac{n^2}{2} + \frac{n}{6}.$$

If n is over 100, and for most applications of interest in our context, $n > 100$, then the first term on the right of this equation is over 200/3 times larger than the second term, which makes the second term irrelevant when only approximate values are needed, and the third term is completely negligible. Hence, we only need the first term when n is large.

Let us now look at how we can use the big O notation. First, we claim

$$\sum_{i=1}^{n} i^2 = O(n^3).$$

To prove this, we set $M = 1$ and find an n_0 such that

$$\sum_{i=1}^{n} i^2 < n^3, \quad n > n_0. \tag{6.2}$$

First, we note that if $\sum_{i=1}^{n} i^2 < n^3$ for some n, it also holds for $n + 1$.

$$\sum_{i=1}^{n+1} i^2 = \sum_{i=1}^{n} i^2 + (n-1)^2 < n^3 + (n+1)^2.$$

Now, $n^3 + (n+1)^2 = n^3 + n^2 + 2n + 1$ is certainly less than $(n+1)^3 = n^3 + 3n^2 + 3n + 1$, which proves that if inequality (6.2) holds for n, it also holds for $n + 1$. Hence, we only need to find an n_0 for which is holds, and $n_0 = 2$ will do. We conclude that $\sum_{i=1}^{n} i^2$ is indeed $O(n^3)$.

The fact that we are free to choose M when establishing the big O notation means that if $f(n) = O(g(n))$, then $f_1(n) = cf(n) = O(g(n))$ where c is a fixed rational number. Thus, when we state that the sum of squares from 1 to n is $O(n^3)$, then we do not know if it is bounded by n^3, $2n^3$ or $n^3/3$. They all are $O(n^3)$. However, from the big O notations, we can give upper bounds for the growth of a faction. In particular, if $f(n) = O(n^3)$, then doubling n will increase $f(n)$ by a factor of 8.

In some cases, we really want to distinguish between $f(n)$ and $cf(n)$. For instance, if $f(n)$ is an execution time, we care about the factor c. To express this, we write

$$\sum_{i=1}^{n} i^2 = \frac{n^3}{3} + O(n^2).$$

If $f(n)$ is a polynomial of degree m, that is, $f(n) = \sum_{i=0}^{m} a_i n^i$, then $f(n) = O(n^m)$. If we want to extract the highest term of the polynomial, then we write

$$f(n) = a_m n^m + O(n^{m-1}).$$

One disadvantage of the big O notation is that it merely gives an upper bound for the growth. Frequently, we want to have more. For instance, when $f(n)$ is a polynomial, then we would like to have its degree. For cases like this, the big Theta notation was introduced. We say that

$$f(n) = \Theta(g(n)),$$

If there are values M_1, M_2 and n_0 such that for $n > n_0$,

$$M_1 g(n) \le f(n) \le M_2 g(n).$$

It follows that if $f(n)$ is a polynomial of degree $m \ge 0$ in terms of n, then $f(n) = \Theta(n^m)$. The Theta notation also allows us to define what we call the *order of a function*. We say that $f(n)$ has *order* $g(n)$ if $f(n) = \Theta(g(n))$. We also say a function $f(n)$ is quadratic if $f(n) = \Theta(n^2)$, and it is cubic if $f(n) = \Theta(n^3)$.

Even though the little o notation is not used widely in complexity analysis, it is used extensively in calculus, and we will use it later. We say that $f(x) = o(g(x))$ if $g(x)/f(x) \to 0$ as $x \to 0$. This implies that if

$$h(x) = f(x) + o(g(x)),$$

then for small enough x, $o(g(x))$ can be neglected.

6.2 Space Complexity

We will measure the space complexity in single precision words, because this is what we are mainly using in our programs. We first show that the transition matrices we are concerned with are large, but sparse, that is, most of their entries are 0. This will be shown first. Next, we will discuss methods of storing the transition matrices such that space requirements are reduced. As it turns out, different algorithms require different storage methods, and this will be discussed in the context of vector-matrix multiplications and solutions of systems of linear equations.

6.2.1 The Sparsity of Transition Matrices

We first give bounds for the density of transition matrices of different types of discrete event systems. The density is expressed as the fraction of non-zero entries of the transition matrix.

In systems with Cartesian state spaces, with state variables X_i, $0 \leq X_i \leq N_i$, $i = 1, 2, \ldots, d$, the number of states is given as

$$N = (N_1 + 1)(N_2 + 1)\ldots(N_d + 1).$$

Usually, this is a large number, and it increases exponentially with d. Thus, if we add a state variable X_{d+1} with $0 \leq X_{d+1} \leq 9$, the number of states increases by a factor of 10. Since the dimension of the transition matrix is equal to the number of states, we are dealing with large transition matrices. The normal method to store transition matrices with dimension N requires N^2 words, and since N is large, this is immense. Also, adding state variable X_{d+1} with $0 \leq X_{d+1} \leq 9$ increases the number of words needed to store the transition matrix by a factor of 100.

Fortunately, the transition matrices are sparse, that is, most of their entries are zero. Consider, for instance, a Poisson event systems with e event types. There, each physical state corresponds to a row in the transition matrix, and each event type creates one entry in this row. It follows that of the N^2 entries of the matrix, at most Ne entries are different from 0. Consequently, of the N^2 entries of the transition matrix, less than Ne entries differ from 0, that is, of the N^2 entries, the fraction of non-zero entries is less than $\frac{e}{N}$. Since N is typically in the thousands, and e below 10, this means the fraction of non-zero elements is small. To exploit sparsity, a sparse matrix representation is introduced. One logical way of storing the non-zero entries is to use three arrays, one used for the row of the entry, one for the column, and one for the value.

For further discussion, we define *transitions* as entries in the transition matrix that are not 0. Every state change that is possible leads to a transition. In addition to that, in CTMCs, the diagonal entries of the transition matrix are non-zero except for absorbing states.

Renewal event systems are also sparse. Our analysis in this case is approximate. In particular, if we ignore the fact that events can be disabled, the number of states in renewal event systems is given by

$$N = \prod_{i=1}^{d}(N_i + 1) \prod_{i=1}^{e} \bar{T}_i,$$

where $\bar{T}_i$ is the upper limit for the time between the events of type Ev_i. We now determine the number of transitions. First, we start with the case where the supplementary state variables represent ages. If $1 \leq Y_i \leq \bar{T}_i - 1$, then each of the e event processes can renew, and it follows that for each state, there are 2^e possible ways to pick the processes that renew, which means that for each state, there are 2^e possible targets. If, however, some Y_i are at their maximum value $\bar{T}_i$, the corresponding process will definitely renew, and in this case, we have fewer than 2^e targets. Still, 2^e can be used as an upper bound, that is, the number of transitions is at most $N2^e$. It follows that the density of the matrix is at most $2^e/N$. Now, $\bar{T}_i > 1$, because otherwise, Y_k would always be 1, which makes it redundant. If $\bar{T}_i \geq 2$, then $2^e \leq \prod_{i=1}^{e} \bar{T}_i$. If D is the number of non-zero entries, one has

$$D \geq \prod_{i=1}^{d}(N_i + 1)2^e.$$

It follows that the density must be less than

$$\frac{1}{\prod_{i=1}^{d}(N_i + 1)}.$$

Consequently, we conclude that if ages are used as supplementary variables, the transition matrices of renewal event systems are sparse.

We now show that the transition matrices of renewal event systems are also sparse when remaining lifetimes are used as supplementary state variables. Indeed, the number of transitions ending in state i, $1 \leq i \leq n$, is at most 2^e. To prove this, we consider the columns instead of the rows. Indeed, each Y_k of a target state can be reached from 2 starting states: Either Y_k was 0, or Y_k was one unit higher, except when $Y_k = \bar{T}_k - 1$, a case we ignore in our approximation. Since this is true for each Y_k, $k = 1, 2, \ldots, e$, we obtain at most 2^k transitions resulting in states with given values Y_k. This shows that the number of transitions is approximately the same for both ages and remaining lifetimes.

In the case of phase-type event systems, we also count changes in phases as events, and in this case, we get the same densities as in Poisson event systems. It follows that in all discrete event systems discussed so far, the transition matrices are sparse.

6.2.2 Storing only the Non-zero Elements of a Matrix

If matrices are sparse, then the matrix can be stored by three arrays, say "row", "col" and "value", where the array "row(k)" stores the row of the kth non-zero element, "col(k)" its column, and "value(k)" its value. This method of storing matrices is implemented in both MATLAB and Octave.

We now apply this method to implement a vector-matrix multiplication. Thus, we have a vector $x = [x_i, i = 1, 2, \ldots, N]$, and we want to find $b = [b_1, b_2, \ldots, b_N] = xA$, where A is a matrix of size N, and it is stored in row, column, value format. The vector-matrix multiplication can now be implemented using the following algorithm, where s is the number non-zero entries.

Algorithm 25. Algorithm for vector-matrix multiplications

b(j) = 0 for j = 1:N
for k = 1 : s
    i = row(k)
    j = col(k)
    b(j) = b(j) + x(i)*value(k)
end for

Such vector-matrix multiplication occur, for instance, when calculating transient solutions for Markov chains according to equation (1.1).

6.2.3 Storage of Banded Matrices

Unfortunately, the row, column, value representation does not work well when solving N equations in N variables by Gaussian elimination, and other storage methods are needed. Fortunately, the transition matrices derived from discrete event systems are also *banded*. By this, we mean that if $P = [p_{i,j}]$ is a transition matrix, the maximum difference between i and j is small when compared to the number of states. Expressed differently, there are constants g and h, such that $i - g \le j \le i + h$.

Consider first Poisson event systems with Cartesian state spaces. If all X_i change at most by 1, and if i is the state number before a change, then the state number after the change must satisfy

$$i - (u_1 + u_2 + \cdots + u_d) \le j \le i + (u_1 + u_2 + \cdots + u_d),$$

where $u_k = \prod_{n=k+1}^{d}(N_k + 1)$. Now $u_2 + u_3 + \cdots + u_d < u_1$, which means that all entries of the matrix P are in a band $i - 2u_1 \le j \le i + 2u_1$. Since $N = u_0 = u_1(N_1 + 1)$, the band occupies a fraction of $\frac{1}{4N}$ the transition matrix. This suggests re-ordering the states such that the state variable with the largest range is made X_1.

Let us now consider systems that require supplementary state variables. For this discussion, we no longer differentiate between physical and supplementary state variables, that is, the state variables are $Z_i, i = 1, 2, \ldots, e$, where Z_i can either be a physical state variable or a supplementary state variable. We now have to consider how much a variable can change. If Z_k is a remaining lifetime, at any renewal, it potentially changes from 0 to its highest value, and if Z_k is age, it can change from its present value to 1. Since Z_k can be at its upper bound, it too can change through its entire range. Generally, other things being equal, state variables that can change by large amounts should be represented by a Z_k with a high value of k.

Essentially, one only wants to store the band, and this can be achieved by using a matrix with N rows and $g + h + 1$ columns. If $Q = [q_{i,j}]$ is the original transition matrix, then $i - g \le j \le i + h$, or $-g \le j - i \le h$. It follows that if we can declare "A" to be an array with N rows and columns ranging from $-g$ to h, the entire transition matrix can be stored in A. The entry $q_{i,j}$ is then in row i, column $j - i$ of the array A. Consequently, we have

$$q_{i,j} = A(i, j - i) \tag{6.3}$$

$$A(i, k) = q_{i,i+k} \tag{6.4}$$

Hence, if an algorithm is formed in terms of $q_{i,j}$, one can use (6.3) to make the conversion. A better way is to create "A" right when the transition matrix is generated. Indeed, if, in a Poisson event system, an event results, for instance, in $j = i + u_1$ then $A(i, u_1)$ is assigned the rate of the event.

6.3 Time Complexity

The question addressed here is how long it takes for a given program, with given input, to complete its task and provide the required output. This time can, of course, be measured by running the program. However, estimates are useful before actually running and often even before programming. Of course, the time complexity depends on the input parameters. For instance, in Algorithm 25, it depends on the number of entries in the transition matrix.

As stated in the introduction of this chapter, we do not need very accurate estimates of the time complexity. Hence, instead of using actual times, it is standard to use flops, or floating point operations. Sometimes, the number of statements is also used to estimate the execution time. Of course, the number of flops depends on the problem, and it is important to find the parameters that determine them. Often one merely selects one parameter at a time, but in other cases, one selects several parameters. For instance, in a vector-matrix multiplication, one often considers only the matrix size N, whereas at other times, one considers the number of non-zero entries as well. The parameters used to estimate the number of flops are referred to as the size of the problem. Most algorithms we discuss are *polynomial in time*, that is, the number of flops can be expressed as a polynomial of its size. In this case, we only retain the dominant term, that is, the term with its highest power. Thus, if an algorithm solving a problem of size N requires $a_3 N^3 + a_2 N^2 + a_1 N + a_0$ flops, we write

$$a_n N^3 + O(N^2),$$

or, if we only need an upper bound for its time complexity, we write $O(N^3)$. Finally, if we want to find the right power, we use the Theta notation and write $\Theta(N^3)$.

Let us first consider the multiplication of two square matrices of dimension N using the formula

$$c_{i,j} = \sum_{k=1}^{N} a_{i,k} b_{k,j}, \ i, j = 1, 2, \dots N.$$

In this case, we have to evaluate N^3 products, and all of these must be added. Consequently, we have $2N^3$ flops, that is, the time complexity is $\Theta(N^3)$. Next, consider a vector-matrix multiplication, such as

$$c_j = \sum_{k=1}^{N} a_k b_{k,j}, \ j = 1, 2, \dots, N.$$

A calculation similar to the one used previously shows that evaluating $c = [c_j, \ j = 1, 2, \dots, N]$ requires $2N^2$ flops. We now apply these results to compare two ways to find transient solutions of DTMCs. Suppose we would like to find $\pi(m)$ for given m, using $\pi(m) = \pi(m-1)P$, which leads to $\pi(m) = \pi(0)P^m$. We could calculate P^m first, and then find $\pi(m)$ as $\pi(0)P^m$. This would require $m-1$ matrix multiplications, with $2N^3$ flops each, plus a vector-matrix multiplication, with $2N^2$ flops, for a total of $2(mN^3 + N^2)$, which is $\Theta(N^3)$. Alternatively, we could evaluate $\pi(n+1) = \pi(n)P$ m times, which requires $2mN^2$ flops, for a complexity of $\Theta(N^2)$, which is much lower. Of course, if m is a power of 2, say $m = 2^k$, then one can find P^m by first calculating P^2, then find P^4 by squaring P^2, and so on. This means that we can find P^m with k matrix multiplications, that is, we need $2(kN^3 + N^2)$ flops. This has to be compared to its alternative, which requires $2mN^2 = 2^{k+1}N^2$ flops. For fixed N, there is always a k such that $2(kN^3 + N^2) < 2^{k+1}N^2$, or $kN + 1 < 2^k$. Consequently, there are cases where this method is preferable. However, in our problems, it is N we need to be concerned with, and for large N, finding P^m first is not a good option.

Note that when rising a sparse matrix to a power, say you rise P to the power of m, then the matrix becomes denser with each step. This not only increases the flop count, it also requires the insertion of additional entries in the matrix. These insertions are called *fill-in*. The fill-in increases program complexity, and consequently the time for programming and testing. In the case of banded matrices, the bandwidth increases with each step, which also increases the flop count. Consequently, we will avoid taking powers of transition matrices.

If vector-matrix multiplications are used to find $\pi(0)P^m$, then P remains unchanged, and the sparsity obviously remains the same. In this case, our best option is to use Algorithm 25. In this algorithm, the outer loop executes s times, and for each s, there is an addition and a multiplication, which adds up to $2s$ flops in total. Since s is less than Ne, the number of flops is reduced from $2N^2$ to $2Ne$, or $\Theta(Ne)$. In conclusion, for discrete-time discrete event systems the number of flops required to find the probabilities to be in the different states at time m is of the order $\Theta(mNe)$.

Before transient solutions can be calculated, the matrix must be generated. When looking at the algorithms for generating transition matrices, we discover that all operations are integer operations, that is, we have no flops. However, in most algorithms, the number of integer operations is proportional to N and e. From this, we conclude that we need $\Theta(N)$ operations to generate a transition matrix of size N.

We now turn to the solution of the equilibrium equations. There, we have N linear equations for N variables, and they can in principle be solved by Gaussian elimination. The problem is that this requires $\Theta(N^3)$ flops, and since N tends to be large, this is a huge number. Also, if N increases by a factor of 10, the number of flops increases by a factor of 1000! The fact that the matrices are sparse does not help to reduce the time complexity significantly, because as the elimination progresses, the matrix becomes denser. Also, the elimination process becomes more complicated when dealing with sparse matrix formats, such as the row, column, value format, and this requires additional programming effort.

When using Gaussian elimination, it is an advantage if the matrix is banded, because the bandwidth does not change as the elimination process progresses. However, the band will fill up. Still, exploiting the bandedness of the matrix can reduce the flop count significantly. We will therefore consider special algorithms for banded matrices in Chapter 9.

Note that the flop count of $\Theta(N^3)$ for solving equilibrium equations exceeds the one for finding transient solutions in DTMCs, which is $\Theta(N^2)$ for each iteration. If the number of iterations required to reach an equilibrium is not too large, then it is easier to find the equilibrium solution as the limit of transient solutions. This idea leads to the so-called i*terative methods* that approach the equilibrium probabilities iteratively. These methods will be discussed in Chapter 9. Almost all iterative methods use vector-matrix multiplications. The question arises how to evaluate the time complexity of iterative algorithms. In this case, we also have to consider the number of iterations, and this depends on the accuracy required. However, the number of iterations may depend on the matrix size, and this has to be taken into account.

To reduce the time complexity, it is recommended to put a great effort to reduce N, the number of states. N increases with the dimensionality of the problem, which is determined by the number of physical variables, as well as the number of supplementary variables. Our first objective should thus be to reduce the number of state variables of the model under consideration. This can lead to thousand-fold savings, even million-fold savings in flops as shown in [41].

6.4 Errors due to Inaccurate Data, Rounding, and Truncation

Often, probabilities and expectations arising in queueing are given with a precision of 6 to 7 digits, but is this really necessary? Clearly, if the input data are inaccurate, even the most accurate calculation cannot deliver accurate results. The problem is thus to identify the main sources of errors, and try to control these. Here, we identify three sources of errors:

1. Errors due to inaccurate data, or errors due to inaccurate models.
2. Errors due to rounding. This error arises because in floating point arithmetic, and this is the arithmetic used here, only a finite number of digits are stored.
3. Errors due to truncation. This error arises if a sum with an infinite number of terms is truncated, or if an improper integral is evaluated numerically. We also consider the error caused by limiting iterative algorithms to a finite number of iterations as truncations.

Many errors cannot be avoided, but frequently, measures are available to reduce them. It is also important to estimate the size of the error. Errors, once committed, also propagate, influencing later results. This is particularly true for subtractions: They can greatly magnify errors committed earlier.

We now discuss the three types of errors, their nature, as well as the methods of estimating them and, if possible, ways to reduce them.

6.4.1 Data Errors

The parameters of our models, such as arrival rates and service rates, are mostly estimated based on past data and/or expert opinion. They are usually not extremely accurate, and estimates that are correct with a precision better than plus or minus 1% are rare. Of course, estimates can be improved by collecting additional data, but this can be costly. To increase the precision of an estimate by one decimal digit, one has to increase the sample size by a factor of 100. This significantly limits the possibility of obtaining accurate estimations at a reasonable cost.

Some parameters have a minor influence on the final results, whereas others are crucial. To find what these crucial parameters are, and increase the effort of estimating them as accurately as possible, one can vary the different parameters of the model, and observe their effect on the final result. This method is called *sensitivity analysis*, and it is widely used in linear programming and in simulation. The senior author also knows from personal experience that practitioners love sensitivity analysis.

In order to be treatable, models must be simplified. This also causes errors. Not all variables can be included in models without making them unmanageable. It is often difficult to determine which variables to include, and which ones to omit. Here, expert opinion can help.

6.4.2 Rounding Errors

In numerical analysis, almost all calculations are done by using the floating point arithmetic. By most computers, standard IEEE 754 is used, with 32 binary digits for single precision, and 64 binary digits for double precision. In single precision, one bit is used for the sign, 8 bits for the exponent, and the remainder for the fraction. In double precision, there is again one bit for the sign, 11 bits for the exponent, and the remainder for the fraction. Without going into details, we merely note that if the result of an operation exceeds the number of bits provided for the fraction, the result is rounded to the nearest number that can be represented internally. Consequently, the value for a number, say for a, is stored in memory as $\mathrm{fl}(a)$, which slightly differs from its true value. We define $\mathrm{fl}(a) - a$ as the *absolute error*, and $\frac{fl(a)-a}{a}$ as the *relative error*. If ρ is the relative error, then the *rounding unit* ϵ_M is defined to be the smallest number such that $|\rho| \leq \epsilon_M$. It follows that $\mathrm{fl}(a) = a(1 + \rho)$ with $|\rho| \leq \epsilon_M$. The rounding unit is approximately 10^{-7} in single precision, and approximately 10^{-16} in double precision. For more details, see [84].

If $\circ$ is any of the four basic operations $+, -, \times$ or $\div$, then the hardware of modern computers is constructed such that

$$\mathrm{fl}(a \circ b) = (a \circ b)(1 + \rho), \ |\rho| \leq \epsilon_M. \tag{6.5}$$

The details of how these arithmetic operations achieve the required precision in their hardware implementation are irrelevant as long as (6.5) is satisfied.

In some cases, no rounding is needed. In particular, if a and b are both positive single precision numbers, then $a - b$ has fewer digits than either a or b. It follows that subtractions do not cause rounding errors. This, of course, seems to be counter intuitive, because subtractions are known to cause major problems when considering rounding, and they do! This will be explained below.

For evaluating the effect of rounding errors, we also have to consider how errors, once committed, affect later calculation. For this purpose, it is convenient to define "calc(expression)" to be the value obtained if "expression" is evaluated in floating point arithmetic. Consider first the multiplication, that is, we consider the expression $a_1 a_2 \ldots a_n$, and we want to determine bounds for $\mathrm{calc}(a_1 a_2 \ldots a_n)$. This is how we proceed. We first form $a_1 a_2$, when $\mathrm{calc}(a_1 a_2) = a_1 a_2 (1 + \epsilon_1)$. To find $a_1 a_2 a_3$, one multiplies $\mathrm{calc}(a_1 a_2) = a_1 a_2 (1 + \epsilon_1)$ with a_3, and the result is

$$\mathrm{calc}(a_1 a_2 a_3) = a_1 a_2 a_3 (1 + \epsilon_1)(1 + \epsilon_2).$$

In this way, one continues, obtaining at last the entire product as

$$\mathrm{calc}(a_1 a_2 \ldots a_n) = a_1 a_2 \ldots a_n (1 + \epsilon_1)(1 + \epsilon_2) \ldots (1 + \epsilon_{n-1}).$$

Since

$$1 - \epsilon_M \leq 1 - \epsilon_i \leq 1 + \epsilon_M, \ i = 1, 2, n - 1$$

we have

$$a_1 a_2 \ldots a_n (1 - \epsilon_M)^{n-1} \leq \mathrm{calc}(a_1 a_2 \ldots a_n) \leq a_1 a_2 \ldots a_n (1 + \epsilon_M)^{n-1}. \tag{6.6}$$

A similar bound can be found for the division.

Let us now consider sums of the form $a_1 + a_2 + \cdots + a_n$. Clearly, $\mathrm{calc}(a_1 + a_2) = (a_1 + a_2)(1 + \epsilon_1)$. Using this value to find $a_1 + a_2 + a_3$ yields

$$\mathrm{calc}(a_1 + a_2 + a_3) = ((a_1 + a_2)(1 + \epsilon_1) + a_3)(1 + \epsilon_2) = (a_1 + a_2)(1 + \epsilon_1)(1 + \epsilon_2) + a_3(1 + \epsilon_2).$$

Continuing in this fashion, one finds

$$\text{calc}(a_1 + a_2 + \cdots + a_n) = (a_1 + a_2)(1 + \epsilon_1)(1 + \epsilon_2)\ldots(1 + \epsilon_{n-1})$$
$$+ a_3(1 + \epsilon_2)(1 + \epsilon_3)\ldots(1 + \epsilon_{n-1}) + \cdots + a_n(1 + \epsilon_{n-1}).$$

This indicates that if $j > i$, the error associated with a_j tends to be less that the one associated with a_i. This would suggest adding the smallest values first.

If all $a_i \geq 0$, the equation above implies

$$(a_1 + a_2 + \cdots + a_n)(1 - \epsilon_M)^{n-1} \leq \text{calc}(a_1 + c_2 + \cdots + a_n) \leq (a_1 + a_2 + \cdots + a_n)(1 + \epsilon_M)^{n-1}.$$
$$(6.7)$$

This relation is not true if some a_i are negative: If $a_i < 0$ for a specific i, then one must take the lower bound rather than the upper bound for the error, and the inequality above is no longer valid. Subtraction does not satisfy this relation either.

The consequence of the inequalities (6.6) and (6.7), and their analog for division implies when there are no subtractions, and if one needs n operations to find the result, the relative error is between $(1 - \epsilon_M)^{n-1}$ and $(1 + \epsilon_M)^{n-1}$. This makes the determination of error bounds easy as long as sums do not involve negative values and as long as there are no subtractions.

If we have to handle negative values, or if we have subtractions, just counting operations is no longer valid. To show this, consider $a + b$: If the calculated value of a is $a(1 + \epsilon)$ instead of a, then we obtain

$$a(1 + \epsilon) + b = (a + b)\frac{a(1 + \epsilon) + b}{a + b} = (a + b)\left(1 + \frac{\epsilon}{1 + b/a}\right).$$

Hence, if a and b have the same sign, then $1 + b/a$ is greater 1, that is, the relative error decreases. Of course, the absolute error stays the same, but it is now taken in relation to a larger number. However, if a and b have different signs, then $1 + b/a < 1$, and the relative error increases. The relative error also increases if instead of $a + b$, we have $a - b$, with $b > 0$. Consequently, subtractions increase relative errors committed earlier. As an example, consider $a - b$, with $a = 1$ and $b = 0.999$, yielding $a - b = 0.001$. If, due to earlier calculations, a has a rounding error of up to $\pm 0.1\%$, that is, $0.999 \leq \text{calc}(a) \leq 1.001$ then $\text{calc}(a - b)$ would be between 0 and 0.002. The relative error can thus reach 100% in this case. Effects like this are called *subtractive cancellation*. They can lead to results that are completely wrong.

Many algorithms involving probabilities do not contain subtractions. For example, Algorithm 25 contains no subtraction, and it is sufficient to count operations. If one needs $n + 1$ operations to find a specific result, the maximum rounding error is bounded by a factor $(1 - \epsilon_M)^n$ and $(1 + \epsilon_M)^n$, or approximately by a factor $(1 - n\epsilon_M)$ and $(1 + n\epsilon_M)$. In Algorithm 25, for instance, the calculated b(j) has a relative error of at most $2s\epsilon_M$ which is $2s10^{-6}$ in single precision arithmetic. Since $s \approx Ne$, this is negligible in most applications. For instance, when the values used in Algorithm 25 are probabilities, it is difficult to provide estimates with a precision of more that 1% for the values used by the algorithm. Moreover, since rounding errors can be positive or negative, some errors cancel each other out, which makes the error bounds derived here pessimistic. If subtractions cannot be avoided, they should be done as early as possible. If the calculations contain no subtractions, then, but only then, one can usually ignore the rounding errors.

When using floating point arithmetic, there is also only a finite space for the exponent, namely 8 bits for single precision, and 11 for double precision. In single precision, this allows to store numbers in the range from approximately 5.910^{-39} to 3.410^{38} in single precision, and 8.910^{-308} to 9.010^{307}. in double precision. We noticed, however, that some implementations have slightly different lower limits. If a number exceeds the upper bound, an exponent overflow results, and if it is below the lower bound, an exponent underflow results. In many languages, both underflow and overflow result in an error message. However, in other languages, such as VBA, the language that comes with Excel$^{\copyright}$, or in Octave, exponent underflows lead to a result of zero, with no error message. Unfortunately, this may lead to completely meaningless numerical results.

6.4.3 Truncation Errors

When evaluating an infinite sum numerically, one must truncate the sum at some point, which can lead to a *truncation error*. The question is now when to truncate. Obviously, one must not truncate before the terms of the sum start to decrease. Once they do, one frequently truncates as soon as a term is encountered with an absolute value less than $\epsilon > 0$, where ϵ is a small positive value. This may not totally safe, but it is often the best method available. In other cases, one can bound the sum of the remaining terms. In particular, the series formed by the remaining terms may be majorized by a geometric series, in which case a bound for all remaining terms is easy to establish.

Of course, truncation is only meaningful if the series converges. In probability theory, this is simplified if all terms are probabilities, and there are no subtractions. In this case, the sum is monotonously increasing, and it converges if there is an upper bound, such as 1 if the sum is a probability. Convergence is important when dealing with double sums. If all terms of the double sum are positive, if there is an upper bound, and if there is no subtraction, then the order of the summation can be changed, that is

$$\sum_i \sum_j a_{i,j} = \sum_j \sum_i a_{i,j}.$$

A similar rule applies to sums of integrals as long as they can be understood as limits of Riemann sums.

When using iterative algorithms, one similarly has to estimate when to stop iterating. In this case, the most frequently used method is to stop iterating as soon as the difference between two iterations is small enough. For instance, one way to find equilibrium distributions of Markov chains is to iterate $\pi(n + 1) = \pi(n)P$ for $n = 0, 1, \ldots$, and stop as soon as $\pi(n + 1)$ is close enough to $\pi(n)$. Of course, since

$\pi(n)$ is a vector, one has to decide what is meant to be "close enough". To measure the closeness of vectors, mathematicians introduced the so-called *vector norms*, and one stops the iteration as soon as the norm of the difference between $\pi(n+1)$ and $\pi(n)$ is below a certain limit. Here, a *vector norm* of a vector v, written as $||v||$, is a mapping from a vector v to a non-negative number x satisfying the following conditions

1. If $x \neq 0$, then $||v|| > 0$.
2. If a is a scalar, then $||av|| = |a|||v||$.
3. If v and w are two vectors, then $||v + w|| \leq ||v|| + ||w||$.

For the purpose of stopping algorithm, one mainly uses the 1-norm and the infinity norm. For the vector $v = [v_1, v_2, \ldots, v_d]$, these two norms are defined as follows:

$$||v||_1 = \sum_{i=1}^{N} |v_i|$$

$$||v||_\infty = \max_i |v_i|.$$

When applied to the iteration $\pi(n + 1) = \pi(n)P$, one stops iterating as soon as $||\pi(n+1) - \pi(n)||$ is less than some prescribed ϵ. When using the 1 norm, this means that one stops iterating as soon

$$\sum_{i=1}^{N} |\pi(n + 1) - \pi(n)| < \epsilon.$$

In the case of the infinity norm, one stops as soon as the maximum difference between $\pi(n + 1)$ and $\pi(n)$ is below some $\epsilon > 0$.

Problems

6.1 Many people believe that with the tremendous increase in computer speed, questions about time complexity will soon become irrelevant. What is your opinion?

6.2 Prove equation (6.1) by complete induction.

6.3 In an $M/M/1$ queue, the expected number in the system is $L = \frac{\rho}{1-\rho}$, where $\rho = \frac{\lambda}{\mu}$. Suppose μ is known, but λ can only be estimated, with an error of $\pm 1\%$ at a 95% confidence interval. Find the corresponding confidence interval for L when $\rho = 0.9$.

6.4 The expressions $a(b - c)$ and $ab - ac$ lead to the same result. If a, b and c are positive, what expression leads to a lower rounding error when using floating point arithmetic? Note that subtractions do not lead to rounding errors, but additions and multiplications do.

Chapter 7
Transient Solutions of Markov Chains

In the previous chapters, we have shown how to generate the transition matrices of discrete-event systems, and we found that these matrices are typically large, but they tend to be sparse. This requires the use of efficient numerical methods for finding transient and equilibrium probabilities. This chapter concentrates on transient solutions, that is, we look for efficient solutions of the transient equations given by equations (1.1) and (1.5) in Chapter 1.

The transient solutions of Markov chains are represented by the $\pi_i(t)$, the probabilities to be in state i at time t. These have to be determined. However, discrete-event systems typically have an immense number of states. and a list of the $\pi_i(t)$ for all i and different values of t is not very informative. It therefore becomes necessary to extract useful information from these data. This is especially true if supplementary variables are used, which may not be of prime interest. What is required is that the data are summarized in terms of marginal distributions and expectations.

Moreover, the $\pi_i(t)$ only reflect the situation at time t. This may not be enough for decision-makers. For instance, in a queueing system, decision-makers may be interested in the time the server is idle in the interval from 0 to T. If the system reflects a business concern, the total profit in the interval from 0 to T is of prime interest. For this reason, we need, beside the $\pi_i(t)$, values $v_i(T)$, which are defined as the expected times the system spends in state i in the interval from 0 to T.

The determination of $\pi_i(t)$ and $v_i(t)$ is straightforward in the case of DTMCs. In the case of CTMCs, one is faced with the solution of differential equations. However, we bypass the solution of these equations by using the so-called randomization or uniformization method, which will be discussed in detail.

In this chapter, we also discuss waiting time problems. To solve waiting time problems, we convert them into Markov chains with an absorbing state, and the waiting time then becomes the time until absorption. If $\pi_0(t)$ is the probability that absorption has occurred before t, then $\pi_0(t)$ is the probability that the waiting time is less than t. By definition, $\pi_0(t)$ is a transient probability, and it can be solved as such.

The outline of this chapter is as follows. We first discuss how to find the appropriate measures in DTMCs, and then we turn to CTMCs. We stress the randomization

W. Grassmann and J. Tavakoli, *Numerical Methods for Solving Discrete Event Systems*, CMS/CAIMS Books in Mathematics 5, https://doi.org/10.1007/978-3-031-10082-6_7

method, also called uniformization. This method is the most popular method for calculating the probabilities and expectations in CTMCs. The main concepts explained in this chapter will be illustrated by an example involving a tandem queue. Finally, we show how to calculate the distribution of waiting times in queues.

7.1 Extracting Information from Data Provided by Transient Solutions

The formulation of systems as Markov chains requires many states, often thousands or hundreds of thousands. For all these states, probabilities are calculated. However, the joint distributions of hundreds of probabilities, as well as hundreds of time averages are confusing, and the problem is to extract the right information from these data. As the first step, one must formulate what information is required for evaluating the performance of the system. For instance, in order to decide the appropriate buffer size M in an $GI/G/1/M$ queue, one would need to know the number of arrivals lost, and this depends on the probability that the number in the system is M. If the cost of waiting is important, and if the cost per customer of waiting increases linearly with the time, then the expected number waiting, averaged over time is also relevant.

For a Markov chain with N states, one concludes from these examples that aside from $\pi(t) = [\pi_1(t), \pi_2(t), \ldots, \pi_N(t)]$, the vector containing the probabilities to be in the different state at time t, one also needs

1. The expected time from 0 to t the system spends in state i for all i.
2. The expected number of events occurring from time 0 to t.

Let $v(T) = [v_1(T), v_2(T), \ldots, v_N(T)]$ be the expected time spent in the different states during the interval from 0 to T. In a DTMC, this number can be found as

$$v(T) = \sum_{t=0}^{T} \pi(t). \tag{7.1}$$

For the CTMC, one has

$$v(T) = \int_0^T \pi(t)dt. \tag{7.2}$$

To prove equation (7.1), note that at time t, the system is either in state i, that is, the actual number of visits at time t is 1, or the system is not in state i, and the actual number of visits is 0. It follows that the expected number of visits at time t is $0 \times (1 - \pi_i(t)) + 1 \times \pi_i(t) = \pi_i(t)$. Adding these expectations over the interval from 0 to T leads to (7.1). Equation (7.2) is just the limit of the appropriate Riemann sums.

To determine the expected number of events from 0 to T in a DTMC, consider the expected number of transitions from state i to state j. There is a transition from state i to state j at t whenever the system is in state i at t, and a transition occurs to j. The probability that these two events occur at t is thus $\pi_i p_{i,j}$, and the expected number of these changes is

$$\text{E(transitions from } i \text{ to } j \text{ in } [0,T]) = \sum_{t=0}^{T} \pi_i(t)p_{i,j} = v_i(T)p_{i,j}.$$

For a CTMC, one has

$$\text{E(transitions from } i \text{ to } j \text{ in } [0,T]) = \int_0^T \pi_i(t)q_{i,j}\,dt = v_i(T)q_{i,j}.$$

Note that $v(t)$ is an expected time, whereas $q_{i,j}v(t)$ is the expected number of occurrences.

Next, the data must be summarized. Of prime interest are joint distributions, marginal distributions, and expectations. Thus, suppose the $\pi_i(t)$, $i = 1, 2, \ldots, N$ are calculated for some t, and the marginal distributions have to be found. The following algorithm shows how to do the accumulation. In this algorithm, "marginal" is a two-dimensional array, and marginal(i, j) accumulates all probabilities in which $X_j = i$, "pv" is the vector of the physical state variables of the state number i, "sv" the vector of the supplementary state variables of state i. The function "decode" finds both "pv" and "sv" corresponding to state i, Of course, decode automatically creates "sv", though it is not needed. The probability of state i is "pit".

Algorithm 26.

```
marginal = 0
for code = 0 : N-1
    [pv(k), sv(k), k=1:d, k=1:e] = decode(code)
    for k = 1 : d
        marginal(pv(k),k) = marginal(pv(k),k) + pit(code)
    end for
end for
```

If the expected duration that state variable X_j has the value i during the time from 0 to t is needed, and the v_i are all calculated, one can modify Algorithm 26 to accomplish that. Essentially, one merely has to change the variable names.

It is often convenient to consider not the Markov process $\{X(t), t \geq 0\}$, but a function of $X(t)$, say $r(X(t))$. Clearly, if $R(t)$ is defined as $r(X(t))$, then $\{R(t), t \geq 0\}$ is a process on its own, but typically not a Markov process. If S is the state space of X, we can find the state space of R as follows: State j is in the state space of R if there is a state $i \in S$ such that $j = r(i)$. We will denote the state space of R by $r(S)$. For instance, if $X = [X_1, X_2]$, with $0 \leq X_1 \leq N_1$, and $0 \leq X_2 \leq N_2$, and $r(X) = X_1$, then $r(S)$ consist of the integers from 0 to N_1.

If $\beta_j(t)$ is the probability that $R(t) = j$, then one has

$$\beta_j(t) = \sum_{i:r(i)=j} \pi_i(t).$$

To calculate this sum, one can first determine $r(i)$ for all $i \in S$, and if $r(i) = j$, include the corresponding probability in the summation. This avoids decoding. Similarly, if

$u_j(T)$ is the expected time $R(t) = j$ in the interval $0 \le t \le T$, one can use

$$u_j(t) = \sum_{i:r(i)=j} v_i(t).$$

If $X = [X_1, X_2, \ldots, X_d]$, then by defining $r(X)$ to be X_k, $\beta_i(t)$ obviously produces the probability for $X_k = i$. The expected time $u_i(T)$ then provides the expected time during the interval from 0 to T that the process $R(t)$ is in state j, $0 \le t \le T$.

The process $R(t)$ is widely used in the theory of *Markov decision processes*, initially described by Howard [50], where it is called a *reward*, and instead of $r(i)$ he used the symbol r_i, where r stands for reward. Actually, he also uses $r_{i,j}$ for the reward of going from i to j, that is, he also associated rewards with state changes.

7.2 Transient Solutions for DTMCs

The transient solutions for a DTMC can be found from (1.1) and in matrix form from (1.2). We repeat (1.2) here for reference.

$$\pi(t) = \pi(t-1)P.$$

Consequently,

$$\pi(t) = \pi(0)P^t. \tag{7.3}$$

The entries of P^t, we call them $p_{i,j}(t)$, give the probabilities to be in state j at time t, given one is in state i at time 0. These probabilities are also called *multi-step transition probabilities*. The vector of the expected number of visits to state i in the interval from 0 to T is given as

$$v(T) = \sum_{t=0}^{T} \pi(t). \tag{7.4}$$

We now provide an algorithms to find $\pi_i(t)$ and $v_i(t)$. We use "pit" for $\pi_i(t)$ and "vit" for $v_i(t)$. We modify Algorithm 25, where the transition matrix is stored in row, column, value entry format. The variables pit(i) and pitnew(i) are used for $\pi_i(t)$ and $\pi_i(t+1)$, respectively. The variable "kbar" represents the number of transitions, and transient solutions $\pi_i(t)$ are obtained for $t \le$ tmax.

Algorithm 27. Algorithm for calculating transient rewards

enter pit(i) for i = 1 : N
vit(i) = pit(i) for i = 1 : N
for t = 1 : tmax
    pitnew(i) = 0 for i = 1 : N
    for k = 1 : kbar
        i = row(k)

```
        j = col(k)
        pitnew(j) = pitnew(j) + pit(i)*prob(k)
    end for
    vit(i) = vit(i) + pitnew(i) for i = 1:N
    pit(i) = pitnew(i) for i = 1 : N
end for
```

Algorithm 27 has two loops, the outer loop executed tmax times, and the inner loop executed kbar times. We conclude that the time complexity is $O(\text{tmax} \times \text{kbar})$. Note that the algorithm contains no subtractions, which makes it numerically stable. At the end of the algorithm, pit(i) yields $\pi_i(t)$, and vit yields $v_i(t)$ with $t = \text{tmax}$.

Some languages, such as MATLAB and Octave, provide a construct for matrix multiplication. These can obviously be used, in particular, if such a language allows sparse matrix representations. For instance, in Octave (MATLAB), the inner loop can be replaced by the statement "pitnew = prob*pitnew". This completes the discussion of transient solutions of DTMCs.

7.3 Transient Solutions for CTMCs

In this section, we discuss how to find $\pi(t)$ given by transient equation (1.6) of Chapter 1, which we repeat here for easy reference

$$\pi'(t) = \pi(t)Q,$$

with $\pi(t) = [\pi_i(t), i = 1, 2, \ldots N]$ and $Q = [q_{i,j}, i, j = 1, 2, \ldots, N]$. In principle, this differential equation can be solved by any of the standard numerical methods available for this purpose (see e.g. [76]). However, as shown in [43], the randomization or uniformization method, also known as *Jensen's method* is more efficient for finding transient solutions of CTMCs.

Equation (1.6) has the following formal solution, where e^{Qt} is defined as $\sum_{n=0}^{\infty}(Qt)^n/n!$:

$$\pi(t) = \pi(0)e^{Qt} = \pi(0)\sum_{n=0}^{\infty}(Qt)^n/n!. \tag{7.5}$$

Here, the elements of the matrix e^{Qt}, say $p_{i,j}(t)$, represent the probabilities to be in state j at time t, given the process started in state i at time 0. To verify that (7.5) satisfies (1.6), one has to take the derivative with respect to t on both sides of (7.5).

Equation (7.5), in its present form, is unsuitable for numerical calculations unless t is small. $(Qt)^n$ increases exponentially with n, and yet the probabilities $\pi_i(t)$ are between 0 and 1. Since all entries of $\pi(t)$ are between 0 and 1, this means that we have to take the difference between large numbers, which leads to intolerable subtractive cancellations. Though (7.5) is not recommended, we can modify it to avoid subtractions.

We set $P = Q/q + I$, where I is the identity matrix and $q \geq -q_{i,i}$, $i = 1, 2, \ldots, N$. With this choice, P becomes a stochastic matrix. Solving $P = Q/q + I$ for Q yields $Q = (P - I)q$, and substituting this value into (7.5) yields

$$\pi(t) = \pi(0)e^{(P-I)qt} = \pi(0)e^{-qt} \sum_{n=0}^{\infty} P^n (qt)^n / n!.$$

Since $e^{-qt}(qt)^n / n! = p(n; qt)$ is the Poisson distribution with parameter qt, one obtains

$$\pi(t) = \pi(0) \sum_{n=0}^{\infty} P^n p(n; qt) \tag{7.6}$$

If $\pi^{(n)} = \pi(0)P^n$, then this equation becomes

$$\pi(t) = \sum_{n=0}^{\infty} \pi^{(n)} p(n; qt). \tag{7.7}$$

For numerical purposes, the sum in this equation must be truncated to some finite value. Methods for achieving this are given later. The vectors $\pi^{(n)}$ can be calculated recursively as

$$\pi^{(n+1)} = \pi^{(n)} P.$$

The same formula was used to find $\pi(n)$ for DTMCs. Consequently, Algorithm 27 can be used here as well.

Equation (7.7) has a probabilistic interpretation. In a DTMC, the state X only changes at times $1, 2, 3, \ldots$ that is, the time interval between changes has a length of 1. If we replace the length of 1 by exponential random variables with parameter q, the discrete-time process is converted into a continuous-time process, while the memory-less property is maintained. Feller [20] calls this method *randomization*. However, the main formula has already been found by Jensen in 1953 [53]. It also has been used extensively by Keilson [55] in order to unify the treatment of continuous-time and discrete-time Markov chains. He calls this method *uniformization*. The method was first used as a numerical tool by one of us in [25]. For details, see [43].

If one is interested in $v_i(T)$, the expected time of being in state i in the interval from 0 to T, one can use

$$v_i(t) = \int_0^T \pi_i(t)dt.$$

The integrals in question can be calculated as follows, using (7.7).

$$\int_0^T \pi_i(t)dt = \sum_{n=0}^{\infty} \pi_i^{(n)} \int_0^T p(n; qt)dt. \tag{7.8}$$

The integral $\int_0^T p(n; qt)dt$ turns out to be (see [28])

$$\int_0^T p(n; qt)dt = \frac{1}{q} \sum_{m=n+1}^\infty p(m, qT). \tag{7.9}$$

To prove this equation, formulate the Poisson process as a CTMC, in which case the transient equations for $m > 0$ become, according to (2.1)

$$p'(m, qt) = -qp(m; qt) + qp(m - 1; qt).$$

Take the sum of this differential equation from $n + 1$ to infinity to obtain:

$$\sum_{m=n+1}^\infty p'(m; qt) = -q \sum_{m=n+1}^\infty p(m; qt) + q \sum_{m=n+1}^\infty p(m - 1; qt).$$

The right side simplifies to $qp(n; qt)$, and we have

$$\sum_{m=n+1}^\infty p'(m; qt) = qp(n; qt).$$

Indefinite integration yields

$$\sum_{m=n+1}^\infty p(m; qt) = q \int p(n; qt)dt.$$

Since for $n > 0$, $p(n; 0) = 0$, the integral from 0 to T can now be found as

$$\int_0^T p(n; qt)dt = \frac{1}{q} \sum_{m=n+1}^\infty p(m; qT).$$

With this, (7.9) is proven.

Using (7.9), equation (7.8) can be written as

$$v(T) = \int_0^T \pi(t)dt = \frac{1}{q} \sum_{m=0}^\infty \pi^{(n)} p^c(n; qt), \tag{7.10}$$

where $p^c(n; qt) = \sum_{m=n+1}^\infty p(m; qt)$.

There is an alternative to (7.10). Instead of $v_i(T)$ the expected time in state i in the interval from 0 to T, we look the average probability, that is, v_i/T. One has

$$\frac{1}{T} \sum_{n=0}^\infty \pi_i^{(n)} \sum_{m=n+1}^\infty p(m; qt) = \sum_{m=0}^\infty p(m; qT) \frac{1}{m+1} \sum_{n=0}^m \pi^{(n)}. \tag{7.11}$$

In other words, to find the average probability for the interval from 0 to T, one can take the average of the $\pi^{(n)}$ from 0 to n, and multiply it by the Poisson distribution. This method was suggested in [43]. Equation (7.11) is proven as follows:

$$\frac{1}{T}\int_0^T \pi(t)dt = \sum_{n=0}^\infty \pi^{(n)} \frac{1}{qT} \sum_{m=n+1}^\infty p(m; qT)$$

$$= \sum_{n=0}^\infty \pi^{(n)} \frac{1}{qT} \sum_{m=n+1}^\infty e^{-qT}(qT)^m/m!$$

$$= \sum_{n=0}^\infty \pi^{(n)} \sum_{m=n}^\infty \frac{1}{m+1} e^{-qT}(qT)^m/m!$$

$$= \sum_{m=0}^\infty e^{-qT}(qT)^m/m! \frac{1}{m+1} \sum_{n=0}^m \pi^{(n)},$$

and (7.11) follows.

Equation (7.11) is interesting on its own. However, as it turns out, its use is less efficient than (7.10) because instead of adding the $p(m; qt)$, one must add vectors.

7.4 Programming Considerations

To find $\pi(t)$ according to (7.7) and $v(t) = [v_i(t)]$ according to (7.10), one must first calculate $\pi^{(n)}$, $p(n; qt)$ and $p^c(n; qt) = \sum_{m=n+1}^\infty p(m; qt)$, $n \geq 0$. This can all be done in a single loop. One also needs an upper limit for this loop.

The $\pi^{(n)}$ can be calculated recursively by using $\pi^{n+1} = \pi^{(n)}P$ as discussed in Section 7.2. To find the $p(n; qt)$, one sets $p(0; qt) = e^{-qt}$, and uses

$$p(n; qt) = p(n-1; qt)qt/n, \ n \geq 1. \tag{7.12}$$

Finally, to obtain $p^c(n; qt)$, one uses

$$p^c(0; qt) = 1 - p(0; qt), \ p^c(n; qt) = p^c(n-1; qt) - p(n; qt).$$

We also need a stopping criterion. The easiest criterion is to stop as soon as $p^c(n; qt)$ is less than $\epsilon > 0$. Let $n = n_u$ at this point. Since $p^c(n_u; qt) = \sum_{m=n_u+1}^\infty p(m; qt)$, all elements of the remaining terms of (7.7) are less than ϵ, and even less than $\pi^{(n_u)}\epsilon$.

We start our discussion with the following algorithm, which will be modified later. In this algorithm "pnqt" stands for the current value of $p(n; qt)$, "pinVec" for the present value of the vector $[\pi_i^{(n)}]$, and "pitVect" is the vector used for accumulating $[\pi_i(t)]$. The variable "ipnqt" is $p^c(n; qt)$, and "ipitVec" accumulates the time integral given by (7.10). The variable "nHigh" is set to some high value to avoid infinite loops.

Algorithm 28. Randomization Algorithm

```
pnqt = exp(-qt)
ipnqt = 1 - pnqt
pinVec = π(0)
```

```
pitVec = pnqt* pinVec
ipitVec = ipnqt* pinVec
for n = 1 : nHigh
    pinVec = pinVec * P
    pnqt = pnqt*(qt)/n
    ipnqt = ipnqt − pnqt
    pitVec = pitVec + pinVec* pnqt
    ipitVec = ipitVec + pinVec* ipnqt
if ipnqt < epsilon then break
end for
ipitVec = ipitVec/q
```

This algorithm breaks down if e^{-qt} results in an exponent underflow. Specifically, in the VBA language, which is included with Excel©, an exponent underflow occurs if $qt > 103$ when using single precision, and $qt > 745$ when using double precision. When an underflow occurs, e^{-qt} is set to 0, and $\pi_i(t) = 0$ for all i, a result that is obviously wrong.

To avoid the underflow, and still use the algorithm above, one can break the interval from 0 to t into several subintervals, say from 0 to t_1, from t_1 to t_2, and so on. At time t_i, the starting probability is not $\pi(0)$, but $\pi(t_i)$, and t is replaced by $t - t_i$. For details of this method, see [86, page 411].

Note that (7.7) has no negative elements, and no subtractions, which assures numerical stability as shown in [30]. There are subtractions when calculating $p^c(n; qt)$, but in this case, subtractive cancellation is a minor factor because the error introduced by subtractive cancellation is small when the values in question are not very close together, and this is the case here. Consequently, when also using double precision, no problems should occur.

There is another way to find an upper limit for the sum of (7.7) that has been used extensively in literature: One approximates the Poisson distribution by the normal distribution [21], [28], [67]. In this way, one can find both a lower limit n_0 and an upper limit n_u for n as follows: $n_0 = qt - z_\alpha \sqrt{qt}$, and $n_u = qt + z_\alpha \sqrt{qt}$, where z_α can be found from normal distribution tables. Of course, if $qt - z_\alpha \sqrt{qt} < 0$, one sets $n_0 = 0$. Note that for large qt, $z_\alpha \sqrt{qt}$ is small compared to qt, and using a higher value of z_α does not significantly increase the computational effort. On the other hand, for small values of qt, the normal distribution does not approximate the Poisson distribution very well, and to counteract the resulting error, one can add a constant value, say c. This gives an upper limit for n in (7.7) of $qt + z_\alpha \sqrt{qt} + c$. In [28], $z_\alpha = 4.26$ and $c = 4$ was chosen. When determining the lower limit for the summation, one can be generous because the effect of the lower limit on the execution time is minor. In [28], the lower limit was set to $n_0 = \max(0, qt - 6\sqrt{qt})$.

Even if there is an upper and lower limit, (7.7) and (7.10) can still be used because for $n < n_0$

$$p(n; qt) \approx 0, \quad p^c(n; qt) = 1 - \sum_{m=1}^{n_0} p(m, qt) \approx 1.$$

The expression $p(n_0; qt) = e^{-qt}(qt)^{n_0}/n_0!$ still contains e^{-qt}, and with it, the danger of an exponent underflow. To avoid underflows, one can calculate a multiple of $p(n; qt)$, say $\hat{p}(n; qt)$ with $\hat{p}(n_0; qt)$ set to an arbitrary value, say $\hat{p}(n_0; qt) = 1$. Now, $\hat{p}(n; qt)$ can be calculated recursively like $p(n; qt)$ for $n > n_0$. The value of $p(n; qt)$ can then be found as $\hat{p}(n; qt)/S$, where $S = \sum_{n=0}^{\infty} \hat{p}(n; qt)$. Since $p(n; qt) = \hat{p}(n; qt)/S$ for all n, one has

$$\pi(t) = \frac{1}{S} \sum_{n=0}^{\infty} \pi^{(n)} \hat{p}(n; qt).$$

Similarly,

$$v(t) = \frac{1}{qS} \sum_{n=0}^{\infty} \pi^{(n)} \left(S - \sum_{m=0}^{n} \hat{p}(m; qt) \right).$$

As we mentioned earlier, one can set $\hat{p}(n_0; qt) = 1$. If the actual value of $p(n_0; qt)$ is very small, then $\hat{p}(n_0; qt)$, $n > n_0$ is much larger than it should be, possibly even leading to an exponent overflow. To obtain a better value for $\hat{p}(n_0; qt)$, one can approximate $n!$ by the Stirling formula

$$n! \approx \sqrt{2\pi n} \left(\frac{n}{e} \right)^n.$$

This leads, after some minor calculations, to the following approximation for $p(n; qt)$

$$p(n; qt) = e^{-qt} \frac{(qt)^n}{n!} \approx e^{n-qt} \left(\frac{qt}{n} \right)^n \frac{1}{\sqrt{2\pi n}}.$$

Assigning this value, with $n = n_0$ to $\hat{p}(n_0; qt)$, will ensure that all $\hat{p}(n; qt)$ calculated using equation (7.12) are close to $p(n; qt)$, which hopefully will avoid any exponent overflow.

We now present an algorithm implementing the theory above. In this algorithm, we use "nLow" for n_0 and "nHigh" for n_u. Note that in this algorithm, one has to calculate S first. Also note that the sums given by (7.7) and (7.10) were broken into two parts, one for $n < n_0$, one for $n \geq n_0$.

Algorithm 29. Randomization Algorithm with Upper and Lower Bound

| Initialization |

qt = qt
nLow = max(0, qt − 5*sqrt(qt))
nHigh = qt + 4*sqrt(qt) + 4.9
If nLow > 0 then
    pn0qt = exp(nLow - qt)(qt/nLow)^nLow /sqrt(2 π nLow)
else pn0qt = exp(−qt)
end if

| Find S |

pnqt = pn0qt; S = pnqt

```
for n = nLow + 1 : nHigh
    pnqt = pnqt*qt/n
    S = S + pnqt
end for
```

Loop for $n < n_0$

```
pinVec = pin
pitVec = 0 ipitVec = 0
for n = 0 : nLow – 1
    ipitVec = ipitVec + pinVec
    pinVec = pinVec * P
end for
```

Loop for $n \geq n_0$

```
pnqt = pn0qt
ipnqt = S -pnqt
for n = nLow:nHigh
    pitVec = pitVec + pinVec* pnqt
    ipitVec = ipitVec + pinVec* ipnqt
    pinVec = pinVec * P
    pnqt = pnqt*qt/(n+1)
    ipnqt = ipnqt – pnqt
end for
pitVec = pitVec/S
ipitVec = ipitVec/(S*q)
```

There are many variations to the algorithm above. For instance, the $p(n; qt)$, given by "pnt", are calculated twice. Instead of recalculating the $p(n; qt)$, one could store them. This requires more memory, yet it hardly saves computer time. Another way to find the $p(n; qt)$ as well as n_0 and n_u has been suggested by Fox and Glynn [21]: They start their calculation from the middle of the distribution, which is qt. The $\hat{p}(n; qt)$ for n less than qt are found recursively as $p(n-1, qt) = p(n; qt)*(n-1)/(qt)$. This almost forces one to store the $p(n; qt)$, and it is questionable if the additional programming effort is worth the trouble, especially since the main programming effort is the vector–matrix multiplication $\pi^{(n)}P$. We therefore believe that the methods implemented in Algorithm 29 are preferable.

7.5 An Example: A Three-Station Queueing System

To illustrate the use of transient solutions, we provide this example: A queueing system in a repair facility has three stations, say checking, fixing, and testing, and any job goes first to checking, then to fixing, and finally to testing. The system opens at 8:00 a.m, and after 5:00 p.m, no arrivals are accepted anymore. However, the facility only closes when all three lines are empty. In this way, the system always starts empty in the morning. The maximum line length is 4 for each station, and if the line is full, the previous line is blocked, or, in the first line, the arrival goes elsewhere. Each line has a single server. Arrivals are Poisson, with a rate of 3 per hour, all service times are exponential, and the service rate of all three servers is 4 per hour. Management requests the following data:

1. How many jobs are lost because line 1 is full? How does this number change if the buffer for line 1 is increased by 1?
2. What is the expected number of completed repairs from 9:00 a.m. to 5:00 p.m?
3. What is the distribution of the three line lengths at 11:00 a.m, 2:00 p.m., and 5:00 p.m?
4. Can you provide a table for the probability that the system becomes empty for times from 5:00 p.m to 8:00 p.m?

We will first describe the general method. To generate the transition matrix, we apply Algorithm 5. Then Algorithm 28 is employed to calculate the transient solution. The resulting transient probabilities for state number i were then obtained as pitVec(i). To find marginal distributions, we essentially used Algorithm 26

Our time is measured in hours, starting at time $t = 0$ at 9:00 a.m. To do the calculation for $t = 3$ (11:00), $t = 6$ (2:00), and $t = 9$ (5:00), we brake the time into three subintervals, for t from 0 to 3, from 3 to 6, and from 6 to 9. The initial probabilities of each subinterval are set to the final probabilities of the previous subinterval, and in each subinterval, the calculation is done for $t = 3$.

The expected line lengths for 9:00, 11:00, and 2:00 are as follows:

Time	line 1	line 2	line 3
11:00	1.541	1.278	1.0309
2:00	1.682	1.508	1.256
5:00	1.717	1.559	1.300

The time averages of the line length for the different 3-hour intervals were also calculated. In particular, the expected time during which the different lines were idle within the different time intervals is as follows:

Time	server 1 idle	server 2 idle	server 3 idle
9:00 to 11:00	1.222	1.584	1.871
11:00 to 2:00	0.844	0.991	1.195
2:00 to 5:00	0.794	0.908	1.108

From these numbers, one concludes that server 3 was idle, on average, for a total time of $1.871 + 1.195 + 1.108 = 4.174$ hours out of 9 hours. Since his service rate is 4 per hour, the expected number served is $4 \times (9 - 4.174) = 19.304$. With an arrival rate of 3 per hour, 27 units arrive on average, of which 19.304 leave until 5:00, while an expected 7.696 units leave later.

Next, consider the expected times in hours the lines are full. These times were found to be:

Time	line 1 full	line 2 full	line 3 full
9:00 to 11:00	0.211	0.099	0.047
11:00 to 2:00	0.395	0.298	0.197
2:00 to 5:00	0.434	0.355	0.243

From these numbers, one concludes that line 1 was full for a total time of 0.211 + 0.395 + 0.434 = 1.04 hours. If line 1 is full, arrivals are lost. With an arrival rate of 3, this yields an expected loss of 3.12, or 16% of all arrivals. Clearly, the expected number of items entering the system, minus the expected number leaving is equal to the number of items in the system at 5:00 p.m. Let us check if this is true for our example. The expected number of units entering is the expected number of arrivals, minus the expected number lost, which is 27 − 3.12 = 23.88. The number of departures until 5:00 is 19.304. If everything is correct, we should find 23.88 − 19.304 = 4.576 items at 5:00 p.m. Indeed, when calculating the expected number of items at 5:00 p.m., we also find 4.576. The fact that there are no noticeable differences between the two results shows that rounding and truncation errors are minimal.

The facility admits no arrivals after 5:00, but it only closes when all lines are empty, which means $X_1 = X_2 = X_3 = 0$. This is state number 0, and the probabilities of being in state 0 are:

Time	5:00	5:30	6:00	6:30	7:00	7:30	8:00
P{empty}	0.0288	0.1132	0.2838	0.5043	0.7105	0.8556	0.9376

From this table, one sees that there is a 28% probability that the system is closed before 6:00. The probability that it is still open at 8:00 is 1 − 0.9376 = 0.0624.

Finally, if the buffer of line 1 is increased by 1, the number of arrivals lost decreases from 3.12 to 2.27, or by 27%. Hence, adding one more space at line 1 should be considered.

One may object that this example has only few states, and more realistic examples would have many more states. To check whether or not larger problems could also be solved by the same method, we increased all buffer sizes to 10, and then there are 1331 states. The results were obtained with no noticeable delay on our computer. We were using a ThinkPad computer, and the language used was Octave.

7.6 Waiting Times

Waiting times have been dealt with extensively in queueing theory, and no wonder. Waiting in queues is a familiar experience for everyone. We join a line, wait for service, and leave after having received service. Technically, we can model waiting times as Markov chains with an absorbing state, a state that is entered as soon as the wait is over. The time until absorption can then be interpreted as a transient solution of a Markov chain, and found by Algorithm 28 or 29.

We sometimes use the word generalized waiting time for the time needed to reach a certain state or group of states. For instance, in Section 7.5, we were interested in the time needed until the system is empty. In this case, the empty state is made into an absorbing state. More generally, we can consider times to absorption as transient solutions. In particular, if X is a PH random variable, then it can be considered as the time to absorption of the process with transition matrix T defining the PH distribution. Consequently, to find the cumulative distribution of X, $F_X(t)$, for several values of t, methods for finding transient solutions can be used.

In queueing theory, one distinguishes between waiting time and sojourn time. The waiting time is the time an entity has to wait until it receives service, whereas the sojourn time is the time until the departure of the entity from the system. In our terms, the sojourn time is a generalized waiting time.

In order to calculate waiting times, we formulate a waiting time process $X(t)$. Here, t represents the time since the beginning of waiting time, that is, $X(0)$ is the state of the waiting time process at the moment the wait starts. We assume that $X(t)$ can only have values between 0 and N, where the state $X(t) = 0$ is the state that the wait is over, an absorbing state. The probability that $X(t) = j$ will be denoted by $\pi_j(t)$, and the vector $\pi(t)$ is defined as $[\pi_0(t), \pi_1(t), \pi_2(t), \ldots, \pi_N(t)]$. To find $\pi(t)$, we need the transition matrix of the process $\{X(t), t \geq 0\}$, as well as $\pi_i(0)$, the probability that $X(0) = i$.

The waiting process starts with a specific arrival, and it ends as soon as a specific event occurs, such as "start of the service". The system for which the waiting time is to be found will be called *parent system* or *parent process*. For instance, in an $M/M/1$ queue, the parent process is given by the number in the system at the time of arrival starting its waiting time. The parent process will be denoted by $\{Z(s), s \geq 0\}$. If the arrival in question occurs at time τ, $Z(\tau)$ determines $X(0)$, that is, $X(0)$ is a function of $Z(\tau)$. Also, let $\phi_i(\tau)$ be the probability that $Z(\tau) = i$, with $\phi(t) = [\phi_j, j = 1, 2, \ldots, N]$.

If the arrival initiating the waiting time occurs at time τ, at which time we set $t = 0$, then we need $Z(\tau)$. The problem is that at time τ, it may be impossible to have an arrival in certain states. For instance, in an $M/M/1/N$ queue, no arrivals are possible when $Z(\tau) = N$, because then, the buffer is full. Similarly, when there are Erlang arrivals, then no arrivals are possible unless the arrival phase is equal to 1. The question now arises of how to account for this. One possibility is to ignore the state and assume arrivals can happen in any state, and it has the same rate in every state. In this approach, one obtains the so-called *virtual waiting time*. The alternative is to consider the probability that arrivals depend on the state of the system. If this is done, we call the resulting waiting time *customer waiting time*.

To find the customer waiting time, one must find the distribution of $Z(\tau)$, under the condition that an arrival occurs immediately after τ. For instance, in an $M/M/1/N$ queue, arrivals only occur when $Z(\tau) < N$. Therefore, we have to determine the probability that $Z(\tau = i)$, given $Z(\tau) < N$, which is:

$$P\{Z(\tau) = i \mid Z(\tau) < N\} = \frac{\phi_i(\tau)}{\sum_{j=1}^{N-1} \phi_j(\tau)}, \quad i < N.$$

Similarly, if arrivals are Erlang, and they can only occur in phase 1 of the arrival process, and if $\phi_{i,j}(\tau)$ is the probability that the number in the system is i, and the arrival phase is j, then we have

$$P\{Z(\tau) = i \mid \text{arrival}\} = \frac{\phi_{i,1}(t)}{\sum_{i=1}^{N_1} \phi_{i,1}(t)}.$$

Here, N_1 is the number of values i can assume.

Next, consider hyper-exponential arrivals, with rate λ_k in arrival phase k, $k = 1, 2$. To obtain the state probabilities at time τ, given there is an arrival, we first calculate the probability that during a small interval from $\tau - h$ to τ, there is an arrival, and then we condition these probabilities such that their sum is 1. Since

$$P\{Z(\tau) \mid \text{arrival in } (\tau - h, \tau)\} = h(\lambda_1 \phi_{i,1}(\tau) + \lambda_2 \phi_{i,2}(\tau)).$$

It follows that

$$P\{Z(\tau) = i \mid \text{arrival}\} = \frac{\lambda_1 \phi_{i,1}(\tau) + \lambda_2 \phi_{i,2}(\tau)}{\sum_{j=1}^{N_1} \lambda_1 \phi_{j,1}(\tau) + \lambda_2 \phi_{j,2}(\tau)}.$$

This result can easily be generalized to arrival with any PH inter-arrival time distribution. If the PH distribution has k phases, and if the arrival rate in phase i is λ_i, one has

$$P\{Z(\tau) = i \mid \text{arrival}\} = \frac{\sum_{n=1}^{k} \lambda_n \phi_{i,n}(\tau)}{\sum_{j=1}^{N_1} \sum_{n=1}^{k} \lambda_n \phi_{j,n}(\tau)}.$$

Note that if arrivals are Poisson, and if they occur at the same rate in every state of the system, then the fact that an arrival is imminent does not allow one to make any conclusions about the state of the system. In this case,

$$P\{Z(\tau) = i \mid \text{arrival}\} = P\{Z(\tau) = i\} = \phi_i(\tau).$$

All the formulas above also hold when the parent process is in a statistical equilibrium. One merely has to change $\phi_i(\tau)$ to ϕ_i. For systems in equilibrium, there is also another way to look at things: Instead of observing the process at any random time τ, one can observe it only when arrivals happen. In other words, the process is $\{Z_n, n = 1, 2, 3, \ldots\}$, where Z_n is the state just before the nth arrival. When arrivals are Poisson and the systems are in equilibrium

$$P\{Z(t) = i \mid \text{arrival}\} = P\{Z_n = i\}.$$

This is the well-know PASTA principle, which stands for "Poisson Arrivals See Time Averages". The term "time averages" indicates that in equilibrium, any τ chosen without regard of the state of the system sees time averages.

The way $Z(\tau)$ determines $X(0)$ depends on the specifics of the model, as well as on the queueing discipline. Here, we consider three queuing disciplines: They are First In, First Out (FIFO), Service In Random Order (SIRO), and Last In First

Out (LIFO). To understand how queueing disciplines and parent models affect the relation between $Z(\tau)$ and $X(0)$, consider the waiting time process of the $E_k/M/c$ queue, first under the FIFO discipline, then under the SIRO discipline. Here, the parent process has two state variables, $Z_1(t)$, which is the number of elements in the system, and $Z_2(t)$, which is the phase of the Erlang process. Under the FIFO discipline, the waiting time process has only one state variable: $X(t)$, which is the number of entities that must be served before the wait of the arrival occurring at τ is over. Consequently, $X(0) = 0$ if $Z_1(\tau) < c$, and $X(0) = Z(\tau) - c + 1$ if $c \geq 0$. However, under the SIRO discipline, $X(t)$ is a vector, that is, $X(t) = [X_1(t), X_2(t)]$, where $X_1(t)$ is the number in the queue, including the arrival, and $X_2(t)$ is the phase of the arrival process. This implies that $X_1(0) = 0$ if $Z(\tau) < c$, and $X_1(0) = Z(\tau) - c + 1$ for $Z(\tau) \geq c$. $X_2(0) = k$, because after the arrival, the arrival process is in phase k.

The waiting time process shares many events of the parent process, but not necessarily all of them. In particular, under the FIFO discipline, the waiting time process does not have arrivals. Moreover, the rates and the event functions in the waiting time process may differ from the ones of the parent process. For instance, under the SIRO discipline, departures, which have a rate of $c\mu$, become stochastic events: With probability $c\mu/X_1(t)$, the result is a transition to state 0, finishing thus the waiting time, and with probability $1 - c\mu/X_1(t)$, $X_1(t)$ decreases by 1. However, the events "next phase" and "arrivals" are not different from the corresponding ones in the parent process.

The waiting time T is the time needed until the waiting time process enters state 0. For CTMCs, we either want to find $F_T(t) = P\{T \leq t\}$ or the density $f_T(t) = F_T'(t)$. For DTMCs, the probability p_t that $T = t$ is of interest. There are three possible ways to find the distribution of T.

1. If the process is in state 0, it must have entered this state at or before t, which implies
$$P\{T \leq t\} = F_T(t) = \pi_0(t).$$

2. If the process is in any state other than 0, absorption has not taken place; consequently
$$P\{T > t\} = \sum_{i \neq 0} \pi_i(t).$$

3. In a discrete system, the probability that absorption occurs at time t is given as:
$$P\{T = t\} = \sum_{i \neq 0} \pi_i(t - 1)p_{i,0}.$$

Similarly, for continuous-time systems, the density for the time to absorption is given as
$$f_T(t) = \sum_{i \neq 0} \pi_i(t)q_{i,0}.$$

For discrete-time Markov chains, the third method is the most efficient one. The first method uses the formula $\pi_0(t) = \pi_0(t - 1) + \sum_{i \neq 0} \pi_i(t - 1)p_{i,0}$, where the sum is equal to the probability of entering state 0. Consequently, finding $\pi_0(t)$ requires one addition more than finding the probability of entering state 0. On the other hand, finding $P\{T > t\}$ by alternative 2 is even less efficient.

7.6.1 Waiting Times in the $M/M/1$ Queue under Different Queuing Disciplines

To illustrate the methods above, we discuss waiting time distributions in $M/M/1$ queues under several different queueing disciplines, including Fist In, First Out (FIFO), Service In Random Order (SIRO), and Last In, First Out (LIFO). The arrival rate will be denoted by λ and the service rate by μ. We also define ρ to be λ/μ. We look at the waiting time of a specific customer, we call her C. Her waiting time is defined as the time from the moment C arrives, until the moment C enters service.

Of great interest is the case where the system is in equilibrium. Let X be the number in the system. It is known that

$$P\{X = n\} = (1 - \rho)\rho^n.$$

As long as the arrival rates are independent of the number in the system, which is true for the $M/M/1$ queue with an infinite buffer, the PASTA principle applies, that is, the equilibrium probabilities also hold immediately before an arrival. Of course, if the system is empty, the waiting time is zero. Let us now discuss the waiting time distributions under the different disciplines, and how they are calculated.

7.6.1.1 FIFO

If the customers are served in the order of their arrival, then all customers in the system at the time when C arrives must be served before C enters service. The probability of serving i customers in time t is $p(i, \mu t)$. If there are i customers in the system when C arrives, and less than i customers are served t units after the arrival of C, then the waiting time of C exceeds t. If T is the waiting time of C, this means

$$P\{T > t \mid i\} = \sum_{n=0}^{i-1} p(n; \mu t). \tag{7.13}$$

Consequently,

$$P\{T \leq t \mid i\} = 1 - \sum_{n=0}^{i-1} p(n; \mu t) = \sum_{n=i}^{\infty} p(n; \mu t).$$

According to equation (7.9), we have

$$\sum_{n=i}^{\infty} p(n; \mu t) = \mu \int_0^t p(i - 1; \mu \tau) d\tau.$$

Consequently, $P\{T \le t \mid i\} = \mu \int_0^t p(i - 1; \mu \tau) d\tau$. Taking the derivative of $P\{T \le t \mid i\}$ yields the density of the waiting time, that is,

$$f_T(t; i) = \mu p(i - 1; \mu t), \ i > 0. \tag{7.14}$$

For $i = 0$, the waiting time is obviously 0.

Equation (7.13) can now be used to find the customer waiting time distribution under FIFO waiting time for the $M/M/1/N$ queue. If Z_n is the number in the system before C joins, then $Z_n < N$ because otherwise, C could not have joined. It follows that

$$P\{Z_n = i\} = \rho^i / c, \ c = \sum_{i=1}^{N-1} \rho^i, \ i = 0, 1, \ldots, N - 1. \tag{7.15}$$

All the customers in the system before C arrived must be served before C. Keeping this in mind, one finds that the unconditional probability that the customer waiting time is greater t is given as:

$$P\{T > t\} = \sum_{i=1}^{N-1} \frac{1}{c} \rho^i \sum_{n=1}^{i-1} p(n; \mu t).$$

The density $f_T(t)$ can be found in a similar way.

In the $M/M/1$ queue with unlimited waiting space, the equilibrium probability is $(1 - \rho)\rho^i$, and according to PASTA, this is the distribution any arrival, including C, will encounter. It follows that the waiting time density is given by the mixture of $(1 - \rho)\rho^i$ and $f_T(t; i)$ given by equation (7.14):

$$\mu(1 - \rho) \sum_{i=1}^{\infty} \rho^i p(i - 1; \mu t) = (\mu - \lambda)\rho e^{-\mu t} \sum_{i=1}^{\infty} \left(\frac{\lambda}{\mu}\right)^{i-1} (\mu t)^{i-1} / (i - 1)!.$$

The sum to the right evaluates to $e^{\lambda t}$, and one has

$$f_T(t) = \rho(\mu - \lambda)e^{-(\mu - \lambda)t}. \tag{7.16}$$

Consequently,

$$P\{T > t\} = \int_t^{\infty} \rho(\mu - \lambda)e^{-(\mu - \lambda)x} dx = \rho e^{-(\mu - \lambda)t}. \tag{7.17}$$

Also

$$P\{T \le t\} = 1 - \rho e^{-(\mu - \lambda)t}.$$

This equation can also be used for $t = 0$. In this situation, $P\{t = 0\} = P\{X = 0\} = 1 - \rho$.

When discussing the other queueing disciplines, we formulate and solve the waiting time problem as a CTMC, and we will do this now for the FIFO discipline with at most N customers in the system. Instead of deriving a formula, we provide the transition matrix of the waiting process $\{X(t)\}$, and the initial probability vector $\pi(0)$. The waiting time distributions can then be obtained by a computer program, like the one based on Algorithm 28 or 29. We define $X(t)$ to be the number in the system ahead of C, with $X(t) = 0$ meaning that the wait of C has ended. Note that the number in the system ahead of C after she joined is equal to Z_n the number in the system before she joined. It follows that $Z_n = X(0)$, and according to equation (7.15), we obtain

$$P\{Z_n = i\} = P\{X(0) = i\} = \pi_i(0) = \rho^i/c.$$

This provides the initial distribution of $X(t)$, and we now provide the transition matrix.

Clearly, when $X(t) = 0$, no event can change $X(t)$ any more. If $X(t) \neq 0$, there is only one event that can change $X(t)$, which is the completion of a service, which reduces $X(t)$ by 1. This leads to the following transition matrix:

	0	1	2	$\ldots$	$N-2$	$N-1$
0	0	0	0	$\ldots$	0	0
1	μ	$-\mu$	0			0
2	0	μ	$-\mu$			$\vdots$
$\vdots$	$\vdots$					$\vdots$
$N-1$	0	0	0	$\ldots$	μ	$-\mu$

One can now find the transient probabilities $\pi_i(t)$, given the initial distribution $\pi_i(0)$ and calculate $\pi_0(t)$, the probability to be in the absorbing state 0 at time t. Clearly,

$$P\{T \leq t\} = \pi_0(t).$$

7.6.1.2 SIRO

Next, we consider the $M/M/1/N$ queue with a SIRO discipline, that is the server selects the customer to be served next randomly from the members in the queue. This system was used in the past for telephone answering systems. As before, we select a particular customer, say C, and find her waiting time.

The SIRO waiting time process, $X(t)$, is given by the number of customers in the queue at time t, including C, except that $X(t) = 0$ means that the waiting time of C has ended. If C is the nth arrival, she finds Z_n customers in the system when she arrives, that is, there are $Z_n - 1$ persons in the queue before C arrives and Z_n after she has joined the queue. It follows that $X(0) = Z_n$, where the distribution Z_n is given by equation (7.15). Consequently,

$$P\{Z_n = i\} = P\{X(0) = i\} = \pi_i(0) = \rho^i/c, \ i < N,$$

where $c = \sum_{i=1}^{N-1} \rho^i$. If the queue is empty, and this has a probability of $1/c$, then C starts service right away. The waiting time is therefore 0 with probability $1/c$.

Next, we look at the events that guide the process $X(t)$. As under FIFO, $X(t) = 0$ is an absorbing state, and no events happen in this state. Otherwise, whenever an arrival occurs, $X(t)$ increases by 1. After a departure, C is chosen with probability $1/X(t)$, and $X(t)$ becomes 0. Hence, with a rate of $\mu/X(t)$, a transition from $X(t)$ to 0 occurs. If C is not chosen, which happens with probability $(X(t) - 1)/X(t)$, $X(t)$ decreases by 1. The event table is therefore as follows.

Event	Event Function	Rate	Condition
Arrival	$X + 1$	λ	$X < N - 1$
C starts service	0	μ/X	$X > 0$
Other starts service	$X - 1$	$(X - 1)\mu/X$	$X > 0$

This leads to the following transition matrix:

	0	1	2	$\ldots$	$N - 2$	$N - 1$
0	0	0	0	$\ldots$	$\ldots$	0
1	μ	$-(\mu + \lambda)$	λ	$\ddots$	$\ddots$	0
2	$\frac{\mu}{2}$	$\frac{\mu}{2}$	$-(\mu + \lambda)$	$\ddots$	$\ddots$	$\vdots$
3	$\frac{\mu}{3}$	0	$\frac{2\mu}{3}$	$\ddots$	$\ddots$	$\vdots$
$\vdots$	$\vdots$	$\ddots$	$\ddots$	$\ddots$	$\ddots$	$\vdots$
$N - 1$	$\frac{\mu}{N-1}$	0	$\ddots$	$\ddots$	$\frac{(N-2)\mu}{N-1}$	$-\mu$

This provides all information to use a program written for finding transient solutions of CTMCs. The probability that the waiting time T is less than t is again $\pi_0(t)$.

7.6.1.3 LIFO

The service discipline LIFO, or Last In, First Out, is obviously unacceptable to human customers. However, it occurs in inventory situations. There may be instances where products, say particle boards, arrive according to a Poisson distribution, and the time between usages is exponential. In this situation, it is natural to place an arriving item at the top of the stack, and when a particle board is needed, it is taken from the top of the stack. This leads to a LIFO queueing discipline.

Though stacks of particle boards do not look like queues of people, the events, their functions, and their rates are the same. It is irrelevant whether or not X is a stack size or a queue length, arrivals increase X by 1, and departures decrease X by 1. There is only a difference between a queue and a stack when only a single element remains. In a queue, this element would be in service, and it will be the next element to leave, whether there are arrivals or not before service is completed. If there is only one element on the stack, this element is only the next element to leave if there is no delivery before the next usage. Also, whereas in human lineups, the time in queue may be more important, in the case of products, it may be the sojourn time, the time the item remains on the stack.

For the model to be discussed, the sojourn time of an entity, we call it A, is the time from the moment A is delivered until the moment it is used. To formulate the sojourn time model, let $X = 0$ represent the state that the waiting time is over. If $X \neq 0$, we need to know the number of entities that were delivered after A, because they are all served before A. However, unless the size of the stack is limited to N, the number of items delivered before A are irrelevant because they are removed after A. Let us first assume that there is no limit to the size of the stack. In this case, we define $X(t)$ to be one more than the number of items delivered after A. In this way, we can reserve $X = 0$ for the absorbing state.

In this model, $X(0) = 1$, because after the delivery of A, there is nothing on top of A. Consequently, $\pi_1(0) = 1$, $\pi_i(0) = 0$, $i > 0$. If $X(t) = 0$, the sojourn time is over, and there are no events. If $X(t) > 0$, then any delivery will increase $X(t)$ by 1, and any usage will decrease $X(t)$ by 1. This leads to the following event table.

Event	Event Function	Rate
Delivery	$X + 1$	λ
Usage	$X - 1$	μ

To solve this model numerically, we need to restrict the state space. We can assume, for instance, that not more than $M - 1$ items are placed on top of A, which means $X(t) \leq M$. In this case, we arrive at the following transition matrix:

	0	1	2	3 ...	$M-1$	M
0	0	0	...	... ...	...	0
1	μ	$-(\mu+\lambda)$	λ	0 ...	...	$\vdots$
2	0	μ	$-(\mu+\lambda)$	λ		$\vdots$
$\vdots$	$\vdots$					$\vdots$
$\vdots$	$\vdots$					$\vdots$
$M-1$	0	...	...	0	μ $-(\mu+\lambda)$	λ
M	0	...	...	... ...	μ	$-\mu$

The distribution of the waiting time T can now be found as

$$P\{T \leq t\} = P\{X(t) = 0\} = \pi_0(t).$$

If the size of the stack is limited to N, then we need to know the stack size S before the arrival of A, which is equal to Z_n. The distribution of Z_n is given by equation (7.15). Let $\pi_{s,i}(t)$ be the probability that at time t, there are i arrivals, including A, while $S = s$. Since at the beginning, $S = Z_n$, and only A is on top of the stack, we have:

$$\pi_{s,1}(0) = P\{Z_n = s\} = \frac{1}{c}\rho^s.$$

Since none of these s items are used before A, s does not change. For given s, the transition matrix is identical to the previous one, with $M = N - s$. We can thus calculate $\pi_{s,0}(t)$ as before for all possible values of s. The overall probability that $T < t$ is then:

$$P\{T < t\} = \sum_{s=0}^{N-1} \frac{1}{c}\rho^s \pi_{s,0}(t).$$

In the next section, we will compare the performance of the different systems. Before doing that, we have to convert the sojourn time distribution to the waiting time distribution. To accomplish this, we only have to set $\pi_0(0) = \frac{1}{c}$ and $\pi_1(0) = 1 - \frac{1}{c}$.

7.6.2 Comparison of the Queueing Disciplines

Note that the expected waiting time does not depend on the queueing discipline because according to Little's theorem, the expected waiting time is $E(Z)/\lambda$, where Z is the line length, and $E(Z)$ is the same for all queueing disciplines. However, there are big differences between their cumulative distributions $F_T(t) = P\{T > t\}$. To highlight the differences, it is convenient to express the waiting time T as a multiple of μ, and to achieve this, we set $\mu = 1$. Consequently, λ must be strictly less than 1 if the system is to reach a statistical equilibrium. We use $\lambda = 0.8$.

Let us now suppose customer C would like to know how much time to reserve for waiting if she has to wait. She would also like to be sure that this time is sufficient with probability 0.95. Hence, she looks for a t such that $P\{T \le t\} > 0.95$ or, equivalently, $P\{T > t\} \le 0.05$. For the $M/M/1$ queue with $\mu = 1$, $\lambda = 0.8$ and FIFO, we can use (7.17), and find the equation

$$P\{T > t\} = 0.8e^{-0.2t} = 0.05.$$

Solving this equation leads to $t = \ln(0.05/0.8)/(-0.2) \approx 14$. This implies that if $\rho = 0.8$, she has to reserve a time equal to 14 times the service time for waiting, if the reserved time is to be sufficient with probability 0.95. This is surprisingly long! Of course, the queueing disciplines SIRO and LIFO are certainly worse. This can be clearly seen from Figure 7.1, which plots $P\{T \ge t\}$, with T being the waiting time, for FIFO, SIRO,s and LIFO. Since we needed numerical methods, we had to bound the number in the system. We chose a bound of 20 for FIFO and SIRO. For this reason, the number for the FIFO disciplines deviates from the one obtainable from (7.17). For LIFO, we allowed M to reach 20. When using the LIFO discipline, the probability of having short waiting times is higher than under FIFO. On the other hand, some customers have very long waiting times under LIFO, much longer than under FIFO. SIRO is somewhere between these extremes (Figure 7.1).

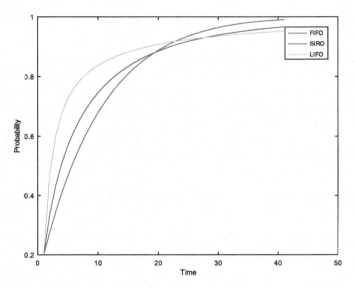

Fig. 7.1 Probability of waiting over t expected service times

7.7 Conclusions

In the real world, systems are changing constantly, and a statistical equilibrium is seldom encountered. This makes the transient solution important. However, the bulk of literature is devoted to equilibrium solutions because, in many situations, the actual system is often close to a statistical equilibrium. Also, in many cases, equilibrium solutions are easier to derive than transient solutions. Equilibrium solutions are also used if one is not given an initial probability vector $\pi(0)$, because, without an initial distribution, no transient distribution can be calculated, and then, the equilibrium solution may be a reasonable approximation.

Some applications of transient solutions may not be obvious. One application is the calculation of waiting time distributions, as they were analyzed in this chapter. Of course, there is a great variety of waiting time problems, and not all could be discussed here. For instance, one can look at waiting time in multiserver queues, or at waiting times in queues with PH inter-arrival times and/or PH service times. Virtual waiting times are also of interest in some applications. However, the methods discussed here can be applied to also solve these more complex waiting time problems.

If a random variable X follows a PH distribution, $P\{X \leq x\}$ can also be cast as a Markov chain problem, a topic that will be explored in the next chapter. Finally, some embedded Markov chains require to find transient solutions, as shown in Chapter 10.

Problems

7.1 Implement Algorithm 29 and run it for the following problem: In a telephone answering service, waiting customers are divided into two groups, with a small group containing only three customers, and the other group containing all other customers. Arrivals to the system are Poisson, with a rate of λ, and each server has a service rate of μ. When a server is free, he or she picks a customer at random from the group of three. At this point, a new waiting customer from outside the group joins the small group. Define the transition matrix Q, as well as the initial probabilities for an $M/M/1$ queue in equilibrium.

7.2 Implement Algorithm 29 and find transient solutions for the $M/M/3/10$ queue.

7.3 Construct a graph for the expected line length at t of the $M/M/1$, the $E_2/M/1$ and the $H_2/M/1$ queue for t running from 0 to 20, with a step size of 1, when $\lambda = 0.8$ and $\mu = 1$.

7.4 In their paper [21], Fox and Glynn suggest to find the Poisson distribution, starting with $p(n; \mu)$, with $n = \lfloor \mu \rfloor$. Change Algorithm 29 to implement this idea.

7.5 Provide formulas to find the FIFO waiting time of the $GI/G/1/N$ queue in equilibrium.

7.6 Find the SIRO actual and virtual waiting time for the $E_2/E_2/1/N$ queue with $\lambda = 0.9$ and $\mu = 1$.

Chapter 8
Moving toward the Statistical Equilibrium

We have repeatedly referred to equilibrum distributions of Markov chains, but we have not yet addressed the question of whether or not they exist, and if they exist, whether or not they depend on the initial state. These are the questions addressed in this chapter.

In the center of our investigation is $p_{i,j}(t)$ which, for the process $\{X(t), t \geq 0\}$, is defined as

$$p_{i,j}(t) = P\{X(t) = j \mid X(0) = i\}.$$

The questions are:

1. Does $p_{i,j}(t)$ have a limit as $t \to \infty$?
2. Does this limit depend on i?
3. How fast is this limit reached?
4. Under what conditions is this limit 0?

To investigate these questions, we need to consider the paths between the states. Based on the existence of such paths, we will introduce a *classification of states*.

The processes we investigate need not necessarily be Markovian. For instance, in a system with several state variables, a single state variable is usually not a Markov chain, but the notion of a path still applies.

To measure convergence, one has recently proposed what is called *mixing time*. Informally, the mixing time of a Markov chain is the minimum time until the transient solution has approached its equilibrium distribution within a certain margin of error ϵ. To determine this, one needs to define a proper norm as defined in Section 6.4.3. Typically, the infinity norm is used [64]. Similar issues will also be addressed when we discuss iterative solutions in Section 9.3.

Aside from the speed of convergence, we look at more behavioral issues, such as

1. How fast does $p_{i,j}$ converge to its limit?
2. Is the approach to the limit monotonic or non-monotonic?

It turns out that these questions are also important to characterize PH distributions, which are really times to absorption of Markov chains. To explore their behavior, we use eigenvalues.

© The Author(s), under exclusive license to Springer Nature Switzerland AG 2022   155
W. Grassmann and J. Tavakoli, *Numerical Methods for Solving Discrete Event Systems*,
CMS/CAIMS Books in Mathematics 5, https://doi.org/10.1007/978-3-031-10082-6_8

Aside from the limits of the $p_{i,j}(t)$, we also look at the limits of time averages, which are defined as follows:

$$\bar{\pi}(T) = \begin{cases} \frac{1}{T+1} \sum_{t=0}^{T} \pi(t) & \text{DTMC} \\ \frac{1}{T} \int_0^T \pi(t)dt & \text{CTMC.} \end{cases}$$

Here, $\bar{\pi}(t)$ represents the probabilities averaged through time. We thus have to distinguish between *standard convergence*, the convergence of $\pi(t)$ and *time-average convergence*, the convergence of $\bar{\pi}(t)$. Generally, standard convergence implies time-average convergence, but the reverse is not necessarily true. Time averages are closely related to time integrals, which have been used in Chapter 7 in some of our models, in particular in the model discussed in Section 7.5.

In the example of Section 7.5, we also used rewards, which were defined as $R(t) = \sum_j r_j \pi_j(t)$, where $\pi_j(t) = \sum_i \pi_i(0)p_{i,j}(t)$. In this definition, marginal distributions, as well as expectations can be interpreted as rewards. Again, one can ask about the behavior of $R(t)$. This, however, can be obtained on a case-by-case basis from the results regarding the $p_{i,j}(t)$.

Let us now assume that $p_{i,j}(t)$ has a limit as $t \to \infty$, and that this limit is independent of i. In this case, the limit is given by the equilibrium probabilities π_j, where π_j is possibly 0. The question is now how quickly this limit is approached. Moreover, one may ask if the approach to the limit is monotonic, or if there are oscillations, that is, if at times the difference $|p_{i,j}(t) - \pi_j|$ first increases and subsequently decreases. Such oscillations make tests for convergence more difficult.

In this chapter, we leave it open as to whether one should consider the absolute or relative error when deciding that $p_{i,j}(t)$ is close enough to its limit. However, the material presented in this chapter should facilitate finding answers to such questions.

Transient solutions, or modifications of transient solutions, are often used to find steady-state solutions. Also, iterative solutions, which will be discussed in Chapter 9, are based on schemes that are similar to the iterations used to find transient solutions of DTMCs. The question then arises of how to find a starting value to reach equilibrium as quickly as possible. This chapter will therefore address the effect of the starting probabilities on the speed of convergence.

The outline of the chapter is as follows: We first discuss the possible paths between states, and their effect on the convergence to steady state. This naturally leads to the classification of states. Next, we show how eigenvalues give important information about the types of convergence, whether the equilibrium is approached exponentially, or as damped oscillation. When discussing eigenvalues, we use Coxian distributions to demonstrate some of these concepts. This not only clarifies many issues about eigenvalues but also is an important topic for its own sake.

8.1 Structure of the Transition Matrix and Convergence toward Equilibrium

The structure of the transition matrix gives many indications if, and how fast, $p_{i,j}(t)$ converges to its limit. Basic to this investigation is the notion of a path between states. The existence of paths between states leads naturally to a partition of the states into what is called *communicating classes*. We also consider partitions into classes that are almost communicating in a sense that will be discussed later.

8.1.1 Paths and Their Effect on the Rate of Convergence

Consider a process $\{X(t), t \geq 0\}$. This may be a DTMC or a CTMC, or even a more general process, as long as the number of states is finite. We now define a path as follows:

Definition 8.1 A path from i to j of length n is a sequence $i_1, i_2, \ldots, i_n, i_n$, with $i_1 = i$ and $i_n = j$, where $p_{i_v, i_{v+1}} > 0$ for $v = 1, 2, \ldots, n$. If there is a path from i to j, j is said to be *reachable* from i.

Note that according to this definition, if $p_{i,i} > 0$, there is a path of length 1 from i to i.

The lengths of paths between i and j certainly influence how fast $p_{i,j}(t)$ converges to its limit as $t \to \infty$. If there is no path between i and j, then $p_{i,j}(t) = 0$ for all t, that is, the limit is trivially 0. If there is a path, then the rate of convergence of $p_{i,j}(t)$ depends on several factors, one of them being the length of the shortest path from i to j. This is the case because $p_{i,j}(t)$ cannot change its value before j is visited at least once when starting from i, that is, the length of the shortest path from i to j is a useful indicator of the rate of convergence of $p_{i,j}(t)$. Let us explore this in the case of discrete-event systems. Specifically, consider the state variable X_k. In many of our examples, X_k could only be increased or decreased by 1. This means that if i is a state where $X_k = i_k$, and j is a state where $X_k = j_k$, $k = 1, 2, \ldots, d$, then one can expect that the speed of convergence of $p_{i,j}(t)$ decreases as $|j_k - i_k|$ increases.

For finding the rate of convergence of $p_{i,j}$ toward its limit, we also have to consider the interactions between state variables. In particular, in tandem queues, starting with an empty system, more events must occur before downstream lines are affected. This reduces the speed of convergence of the X_k that represents the downstream waiting lines.

In Markov chains with stochastic transition matrices, respectively, with proper infinitesimal generators, the convergence of $p_{i,j}(t)$ toward its limit π_j also depends on the speed of convergence of $\pi_{i,k}(t)$ toward its limit π_k, where $k \neq j$. For instance, if the equilibrium probability π_k is high, say it is 0.5, then it is impossible that $p_{i,j}(t)$, can approach its equilibrium probability π_j before $\pi_{i,k}(t)$ is reasonably close to its equilibrium probability of 0.5 because $\sum_{j \neq k} \pi_{i,j}(t) = 1 - p_{i,k}(t)$.

The $p_{i,j}(t)$ in large, but sparse transition matrices often converge slowly toward their equilibria. First of all, in large, but parse matrices, the paths between states can be long. If the bulk of the probability mass is concentrated in states that can only be reached from i by long paths, then convergence is slowed down even more.

The existence of paths can be used to determine if $p_{i,j}(t)$ converges to 0 as $t \to \infty$. If there is a path from j to any state k, but no path from k to j, then $\lim_{t \to \infty} p_{i,j}(t) = 0$. This situation occurs in the case of PH distributions and waiting time distributions. In both cases, there were absorbing states, and once in an absorbing state, there is no path back. In these models, the length of the paths to the absorbing state can still be used as an indicator for the speed of convergence.

8.1.2 Communicating Classes

By definition, state j is *reachable* from i if there is a path from i to j. Clearly, $p_{i,j}(t) = 0$ for all t if j is not reachable from i. Two states are said to be *communicating*, or $i \leftrightarrow j$ if i is reachable from j and j is reachable from i. The relation $\leftrightarrow$ is an equivalence relation, that is, it is reflexive, symmetric, and transitive. Like any equivalence relation, this relation divides the set of states into equivalence classes, called *communicating classes*, and we denote them by $C_1, C_2, \ldots, C_m$, where m is the number of communicating classes. If C_i and C_j, $i \neq j$ are two communicating classes, then if there is a path from a state in C_i to a state in C_j, there cannot be a path from any state in C_j back to C_i, because otherwise, C_i and C_j would form a single communicating class.

A state can either be *recurrent* or *transient*. These terms are defined as follows:

Definition 8.2 A state i is *transient* if the probability of returning to i when starting at i is less than 1. A state is *recurrent* if, when starting at i, the probability of returning to i is 1.

It can be shown [54, page 66] that a recurrent state is visited infinitely often. In a DTMC, one can prove that state i is recurrent by showing that the expected number of visits to i, when starting from i diverges, that is $\sum_{t=0}^{\tau} p_{j,j}(t)$ diverges as $\tau \to \infty$. If the expected number of visits stays finite as $\tau \to \infty$, i is transient. Similar definitions can be formulated for CTMCs.

The states of a communicating class are either all transient or all recurrent. If all states in a communicating class are recurrent, we say that the class is recurrent. The term transient communicating class is to be understood in a similar way.

Transient communicating classes have at least one state with a path to at least one state of another communicating class, which implies that they will eventually be left. A class is recurrent if there is no such path from any state within the class to any state outside the class. If i is any state with a path to j, and j is within a transient class, then the limit of $p_{i,j}(t)$ converges to 0 as $t \to \infty$. If, on the other hand, j is in a recurrent class, then $p_{i,j}(t)$ must converge to some positive limit c if it converges at all. Otherwise, the expected number of visits to state j (or the expected time in state j) cannot diverge.

Markov chains can have different recurrent classes. Once a recurrent class is entered, say it has reached j at time t, then we do not need any information about any transition probability or rate for states outside the recurrent class in question to calculate $p_{j,k}(t + \tau)$ for any k and τ. The recurrent class therefore forms a Markov chain on its own.

Processes that have several recurrent classes, but no transient class, lead to *decomposable* transition matrices, that is, one can decompose them into several transition matrices, one for each communicating class. Each class has its own set of states, and its own transient and steady-state solution. This simplifies the solution of such systems.

We already have used transient communicating classes when we investigated Markov chains with transient states, including waiting time problems. In these Markov chains, we had an absorbing state, which forms a one-state communicating class. Of interest was the time spent in the different states. As it turns out, the expected time spent in each state from 0 to T also has a limit. These limits can be found from systems of linear equations, which are similar to the equilibrium equations of Markov chains. We will discuss these in Chapter 9.

The existence of multiple communicating classes can lead to numerical difficulties when calculating equilibrium probabilities. To avoid problems, one should first identify and remove transient commuting classes. Each recurrent class can then be solved separately. To identify the communicating classes, one can use algorithms from graph theory [14] to find *strongly connected components* in a graph. Indeed, a Markov chain be represented by a directed graph, with an edge from i to j if $p_{i,j} > 0$, respectively, $q_{i,j} > 0$. Once the communicating classes are identified, each communicating class can then be handled individually.

It is also possible to identify communicating classes when enumerating the states. Of particular interest in this respect is the reachability method discussed in Section 3.6. This method follows all paths in order to create the transition matrix, and while doing this, it visits all states that can be reached from the starting state(s).

We still have to address the question of whether $\lim_{t \to \infty} p_{i,j}(t)$ depends on i. Any limit would have to satisfy the equilibrium equations of the process, and if the limit depends on i, there would have to be different solutions of the equilibrium equations. This is true if there are different recurrent classes, because each recurrent class can be considered on its own as a Markov chain, with its own equilibrium probabilities. Consequently, if i and j are in the same communicating class, $p_{i,j}(t)$ converges to the equilibrium probabilities of the communicating class of i, that is, if j is in the same communication class as i, we obtain the equilibrium probability for their common class. If i and j are not in the same communicating class, then $p_{i,j}(t) = 0$ for all t. Hence, the limit of $p_{i,j}(t)$ depends on the class i is in, and consequently on i. The same is true if i belongs to a transient class, because then, the probability of ending in one of the recurrent classes depends on the starting state i. On the other hand, if there is a path between any two states of the process, then i does not affect the limit of $p_{i,j}(t)$ as $t \to \infty$.

We already discussed a model with several recurrent communicating classes in Chapter 2, Section 2.5.3: In this model, there were two state variables, X_1 and X_2, and in one communicating class, $X_1 - X_2$ was even, and in the other class, $X_1 - X_2$ was odd. Communicating classes also arise in models with state variables X_1 and X_2 and two events, which have event function $X_1 + 1, X_2 - 1$ and $X_1 - 1, X_2 + 1$. In this case, $X_1 + X_2$ always has the same value, that is, there is a separate recurrent communicating class for each value of $X_1 + X_2$. Such situations occur in closed queueing networks, which will be discussed in Chapter 11.

Often, one has groups of states that behave almost like communicating classes, but not quite. Specifically, there may be two sets of states, say S_1 and S_2, in the sense that there are transition with rates $q_{i,j} > 0$ and $q_{j,i} > 0, i \in S_1, j \in S_2$, but these rates are small when compared to the rates within the class. For example, consider a model with two state variables X_1 and X_2. This model has initially only two events, with the respective event functions $X_1 + 1, X_2 - 1$ and $X_1 - 1, X_2 + 1$. In this model, there is a communicating class corresponding to every possible value of $M = X_1 + X_2$. If two additional events are added, one with event function $X_1 + 1, X_2$ and the other one with $X_1 - 1, X_2$, both having small event rates, these classes are no longer communicating, but for small rates of the added events, they almost are. The effect is that $p_{i,j}(t)$ converges slowly if $X_1 + X_2$ is different in state i and j. The convergence toward an equilibrium may be improved by giving initial probabilities to at least one state of each of the sets that almost behave like communicating classes. The transition matrices of such models are known as *almost completely decomposable* matrices.

8.1.3 Periodic DTMCs

The transient probabilities of a finite CTMC always converge to a limit, possibly a limit of zero. However, the transient probabilities of some DTMCs do not converge. For instance, the DTMC with states 1 and 2 having the following structure will never reach an equilibrium:

$$P = \begin{bmatrix} 0 & 1 \\ 1 & 0 \end{bmatrix}.$$

Here, the transient equations are

$$\pi_1(t + 1) = \pi_2(t) \tag{8.1}$$

$$\pi_2(t + 1) = \pi_1(t). \tag{8.2}$$

If we start at time 0 in state 1, $\pi_1(1) = 0, \pi_2(1) = 1, \pi_1(2) = 1, \pi_2(2) = 0, \pi_1(3) = 0,$ $\pi_2(3) = 1$, and so on. Markov chains like this are said to be *periodic*. We define:

Definition 8.3 State i is said to be periodic with periodicity $r > 1$ if $p_{i,i}(t)$ is greater 0 only for t divisible by r. If $r = 1$, the state is said to be *aperiodic*.

Hence, in our introductory example, state 1 has a periodicity $r = 2$, and state 2 also has a periodicity of 2.

Later, we will use the term *periodic DTMCs*. This is allowed because of the following theorem:

Theorem 8.1 *If any state in a communicating class is periodic with period r, all other states of this communicating class are periodic with the same period r.*

Proof Since i is communicating with any state j of the communicating class of i, there must be a path going from i to j and back to i. Let w_1 and w_2 be the lengths of these two paths. If i is a periodic state with periodicity r, $w_1 + w_2$ must be divisible by r. Now, if we add a path from j to j of length w_3, resulting in a path first from i to j of length w_1, next a path from j to j of length w_3, and finally a path from j to i of length w_2. Now, $w_1 + w_3 + w_2$ must still be divisible by r, and this is only possible if w_3 is divisible by r. Hence, if there is any path from j to j that is not divisible by r, i cannot be periodic with periodicity r. If follows that the length of any path from j to j must be divisible by r, that is, j must also be periodic with periodicity r. To complete the proof, note that j can be any state of the communicating class of i. □

The theorem implies that periodicity is a property of the communicating class. A DTMC with only one communicating class is called periodic with periodicity r if all states have a periodicity of $r > 1$. A DTMC that is not period is called *aperiodic*.

Note that periodic Markov chains still may have equilibrium probabilities in the sense that they satisfy the equilibrium equations $\pi = \pi P$. In the example given by equations (8.1) and (8.2), for instance, we obtain the following two equilibrium equations, which are identical:

$$\pi_1 = \pi_2$$
$$\pi_2 = \pi_1.$$

Hence, $\pi_1 = \pi_2$, and since their sum is 1, $\pi_1 = \pi_2 = 0.5$. This means that if $\pi(0) = [0.5, 0.5]$, then $\pi(t) = [0.5, 0.5]$ forever. Also note that the vector of the time averages for states 1 and 2 converges to $[0.5, 0.5]$.

The simplest example of a periodic Markov chain with periodicity r is a Markov chain with r rows of the following structure:

$$P = \begin{bmatrix} 0 & 1 & 0 & 0 & \dots & 0 \\ 0 & 0 & 1 & 0 & \dots & 0 & 0 \\ \vdots & \vdots & \ddots & \ddots & \dots & \vdots & \vdots \\ 0 & 0 & 0 & 0 & \dots & 0 & 1 \\ 1 & 0 & 0 & 0 & \dots & 0 & 0 \end{bmatrix}. \tag{8.3}$$

The Markov chain retains the periodicity if the non-zero entries are replaced by matrices. The states of each matrix will form a periodicity class. The sum of all equilibrium probabilities of each periodicity class will be $1/r$.

To form the periodicity classes, say $R_1, R_2, \dots, R_p$, and to put the states into the correct class, we start in some initial state, say state 1, and put it into periodicity class R_1. All states that can be reached from state 1 in one step are placed into R_2, all states that can be reached from any of the states we have placed into R_2 go into periodicity class R_3, and so on, until we reach periodicity class r. Any state that can be reached from R_r goes into R_1, and this continues until all states have been inserted into one of the periodicity classes. If any state i placed into the set R_n, $n < r$ can reach any state in R_m, $m < n$, then the DTMC is not periodic.

Note that even though the $p_{i,j}(t)$ of a periodic Markov chain do not converge, the time average $\frac{1}{T} \sum_{t=0}^{T} p_{i,j}(t)$ converges as $T \to \infty$. Moreover, if the length of the period is r, $p_{i,j}(rt)$ also converges. This implies that if we select one state from each periodicity class, and assign it a probability of $1/r$, then $\pi(t)$ converges quickly toward its equilibrium. This approach is appropriate if a Markov chain is almost periodic, and one is interested in initializing such that the equilibrium is reached as quickly as possible. More details will be provided in Section 9.3.4.1 when we discuss iterative methods, because, in these methods, periodicity can cause problems, which makes the theory of periodic Markov chains important.

Though CTMCs cannot be periodic, their behavior can resemble the one of an almost periodic DTMC. Moreover, when using uniformization, with $P = Q/q + I$, then one should avoid making the Markov chain periodic. This can be done by selecting $q < \max(|q_{i,i}|)$, because then, all $p_{i,i} > 0$. This makes paths from i to i possible, and these paths have a length of 1, making P aperiodic.

Markov chains with only one aperiodic recurrent class are called *completely ergodic*. The word *ergodic* is used for Markov chains with only one recurrent class, irrespective of whether or not this class is periodic.

8.2 Transient Solutions using Eigenvalues

Eigenvalues and eigenvectors, the topic of this section, provide important insights into the nature of transient solutions, and they can be used to find transient solutions numerically. Eigenvalues are also important in Chapter 9 for analyzing iterative methods to determining equilibrium probabilities. Moreover, the theory of eigenvalues will be used in Chapter 12 to analyze Markov chains with block-structured transition matrices with repeating rows.

Since numerically determining eigenvalues is highly technical, and since there are many textbooks on this subject, such as [85] and [95], these methods will not be discussed here. Instead, we suggest the reader uses canned programs or languages that contain functions that can find eigenvalues and eigenvectors, such as Maple, Mathematica, MATLAB, and Octave. Here, we use Octave.

8.2.1 Basic Theorems

We begin with the following theorem:

Theorem 8.2 *Let P be the stochastic or substochastic transition matrix of a DTMC with N states. If one can find N different independent solutions of the equation*

$$\phi(P - \gamma I) = 0, \tag{8.4}$$

where ϕ is a non-zero row vector, and γ is a scalar, then

$$\pi(t) = \sum_{k=1}^{N} c_k \phi_k \gamma_k^t. \tag{8.5}$$

Here, (ϕ_k, γ_k) represents the kth solution of the (8.4), and the c_k are given by the equations:

$$\pi(0) = \sum_{k=1}^{N} c_k \phi_k. \tag{8.6}$$

The solutions (ϕ_k, γ_k) are said to be independent if the ϕ_k are mutually independent, that is, if the ϕ_k define a vector space of dimension N.

Proof To find a solution of the equation

$$\pi(t + 1) = \pi(t)P, \tag{8.7}$$

let us try

$$\pi(t) = \phi \gamma^t.$$

Substituting this into (8.7) and dividing by γ^t yields (8.4). If we have k independent solutions (ϕ_k, γ_k), $k = 1, 2, \ldots, N$, then it is easily verified that $\sum_{k=1}^{N} c_k \phi_k \gamma_k^t$ also satisfies (8.4). Setting $t = 0$ yields the right side of (8.6). If we can solve this system of equations for $c_1, c_2, \ldots, c_n$, then the problem is solved. A solution exists because of the assumption that the ϕ_k are mutually independent vectors.

The theorem has the following counterpart for CTMCs:

Theorem 8.3 *Let Q be the infinitesimal generator or subgenerator of a CTMC with N states. If one can find N different independent solutions of the equation*

$$\phi(Q - \gamma \mathbf{I}) = 0, \tag{8.8}$$

where ϕ is a non-zero row vector, and γ is a scalar, then

$$\pi(t) = \sum_{k=1}^{N} c_k \phi_k \exp(\gamma_k t), \tag{8.9}$$

As before, (ϕ_k, γ_k) represents the kth solution of the (8.8), and the c_k are given by the equations

$$\pi(0) = \sum_{k=1}^{N} c_k \phi_k. \tag{8.10}$$

The solutions (ϕ_k, γ_k) are said to be independent if the ϕ_k are mutually independent, that is, if the ϕ_k define a vector space of dimension N.

Proof As a solution of the equation

$$\pi'(t) = \pi(t)Q, \tag{8.11}$$

let us try

$$\pi(t) = \phi e^{\gamma t}.$$

Substituting this into (8.11) and dividing by $e^{\gamma t}$ yields (8.8). If we have N such solutions, we can determine the c_k, $k = 1, 2, \ldots, N$ such that the initial probability vector $\pi(0)$ is correct. This is possible as long as the ϕ_i are independent. Once this is established, (8.9) is satisfied, which completes the proof.   □

In the discussion above, we actually used eigenvalues and eigenvectors. In fact, if A is any square matrix of dimension N, and

$$\phi(A - \gamma I) = 0, \quad \phi \neq 0,$$

then γ is said to be an *eigenvalue* of A, and ϕ is its corresponding *eigenvector*. The pair (γ, ϕ) will be called *eigenpair*. Consequently, the solutions based on equations (8.5) and (8.9) are eigenvalue solutions.

We should point out that in the literature of eigenvalues, such as [85], λ is used instead of γ. Also, when $Ax = \lambda x$, then x is an eigenvector. The vector ϕ as defined above is called *left eigenvector*. We will call x the *right eigenvector*.

Unless the determinant of the matrix $A - \gamma I$ is zero, the only solution of $\phi(A - \gamma I)$ is the trivial solution $\phi = 0$, which is excluded. The expression $\det(A - \gamma I)$ is a polynomial of degree N in terms of γ, and it is called the *characteristic polynomial* of A. The characteristic polynomial must always have N zeros, counting multiplicities. If all zeros are simple, then we can find N independent eigenvectors, which are determined up to a factor, and equations (8.5) and (8.9) can be applied without any problem. Even when γ_k is a zero with a multiplicity greater than one, these equations can be used as long as there are enough independent eigenvectors associated with γ_k.

Now, let m_k be the multiplicity of the zero γ_k of the characteristic polynomial, and let $\hat{m}_k$ be the number of independent solution vectors of $\phi(A - \gamma_k I) = 0$. In literature, m_k is called the *algebraic multiplicity* of γ_k, and $\hat{m}_k$ is called the *geometric multiplicity* of γ_k. It can be shown that the algebraic multiplicity of a zero can never be lower than the geometric multiplicity. If the geometric multiplicity equals the algebraic multiplicity, then equations (8.5) and (8.9) can still be applied. If not, there will be problems, and methods of overcoming these problems are shown in Section 8.2.7.

The characteristic polynomial is especially easy to find if the transition matrix is triangular as in Coxian distributions and FIFO waiting time problems. Generally, for any triangular matrix A of dimension N, with diagonal elements $a_{i,i}$, the characteristic polynomial is $\prod_{i=1}^{N}(a_{i,i} - \gamma)$, which has the N solutions $\gamma_i = a_{i,i}$, $i = 1, 2, \ldots, N$.

8.2.2 Matrix Power and Matrix Exponential

One can express A^n, the nth power of the matrix A with dimension N in terms of its eigenvalues and eigenvectors, provided one has a complete set of N independent eigenvectors. If Φ is the matrix $[\phi_k, k = 1, 2, \ldots, N]^T$, and if Γ is the diagonal matrix with the diagonal elements $\gamma_1, \gamma_2, \ldots, \gamma_N$, then one has

$$A^n = \Phi^{-1}\Gamma^n\Phi. \tag{8.12}$$

To prove this equation, note that $\phi_k A = \nu_k \phi_k$ implies $\Phi A = \Gamma\Phi$, which leads to

$$A = \Phi^{-1}\Gamma\Phi.$$

Now

$$A^n = (\Phi^{-1}\Gamma\Phi)(\Phi^{-1}\Gamma\Phi)\cdots(\Phi^{-1}\Gamma\Phi).$$

Since $\Phi\Phi^{-1} = I$, (8.12) follows.

Note that (8.12) implies that A^n has the same eigenvectors as A, and its eigenvalues are γ^n.

The matrix exponential $\exp(A)$ is defined as

$$\exp(A) = \sum_{n=0}^{\infty} A^n/n!.$$

Using (8.12), this can be written as

$$\exp(A) = \sum_{n=0}^{\infty} \frac{1}{n!}\Phi^{-1}\Gamma^n\Phi = \Phi^{-1}\left(\sum_{n=0}^{\infty}\frac{1}{n!}\Gamma^n\right)\Phi.$$

The sum in parenthesis is a diagonal matrix, with the diagonal elements $\exp(\gamma_k)$. Consequently,

$$\exp(A) = \Phi^{-1}\mathrm{diag}(\exp(\gamma_k))\Phi.$$

This result can be generalized to any function that has a Taylor expansion, that is, if $f(\gamma)$ is such a function, we have

$$f(A) = \Phi^{-1}\mathrm{diag}(f(\gamma_k))\Phi. \tag{8.13}$$

There is an alternative to find Φ^{-1}, as indicated by the following theorem.

Theorem 8.4 *Let ψ be a solution of $(A - \gamma_k I)\psi_k = 0$. If d_k is the scalar product of ϕ_k with ψ_k, or $d_k = \phi_k\psi_k$, and $\Psi = [\psi_k/d_k]$, then $\Psi = \Phi^{-1}$.*

Proof If γ is an eigenvalue, $A - \gamma I$ is singular. Consequently, there must be a vector $\psi \neq 0$ such that $(A - \gamma I)\psi = 0$. The column vector ψ is called *right eigenvector* of γ. We now prove that $\phi A = \gamma\phi$ and $A\psi = \gamma\psi$ implies $\phi_k\psi_\ell = 0$ for $\gamma_k \neq \gamma_\ell$. To show this, consider the expression $\phi_k A\psi_\ell$. If, in this expression, $\phi_k A$ is replaced by $\phi_k\gamma_k$, we get $\phi_k A\psi_\ell = \phi_k\gamma_k\psi_\ell$, and if $A\psi_\ell$ is replaced by $\psi_\ell\gamma_\ell$, we get $\phi_k A\psi_\ell = \phi_k\gamma_\ell\psi_\ell$.

Consequently

$$\phi_k \gamma_k \psi_\ell = \phi_k \gamma_\ell \psi_\ell,$$

and this is only possible if either $\phi_k \psi_\ell = 0$ or $\gamma_k = \gamma_\ell$. It follows that $\Phi\Psi$ is a diagonal matrix, and by dividing each diagonal entry by d_k, one obtains $\Phi\Psi = I$.

As a consequence of Theorem 8.4, together with (8.13), we have

$$f(A) = \Psi \mathrm{diag}(f(\gamma_k))\Phi. \tag{8.14}$$

The elements of Ψ will be denoted by $\psi_{i,j}$, such that $\psi_j = [\psi_{1,j}, \psi_{2,j}, \ldots, \psi_{N,j}]^T$. Of course, to find Ψ, we divide all ψ_k by d_k, but since ψ_k can only be determined by a factor, ψ_k/d_k is still a right eigenvector.

Equation (8.12) indicates that as n increases, the importance of the largest eigenvalues increases when evaluating A^n, and the other eigenvalues become negligible for high enough n. If γ_1 is the largest eigenvalue, and γ_1 is strictly greater than all other eigenvalues, then A^n approaches $\gamma_1^n \psi_1 \phi_1$.

All the results of this section can only be applied to matrices where the geometric multiplicity is equal to the algebraic multiplicity. Such matrices are called *diagonalizable*. Matrices that are not diagonalizable will be discussed later.

Let us now return to Markov chains, where $A = P$ for DTMCs, and $A = Q$ for CTMCs. Consider now the c_k of equation (8.5), which leads to, if $\phi_{k,j}$ is the jth entry of ϕ_k.

$$\pi_j(t) = \sum_{k=1}^{N} c_k \phi_{k,j} \gamma_k^t.$$

On the other hand, from $\Psi\Gamma\Phi$, we find, if $\psi_{i,k}$ is the ith entry of ψ_k

$$p_{i,j}(t) = \sum_{k=1}^{N} \psi_{i,k} \phi_{k,j} \gamma_k^t. \tag{8.15}$$

Now $\pi_j(t) = p_{i,j}(t)$ if $\pi_i(0) = 1$, $\pi_n(0) = 0$, $n \neq i$. If follows that when starting in state i, $c_k = \psi_{i,k}$.

A result similar to (8.15) holds for CTMCs:

$$p_{i,j}(t) = \sum_{k=1}^{N} \psi_{i,k} \phi_{k,j} \exp(\gamma_k t). \tag{8.16}$$

Again, if the process starts in state i, then $c_k = \psi_{i,k}$.

Note that the equilibrium probabilities are really eigenvectors for the eigenvalue $\gamma_1 = 1$ for DTMCs, and $\gamma_1 = 0$ for CTMCs. Indeed, for the DTMC, $\pi = \pi P$ can be written as

$$\pi(P - \gamma_1 I) = \pi(P - I) = 0,$$

and the result is obvious. Also, the right eigenvector of γ_1 is $\mathbf{e}$, the column vector with all its elements equal to 1. This follows from $P\psi = \gamma\psi$, which becomes, if

$\gamma = 1$ and $\psi = \mathbf{e}$, $P\mathbf{e} = \mathbf{e}$, which is true if all row sums are 1. A similar result holds for the CTMC, except that $\gamma = 0$ is an eigenvalue.

8.2.3 An Example of a Transient Solution using Eigenvalues

To demonstrate the use of eigenvalues, we use Octave to solve a problem discussed earlier. However, though eigenvalues can give useful insights into the behavior of Markov chains, it is typically not the first choice when solving problems numerically.

The problem to be solved is the repair problem given in Table 2.9 of Chapter 2. In this model, we have two communicating classes, depending on whether $X_1 - X_2$ is even or odd, We consider only the case where $X_1 - X_2$ is even. The transition matrix for this model is as follows:

$$Q = \begin{bmatrix} -(\lambda_1 + \lambda_2) & \lambda_2 & 0 & 0 & \lambda_1 & 0 & 0 & 0 \\ 0 & -\lambda_1 & 0 & 0 & 0 & \lambda_1 & 0 & 0 \\ \mu & 0 & -(\mu + \lambda_1 + \lambda_2) & \lambda_2 & 0 & 0 & \lambda_1 & 0 \\ 0 & \mu & 0 & -(\mu + \lambda_1) & 0 & 0 & 0 & \lambda_1 \\ 0 & 0 & 0 & 0 & -\lambda_2 & \lambda_2 & 0 & 0 \\ 0 & 0 & \mu & 0 & 0 & -\mu & 0 & 0 \\ 0 & 0 & 0 & 0 & \mu & 0 & -(\lambda_2 + \mu) & \lambda_2 \\ 0 & 0 & 0 & 0 & 0 & \mu & 0 & -\mu \end{bmatrix}.$$

We first entered Q with $\lambda_1 = \lambda_2 = 1.5$ and $\mu = 1$ into Octave. To find a transient solution, we used the following statements, where t was given different values

```
[V, D] = eig[Q];
Pt = V*expm(t*D)*inv(V);
Pt = real(Pt)
```

Notice that V is really Ψ, the matrix of the right eigenvector. Typically, programs find the right eigenvector because this matches with the typical way equations are written in linear algebra, which solve $Ax = b$, where A is a matrix, and x is the column vector of the unknowns. In the theory of Markov chains, however, we write, $0 = \pi Q$, making π the left eigenvector of $\gamma_1 = 0$.

When running the above program, one obtains the following real eigenvalues: $0, -1.189, -1, 5, -2.5, -2.83$, and -4.606. There is also a pair of conjugate complex eigenvalues, namely $-2.192 \pm 1.537i$. The eigenvalue 0 is, of course, the eigenvalue corresponding to the equilibrium vector. Since V is the matrix of the right eigenvectors, and the equilibrium vector π is the left eigenvector of $\gamma_1 = 0$, the equilibrium probabilities are found, up to a factor, as a row in matrix V^{-1}. One has to identify this row, which is really ϕ_1, and divide each entry by $\sum_{j=1}^{N} \phi_{1,j}$ such that the sum of all equilibrium probabilities is equal to 1. The resulting equilibrium probabilities are

0.0340 0.0748 0.1020 0.0612 0.0748 0.4082 0.0612 0.1837.

The i, j entry of the matrix Pt of the MATLAB program yields the transient probabilities $p_{i,j}(t)$. The first row thus shows $p_{1,j}(t)$, the probabilities to be in state j at t, given the process starts in state 1, which is $X_1 = X_2 = 0$. For the times $t = 1, 2, 4$ we got the following values for $p_{1,j}(t)$:

	$j = 1$	$j = 2$	$j = 3$	$j = 4$	$j = 5$	$j = 6$	$j = 7$	$j = 8$
$t = 1$	0.0670	0.1857	0.0813	0.0282	0.1857	0.4028	0.0282	0.0211
$t = 2$	0.0360	0.1135	0.1075	0.0641	0.0637	0.4217	0.0641	0.1294
$t = 4$	0.0340	0.0751	0.1019	0.0611	0.0727	0.4077	0.0611	0.1865

One sees how the transient solutions approach their equilibrium. This completes the solution of the above problem.

8.2.4 The Theorem of Perron-Frobenious

Crucial to our further discussion is the following theorem about positive matrices. We will not prove this theorem, but a proof can be found, for instance, in [48].

Theorem 8.5 *If P is a matrix where all entries are strictly positive, and if γ_1 is the largest eigenvalue in modulus, then γ_1 is simple and positive. Moreover, both the left and the right eigenvectors of γ_1 contain strictly positive values. Also, there is no other eigenvalue whose modulus equals γ_1.*

We call the unique largest eigenvalue the *Perron–Frobenius* or PF eigenvalue, and the corresponding eigenvector the PF eigenvector. Often, the modulus of the largest eigenvalue is called *spectral radius*. We can thus say that in strictly positive matrices, the largest eigenvalue is equal to the spectral radius.

At first glance, it seems that this theorem is only applicable to DTMCs, and even there, it may not apply because the matrices arising from discrete-event systems have many zero entries, and they are therefore not strictly positive. However, as long as the transition matrices we are looking at are completely ergodic, that is, they are aperiodic, and they only contain one communicating class, then we can always find an n such that P^n is strictly positive, and then, the theorem of Perron–Frobenius applies as well for P [48]. Of course, the eigenvalues of P^n are equal to the ones of P, raised to the nth power. Also, the eigenvectors of γ_1 of P are identical to the ones of γ_1^n of P^n, and they must be strictly positive. In short, the theorem is applicable to any completely ergodic DTMC.

If P is a stochastic matrix, then one eigenvalue of P is $\gamma_1 = 1$, and its left eigenvector is equal to its equilibrium probabilities. In fact, if $\gamma_1 = 1$, then $\pi\gamma_1 = \pi P$ is exactly the equilibrium equation. Moreover, if P is the transition matrix of a completely ergodic Markov chain, γ_1 is the PF eigenvalue. It follows from the theorem of Perron–Frobenius that all other eigenvalues must have absolute values less than 1, that is

$$0 < |\gamma_k| < 1, \ k = 2, 3, \ldots, N.$$

The Perron–Frobenius theorem is also valid for substochastic transition matrices. These matrices have a positive eigenvalue γ_1, which must be less than 1, because otherwise, some $p_{i,j}(t)$ would eventually increase beyond any bound as t increases. The theorem of Perron–Frobenius further implies that the eigenvector ϕ_1 would have to be positive.

To work with a CTMCs with the transition matrix Q, we use uniformization as described in Chapter 7, that is, we create a matrix P related to Q as $P = Q/q + I$, where q must be chosen such that P has no negative entries. If $P = Q/q + I$, then $Q = (P - I)q$, which makes Q a function of P. This allows us to apply equation (8.14), with $f(P) = (P - I)q$, which implies that Q has the same eigenvectors as P, and its eigenvalues are in the form of $\beta = (\gamma - 1)q$, where γ is an eigenvalue of P. If Q has only one communicating class, the same is true for P. In addition, P can be made aperiodic by choosing q such that at least one $p_{i,i} > 0$. Now, the PF theorem can be applied to P, and indirectly to Q. If γ is an eigenvalue of P, then the corresponding eigenvalue of Q is obtained by subtracting 1 from γ, and multiplying this result by q, that is, $\beta_k = (\gamma_k - 1)q$. Now $|\gamma_k| \leq 1$ implies that the real part of γ_1 is between 1 and -1:

$$-1 \leq \Re\gamma_k \leq 1,$$

or, since $q = \max(-q_{i,i})$ is the smallest value for q:

$$-2\max(-q_{i,i}) \leq \Re\beta_k \leq 0. \tag{8.17}$$

It P is stochastic, then $\gamma_1 = 1$, and β_1 must therefore be 0. In this case, $\pi Q = \pi\beta_1$ with $\beta_1 = 0$ yields the equilibrium equations of CTMCs. If Q is transient, then the real part of γ_1 must be negative.

8.2.5 Perron–Frobenius and Non-Negative Matrices

The theorem of Perron–Frobenius can only be applied when there is an m such that all entries of P^m are positive, that is, there are no entries of 0 in P^m. In the case of DTMCs, there are two types of matrices where this is false:

1. The Markov chain has more than one communicating class. This is also true for any triangular matrix.
2. The Markov chain is periodic.

Let us now consider case 1, where there are several communicating classes. In this case, all communicating classes with more than one state can be analyzed independently. In particular, all recurrent communicating classes have an eigenvalue of 1, that is, the eigenvalue 1 has an algebraic multiplicity greater 1. Hence, even though the Perron–Frobenious theorem is applicable for each recurrent communicating class of a Markov chain, it is not applicable for transition matrices containing several communicating classes. In particular, the largest eigenvalue needs no longer be simple.

To see this, consider a Markov chain with several recurrent communicating classes. There, each class has an eigenvalue of 1, that is, γ_1 is no longer simple.

Consider now periodic Markov chains. We start with the simplest case of a periodic Markov chain with periodicity r, which is the r state Markov chain with the transition matrix given by (8.3). In this case, the determinant of $P - I\gamma$ is $1 - \gamma^r$, which is a polynomial of degree r in γ therefore has r zeros. The zeros of $1 - \gamma^r$ are given by

$$\gamma_k = \cos\left(\frac{2\pi(k-1)}{r}\right) + i\sin\left(\frac{2\pi(k-1)}{r}\right), \quad k = 1, 2, \ldots, r.$$

To verify that γ_k is indeed a zero of the characteristic polynomial, note that, for any complex number $z = (\cos\alpha + i\sin\alpha)$, the value of z^n is $z^n = \cos n\alpha + i\sin n\alpha$, which implies

$$\gamma_k^r = \cos(2(k-1)\pi) + i\sin(2(k-1)\pi) = 1.$$

All of these zeros have a modulus of 1. This means that there are several eigenvalues with an absolute value of 1, one of which, say γ_1 is simple and positive. In periodic transition matrices with a single communicating class, we will still call the largest positive eigenvalue the PF eigenvalue. In contrast to the theorem of Perron Frobenius, for positive matrices, γ_1 is not necessarily greater in modulus than all other eigenvalues. The theorem of Perron–Frobenius must therefore be weakened when it is applied to non-negative matrices. Let $s_i = \sum_{j=1}^{N} p_{i,j}$, then we have [48]:

$$\min_i s_i \leq \gamma_1 \leq \max_i s_i.$$

In particular, if the transition matrix is stochastic, then all row sums are 1, then there is an eigenvalue $\gamma_1 = 1$. If the transition matrix is substochastic, the spectral radius is less than 1.

In the case of CTMCs, we again use the matrix $P = Q/q + I$. If P happens to be periodic, one can increase q, and by doing this, all $p_{i,i}$ become positive, which prevents the Markov chain from being periodic. Moreover, the communicating classes of P are the same as the ones of Q, which means that if there are several recurrent communicating classes, each class has an eigenvalue of 0, together with the corresponding eigenvector.

8.2.6 Characterization of Transient Solutions

In this section, we will discuss the general form of eigenvalue solutions for diagonalizable transition matrices of DTMCs and CTMCs. We begin with DTMCs.

In a DTMC with a transition matrix P, the transient solution is given by equation (8.15), that is

$$p_{i,j}(t) = \sum_{k=1}^{N} \psi_{i,k}\phi_{k,j}\gamma_k^t.$$

If P is stochastic, and the DTMC is recurrent, the largest eigenvalue is always 1, say $\gamma_1 = 1$. In this case, (8.4) reduces to the equilibrium equation with $\phi_1 = \pi$. Since all eigenvalues can only be determined up to a factor, we can set $c_1 = 1$ to obtain

$$p_{i,j}(t) = \pi + \sum_{k=2}^{N} \psi_{i,k} \phi_{k,j} \gamma_k^t. \tag{8.18}$$

This splits the transient solution into a permanent part and a transient part. $p_{i,j}(t)$ approaches an equilibrium as the transient part approaches 0.

The split given by equation (8.18) remains valid for periodic Markov chain. If the Markov chain has more than 1 recurrent communication class, then equation (8.18) is valid for each communicating class.

For substochastic matrices, one has, so to say, only the transient part of equation (8.18). If all states communicate, the theorem of Perron–Frobenius still applies, and the largest eigenvalue is equal to the spectral radius.

Consider now the individual terms of equation (8.18), that is, $\psi_{i,k} \phi_{k,j} \gamma_k^t$. If the γ_k are ordered such that $|\gamma_1| \geq |\gamma_2| \geq |\gamma_3| \geq \ldots \geq |\gamma_N|$, then for large enough t, the eigenvalue γ_2, the *subdominant* eigenvalue, dominates the transient part of (8.18). The value of t when the subdominant eigenvalue determines the deviation from the equilibrium depends, of course, on its coefficient $\psi_{i,2} \phi_{2,j}$. As soon as γ_2 dominates the transient part, the behavior of the system depends on whether the γ_2 is positive, negative, or complex. If it is positive, then $p_{i,j}(t)$ approaches the equilibrium in an exponential fashion. The approach is from below if $\psi_{i,2} \phi_{2,j} < 0$ and from above if $\psi_{i,2} \phi_{2,j} > 0$. Of course, if $\psi_{i,2} \phi_{2,j} = 0$, then γ_2 has no effect on $p_{i,j}(t)$.

If γ_2 is complex, say $\gamma_2 = a + ib$, or $r(\cos(\alpha) + i \sin(\alpha))$, then $p_{i,j}(t)$ is repeatedly above and below the equilibrium. In fact, given α, γ_2^t is 0 if $\alpha t = 2n\pi$, where n is an integer. If follows that the time between two crossings of zero is $2\pi/\alpha$. In other words, there is some kind of periodicity. If γ_2 is negative, one has an oscillation with a periodicity of 2.

Before the time where γ_2 dominates one also has to consider the other eigenvalues, and their coefficients $\psi_{i,k} \phi_{k,j}$. Since these coefficients may be positive or negative, in the case of recurrent Markov chains, the process may very well be above the equilibrium at one time and below at another time.

Consider now ergodic CTMCs that have transition matrices with row sums of 0. In this situation, the largest eigenvalue is always 0, say $\gamma_1 = 0$. All other eigenvalues have a negative real part. As in the case of DTMCs, (8.4) reduces to the equilibrium equation. As before, we can set $c_1 = 1$ to get

$$p_{i,j}(t) = \pi_j + \sum_{k=2}^{N} \psi_{i,k} \phi_{k,j} \exp(\gamma_k t). \tag{8.19}$$

As before, this splits the transient solution into a permanent part and a transient part. $p_{i,j}(t)$ reaches an equilibrium when the transient part approaches 0.

For subgenerator matrices, one has, so to speak, only the transient part of equation (8.19). If all states communicate, the theorem of Perron–Frobenius implies that the largest eigenvalue is real, but it must be less than 0 because otherwise, the probabilities to be in state j would eventually exceed 1.

Consider now the individual terms of equation (8.19), that is, $\psi_{i,k} \phi_{k,j} \exp(\gamma_k t)$. If γ_k is real, then it must be negative, that is, $\exp(\gamma_k t)$, $k \neq 1$, converges monotonically to 0. For complex eigenvalues, where $\gamma = a + ib$, one has

$$\exp(\gamma t) = \exp(at + ibt) = e^{at}(\cos(bt) + i\sin(bt)). \tag{8.20}$$

Here, a is negative. It follows that the closer a is to zero, the slower $\exp(\gamma t)$ converges to zero. We arrange the eigenvalues according to their real parts, that is,

$$\Re\gamma_1 \geq \Re\gamma_2 \geq \ldots \geq \Re\gamma_N.$$

For recurrent CTMCs, $\Re\gamma_1 = \gamma_1 = 0$. Also, γ_2 is the subdominant eigenvalue, and for large t, it determines how fast the transient part of equation (8.19) vanishes. To begin with, let us assume that the imaginary part of γ_2 is zero, that is, γ_2 is real. In this case, the transient part will disappear exponentially at a rate given by the real part of γ_2. In fact, as we will show in Section 8.2.11, in CTMCs with bidiagonal matrices, all eigenvalues are real, including γ_2.

If the complex part of γ_k is not 0, we have damped oscillations as shown by equation (8.20). This expression is zero whenever $bt = 2\pi$. If γ_2 is complex, then if t is high enough, $p_{i,j}(t)$ will be periodically above and below its equilibrium π_j, with the length of the period given by $t = 2\pi/b$.

For transient CTMCs with an absorbing state, which arise in connection with PH distributions or waiting time problems, the eigenvalue with the largest real part cannot be complex because otherwise, negative probabilities would appear. This means that $p_{i,j}(t)$ approaches 0 in an exponential fashion. From this, it follows that all Coxian PH distributions have exponential tails. Of course, for t not that large, the other eigenvalues γ_k, $k > 1$ also play a role. In Coxian distributions, all eigenvalues must be real, and unless the arithmetic multiplicity exceeds the geometric one, the Coxian distributions have the form $\sum_{k=1}^{N} b_k \exp(\gamma_k t)$, where the b_k could be positive or negative.

The question is now what happens if the transition matrix is not diagonalizable. This question will be addressed in the next section. As it turns out, in nondiagonalizable stochastic matrices, one finds, instead of γ^t, expressions of the form $t\gamma^t$, or, more generally, terms of the form $\binom{t}{m}\gamma^{t-m}$ for some m. In the case of CTMCs, instead of $\exp(\gamma t)$, one has expressions like $t^m \exp(\gamma t)$. For instance, the Erlang distribution, which, as a phase-type distribution, leads to a CTMC, and its transition matrix is not diagonalizable. As a result, we find terms like $t^k \exp(\lambda t)$, something that cannot occur for diagonalizable transition matrices.

8.2.7 Eigenvalues with Multiplicities Greater than One

The eigenvalues are the zeros of the characteristic polynomial, and multiplicities of zeros of this polynomial lead to the corresponding multiplicities of the eigenvalues. However, as long as there are enough eigenvectors, this does not cause major theoretical problems. Consider, for instance, the eigenvalues of a diagonal matrix where the eigenvalues are given by the diagonal entries. It follows that repeating diagonal entries leads to multiplicities of the eigenvalues. In this case, it is possible that there are not enough eigenvectors. For example, consider the matrix T of the Erlang 3 distribution, which is

$$T = \begin{bmatrix} -\lambda & 0 & 0 \\ \lambda & -\lambda & 0 \\ 0 & \lambda & -\lambda \end{bmatrix}.$$

In this matrix, the eigenvalue of $-\lambda$ has a multiplicity of 3. However, the null space of this matrix is only 1, that is, there can only be one eigenvector. except for a factor, but we need three eigenvectors in order to apply equation (8.9).

To handle the case where there are not enough eigenvectors, we have to introduce *similarity transforms*.

Definition 8.4 Two matrices A and B are *similar* if there is an invertible matrix U such that

$$B = U^{-1}AU,$$

The transformation $U^{-1}AU$ is called a *similarity transformation*.

It follows from equation (8.12) that diagonizable matrices are similar to the diagonal matrix Λ.

Generally, we have:

Theorem 8.6 *If A and B are similar, then they have the same eigenvalues, the same eigenvectors, and the same characteristic polynomial.*

Proof Let $B = U^{-1}AU$. If γ is an eigenvalue of B, then the equation $X(B - I\gamma) = 0$ must have a non-trivial solution X. Now

$$B - \gamma I = U^{-1}AU - U^{-1}\gamma U$$
$$= U^{-1}(A - \gamma I)U.$$

It follows that if $X(B - \gamma I) = 0$, then $XU^{-1}(A - \gamma I)U = 0$, or

$$X(B - \gamma I) = 0 \text{ if and only if } XU^{-1}(A - \gamma I) = 0.$$

Consequently, if γ is an eigenvalue of B, with the eigenvector X, then γ is an eigenvalue of A with the eigenvector of XU^{-1}. Also, if ϕ is an eigenvector of A, ϕU is an eigenvector of B. Consequently, each eigenvector of A corresponds to an eigenvector of B, and vice versa. It follows that if A and B are similar then they have the same eigenvalues and the eigenvectors are linear transformations of each other. Moreover, they share the characteristic polynomial which can be shown as follows:

$$\det(B - \gamma I) = \det(U^{-1}(A - \gamma I)U) = \det(U^{-1})\det(A - \gamma I)\det(U).$$

Since products of determinants are commutative, we have

$$\det(B - \gamma I) = \det(U^{-1})\det(U)\det(A - \gamma I) = \det(A - \gamma I).$$

Consequently, A and B have the same characteristic polynomial as stated.   □

If every eigenvector of A is in a one-to-one relation to an eigenvalue of B, A and B have the same geometric multiplicities. Since they have the same characteristic polynomial, they also have the same algebraic multiplicities. It follows that there cannot be any similarity transformation that changes either the geometric or the algebraic multiplicities. If the algebraic multiplicity exceeds the geometric one, no similarity transform can change this. In any diagonal matrix, the algebraic multiplicity is equal to the geometric multiplicity. This implies that a matrix where the geometric and the algebraic multiplicities are different cannot be similar to a diagonal matrix, that is, such a matrix is not diagonalizable.

To handle matrices that are not diagonalizable, we are looking at block-diagonal matrices. To obtain the block structure of the resulting matrix, let m_k be the algebraic multiplicity of γ_k and $\hat{m}_k$ its geometric multiplicity. In this case, one needs $\hat{m}_k$ square blocks, with a combined dimension of m_k. Aside from this, the dimension of the blocks cannot be determined. For instance, if $m_k = 5$, and $\hat{m}_k = 2$, one needs two blocks, and their dimensions are unknown. It is possible that one block has a dimension of 1, and the other block a dimension of 4, but it is also possible that one block has the dimension of 2, and the other one has the dimension of 3. In either case, the sum of the dimensions is 5. In cases like this, one has to consider all alternatives and determine by other means which alternative is the correct one.

The types of blocks used here are called *Jordan blocks* in literature. In a Jordan block for γ_k, the diagonal elements are all set to γ_k, and the superdiagonal is 1. For our purposes, we change this convention slightly, and set the subdiagonal rather than the superdiagonal to 1. Hence, let J_k be a Jordan block of γ_k, that is,

$$J_k = \begin{bmatrix} \gamma_k & 0 & 0 & \dots & 0 & 0 \\ 1 & \gamma_k & 0 & \dots & 0 & 0 \\ 0 & 1 & \gamma_k & \ddots & 0 & 0 \\ \vdots & \ddots & \ddots & \ddots & \vdots & \vdots \\ 0 & \dots & 0 & 1 & \gamma_k & 0 \\ 0 & 0 & \dots & 0 & 1 & \gamma_k \end{bmatrix}.$$

Given the eigenvalues, we still have to determine the appropriate similarity transform to obtain such a block-diagonal matrix. It is best for our purposes to use the approach already used in Theorem 8.2. Suppose that we have a Jordan block J_k associated with the eigenvalue γ_k, and its dimension is v_k. We try a particular solution of equation (8.5), except that instead of a vector ϕ of dimension N, we use a matrix $\hat{\phi}$ of dimensions $v_k \times N$, and instead of γ we use the Jordan block J, that is, our trial solution is $\pi(t) = J^t \hat{\phi}$. With these values, the transient equation $\pi(t + 1) = \pi(t)P$

then becomes

$$J^{t+1}\hat{\phi} = J^t \hat{\phi} P. \tag{8.21}$$

To satisfy this equation, it is sufficient to satisfy

$$J\hat{\phi} = \hat{\phi}P. \tag{8.22}$$

This is true because (8.21) can be obtained by pre-multiplying both sides of $J\hat{\phi} = \hat{\phi}P$ by J^{-t}. For simplicity, we now treat Jordan blocks of size 3, but the result can easily be extended to blocks of any size. If we expand (8.22), we obtain, where $\hat{\phi}^{(n)}$ is the nth row of the matrix $\hat{\phi}$:

$$\begin{bmatrix} \gamma & 0 & 0 \\ 1 & \gamma & 0 \\ 0 & 1 & \gamma \end{bmatrix} \begin{bmatrix} \hat{\phi}^{(1)} \\ \hat{\phi}^{(2)} \\ \hat{\phi}^{(3)} \end{bmatrix} = \begin{bmatrix} \hat{\phi}^{(1)}P \\ \hat{\phi}^{(2)}P \\ \hat{\phi}^{(3)}P \end{bmatrix},$$

that is

$$\gamma\hat{\phi}^{(1)} = \hat{\phi}^{(1)}P$$
$$\hat{\phi}^{(1)} + \gamma\hat{\phi}^{(2)} = \hat{\phi}^{(2)}P$$
$$\hat{\phi}^{(2)} + \gamma\hat{\phi}^{(3)} = \hat{\phi}^{(3)}P.$$

The first equation is the normal eigenvalue equation, that is, $\hat{\phi}^{(1)}$ can be obtained up to a factor. One can then find $\hat{\phi}^{(2)}$ from the second equation, and with $\hat{\phi}^{(2)}$, one can obtain $\hat{\phi}^{(3)}$ from the third equation, and the problem is solved. For the detailed analysis of eigenvalues with multiplicities greater than one, see [95].

To find $p_{i,j}(t)$, one needs a method to find the nth power of P, and this requires to form the nth power of Jordan blocks. To accomplish this, we write

$$J = S^{-1} + \gamma I,$$

where S^{-1} is the matrix with ones on the subdiagonal, and zeros everywhere else. Note that $(S^{-1})^i$ is the matrix with one i placed to the left of the diagonal. We will denote this matrix as S^{-i}. Clearly, $S^{-i} = 0$ for $i \geq \nu - 1$, where ν is the dimension of J_k. To find J_k^n, we use the binomial theorem with $(S^{-1})^i = S^{-i}$ to obtain

$$J_k^n(S^{-1} + \gamma I)^n = \sum_{j=0}^{n} \binom{n}{j} \gamma^{n-j} S^{-j} = \sum_{j=0}^{\nu-1} \binom{n}{j} \gamma^{n-j} S^{-j}.$$

For instance, if J_k has a dimension of 4, and $n > 4$, we obtain

$$J_k^n = \begin{bmatrix} \gamma_k^n & 0 & 0 & 0 \\ n\gamma_k^{n-1} & \gamma_k^n & 0 & 0 \\ \binom{n}{2}\gamma^{n-2} & n\gamma_k^{n-1} & \gamma_k^n & 0 \\ \binom{n}{3}\gamma^{n-3} & \binom{n}{2}\gamma^{n-2} & n\gamma_k^{n-1} & \gamma_k^n \end{bmatrix}. \tag{8.23}$$

If no Jordan blocks are needed, we have, according to equation (8.15),

$$p_{i,j}(t) = \sum_{k=1}^{N} r_{i,j,k} \gamma_k^t,$$

where $r_{i,j,k} = \psi_{i,k} \phi_{k,j}$. In this expression, only γ_k^t depends on t. On the other hand, if there are Jordan blocks of size $v > 1$, there are also terms of the form $\binom{t}{\ell} \gamma^{t-\ell}$, $\ell \le v$, where $\binom{t}{\ell}$ depends on t. It follows that in (8.5), we have terms $\binom{t}{\ell} \gamma^{t-\ell}$ instead of γ^t.

For the continuous case, we have to form the matrix exponential of Jt in order to find $p_{i,j}(t)$.

$$\exp(Jt) = \sum_{n=0}^{\infty} \frac{1}{n!} (Jt)^n.$$

Now

$$\exp(Jt) = \sum_{n=0}^{\infty} \frac{t^n}{n!} (S^{-1} + \gamma I)^n.$$

If we expand $(S^{-1} + \gamma I)^n$ by using the binomial theorem, we obtain

$$\begin{aligned}
\exp(Jt) &= \sum_{n=0}^{\infty} \frac{t^n}{n!} \sum_{k=0}^{n} \frac{n!}{(n-k)!k!} \gamma^{n-k} S^{-k} \\
&= \sum_{k=0}^{n} \frac{t^k}{k!} S^{-k} \sum_{n=k}^{\infty} \frac{t^{n-k}}{(n-k)!} \gamma^{n-k} \\
&= \sum_{k=0}^{n} \frac{t^k}{k!} S^{-k} \exp(\gamma t).
\end{aligned} \tag{8.24}$$

We see that if $n > 1$, then we have to add terms of $t^k \exp \gamma t$ to equation (8.16). It follows that in (8.9), we have terms $t^k \exp(\gamma t)$ instead of simply $\exp(\gamma t)$. All these new terms have their own eigenvectors ϕ. This result will be needed to characterize Coxian distributions which we will discuss next.

8.2.8 Coxian Distributions Characterized by Eigenvalues

In Section 5.1.3, we discussed Coxian distributions, and their representation as CTMCs. Here, we use these distributions to illustrate the theory of eigenvalues presented above. It is very easy to find the eigenvalues and eigenvectors of Coxian distributions because they are bidiagonal, and the eigenvalues are given by the diagonal entries. Because of the extensive use of Coxian distributions, the results derived in this section are important for their own sake.

The Coxian distribution is based on a CTMC with the following transition matrix:

$$Q = \begin{bmatrix} -\lambda_1 & 0 & \cdots & \cdots & \cdots & 0 \\ \mu_2 & -\lambda_2 & 0 & \cdots & \cdots & 0 \\ 0 & \mu_3 & -\lambda_3 & 0 & \cdots & 0 \\ \vdots & \ddots & \ddots & \ddots & \ddots & \vdots \\ 0 & \ddots & \cdots & 0 & \mu_N & -\lambda_N \end{bmatrix}.$$

In lower triangular matrices, the eigenvalues are given by the diagonal entries, and consequently, we have $\gamma_i = -\lambda_i$. The left eigenvectors can now be obtained from the following equation:

$$\phi(Q - I\gamma) = 0.$$

For the eigenvector $\phi_i = [\phi_{i,j}]$, and the eigenvalue $\gamma = \gamma_i$, this equation expands to

$$0 = -\phi_{i,j}(\lambda_j + \gamma_i) + \phi_{i,j+1}\mu_{j+1}, \; j < N \qquad (8.25)$$
$$0 = -\phi_{i,N}(\lambda_N + \gamma_i). \qquad (8.26)$$

Since Q is a transition matrix, $0 \le \mu_i \le \lambda_i, i = 1, 2, \ldots, N$. We also assume that no μ_i is 0. To find the values $\phi_{i,j}, j = 1, 2, \ldots, N$, we set $\phi_{i,1} = 1$ and write (8.25) as

$$\phi_{i,j+1} = \phi_{i,j}\frac{\lambda_j - \lambda_i}{\mu_{j+1}}.$$

This allows us to find all the $\phi_{i,j}$ recursively. Note that as soon as any λ_j has the value λ_i, all $\phi_{i,k}, k > j$ are zero. It follows that no matter how many λ_j have the same value, we only find one solution for this system of equations. The geometric multiplicity is therefore always 1. However, when m different λ_j have the same value as λ_i, then the algebraic multiplicity is m. As $m > 1$, there are, therefore, not enough eigenvectors.

If all λ_j are different, we can apply Theorem 8.3, and to this end, we set $\pi(0)$ to α and calculate the c_k from equation (8.10), and $\pi(t) = \sum_k c_k\phi_k \exp(-\lambda_k t)$. Alternatively, we can determine the right eigenvectors ψ_k, and apply equation (8.16).

The right eigenvectors ψ_i of $\gamma_i = -\lambda_i$ are given by the following equations, where $\psi_{j,i}$ is the jth entry of ψ_i:

$$0 = -(\lambda_1 + \gamma_i)\psi_{1,i}$$
$$0 = \mu_j\psi_{j-1,i} - (\lambda_j + \gamma_i)\psi_{j,i}, \; j = 2, 3, \ldots, N. \qquad (8.27)$$

To find the $\psi_{j,i}$, we set $\psi_{N,i} = 1$ and solve equation (8.27) for $\psi_{j-1,i}$:

$$\psi_{j-1,i} = \frac{\lambda_j - \lambda_i}{\mu_j}\psi_{j,i}.$$

This equation allows us to find all $\psi_{j,i}$ recursively. Note that as soon as $\lambda_j = \lambda_i$, $\psi_{i,k} = 0$ for $k < j$. Again, we have only one eigenvector, no matter how many λ_j are equal to λ_i.

If m values of λ_i are identical, and no μ_i vanishes, then the algebraic multiplicity is m, yet, as shown earlier, the geometric multiplicity is 1. The result is that there may be one or more Jordan blocks with a combined size of m. It follows from (8.24) that Coxian distributions can have terms of the form $t^\ell \exp(\gamma t)$, where $\ell < m$. The main application of this is the Erlang distribution. We conclude that every Coxian distribution can have only real eigenvalues and that its density can be expressed as $\sum_k d_k f_k(x)$, where the $f_k(x)$ is either an exponential density or, if some λ_i repeat, an Erlang-like density.

8.2.9 Further Insights about PH Distributions Gained through Eigenvalues

Using the theory of eigenvalues, we now show that different transition matrices can lead to the same distribution. Consider the complementary cumulative distribution of T with

$$P\{T > t\} = b_1 e^{-\lambda_1 t} + b_2 e^{-\lambda_2 t}. \tag{8.28}$$

We now want to find any two-state Coxian distribution, and the corresponding α, leading to this distribution.

We start with the transition matrix

$$Q = \begin{bmatrix} -\lambda_1 & 0 \\ \mu & -\lambda_2 \end{bmatrix}, \tag{8.29}$$

and try to find the initial vector α and the rate μ such that $P\{T > t\} = \pi_1(t) + \pi_2(t)$ is equal to (8.28). The eigenvalues of Q are λ_1 and λ_2, which ensures that the exponents of (8.28) are correct. Also, the eigenvector corresponding to $-\lambda_1$ is $\phi_1 = [1, 0]$, and the eigenvector of $-\lambda_2$ is $\phi_2 = [\frac{\mu}{\lambda_1 - \lambda_2}, 1]$. This yields the solutions $[1, 0]e^{-\lambda_1 t}$ and $[\frac{\mu}{\lambda_1 - \lambda_2}, 1]e^{-\lambda_2 t}$, and the general solution, which is according to equation (8.9)

$$\pi_1(t) = c_1 e^{-\lambda_1 t} + c_2 \frac{\mu}{\lambda_1 - \lambda_2} e^{-\lambda_2 t} \tag{8.30}$$

$$\pi_2(t) = e^{-\lambda_2 t}. \tag{8.31}$$

Therefore,

$$P\{T > t\} = \pi_1(t) + \pi_2(t) = c_1 e^{-\lambda_1 t} + c_2 \left(\frac{\mu}{\lambda_1 - \lambda_2} + 1 \right) e^{-\lambda_2 t}.$$

It follows that

$$b_1 = c_1 \tag{8.32}$$

$$b_2 = c_2 \left(\frac{\mu}{\lambda_1 - \lambda_2} + 1 \right). \tag{8.33}$$

This gives two equations for the three variables c_1, c_2, and μ, that is, we are free to set one variable to any value we like, under the condition that all variables that represent probabilities must be between 0 and 1. No matter how we choose our parameters, we have to find the initial probabilities α_1 and α_2. We set $t = 0$ in equation (8.30) and (8.31) to obtain:

$$\alpha_1 = c_1 + c_2 \frac{\mu}{\lambda_1 - \lambda_2}, \quad \alpha_2 = c_2.$$

By equation (8.32), $c_1 = b_1$, and by equation (8.33), $c_2 = \frac{b_2(\lambda_1-\lambda_2)}{\mu+\lambda_1-\lambda_2}$. Consequently

$$\alpha_1 = b_1 + b_2 \frac{\mu}{\mu + \lambda_1 - \lambda_2}, \quad \alpha_2 = b_2 \frac{(\lambda_1 - \lambda_2)}{\mu + \lambda_1 - \lambda_2}.$$

We are now free to choose μ, subject to the condition that $\alpha_1 \geq 0$ and $\alpha_2 \geq 0$. In particular, if b_1 and b_2 are both non-negative, we can set $\mu = 0$, in which case we obtain the hyper-exponential distribution, and $\alpha_1 = b_1$, $\alpha_2 = b_2$. Alternatively, we can set $\mu = \lambda_2$, in which case we obtain a strong Coxian distribution as defined in Section 5.1.3, with $\alpha_1 = b_1 + b_2 \frac{\lambda_2}{\lambda_1}$, $\alpha_2 = b_2 \frac{\lambda_1-\lambda_2}{\lambda_1}$. Also, all values of μ between these extremes yield the same distribution.

Earlier, we showed that no Coxian distribution can have complex eigenvalues. However, there are phase-type distributions that do have complex eigenvalues. An example is the PH distribution resulting in the following transition matrix:

$$\mathbf{T} = \begin{bmatrix} -\lambda & 0 & p\lambda \\ \lambda & -\lambda & 0 \\ 0 & \lambda & -\lambda \end{bmatrix}.$$

This is the transition matrix of an Erlang distribution that has to be repeated with probability p. To find the eigenvalues of $\mathbf{T}$, we find the determinant of $\mathbf{T} - \gamma I$, which is

$$\det(\mathbf{T} - \gamma I) = -(\lambda + \gamma)^3 + p\lambda^3.$$

This is a polynomial in γ, and its zeros are:

$$\gamma_1 = \lambda \left(\sqrt[3]{p} - 1 \right)$$
$$\gamma_2 = \frac{1}{2}\lambda \left(\sqrt[3]{p} - 1 \right) \left(-1 + i\sqrt{3} \right)$$
$$\gamma_3 = \frac{1}{2}\lambda \left(\sqrt[3]{p} - 1 \right) \left(-1 - i\sqrt{3} \right).$$

Here, γ_1 is the PF eigenvalue, and γ_2 and γ_3 are conjugate complexes. Hence, we have damped oscillation, that is, the equilibrium is approached in a non-monotonic way.

8.2.10 Eigenvalues of Zero

Let P be an N state stochastic or substochastic matrix, which has an eigenvalue of zero. In this case, at least one state can be eliminated. To prove this, let ϕ_0 be an eigenvector corresponding to $\gamma = 0$. Since 0 is an eigenvalue of P, it is also an eigenvalue of P^m, that is, $\phi_i P^m = 0$, a result that can also be obtained by multiplying $\phi_i P = 0$ by P^{m-1}. Moreover, the right and left eigenvectors of $\gamma = 0$ are eigenvectors of P^m. Let $\psi_0 = [\psi_{j,0}]$ be an eigenvector of $\gamma = 0$, and suppose $\psi_{N,0} \neq 0$. Since the eigenvalues can only be determined up to a factor, we can assume, without loss of generality, that $\psi_{N,0} = 1$. Now $P^t \psi_0 = 0$ implies for $i = 1, 2, \ldots, N$

$$\sum_{j=1}^{N} p_{i,j}(t)\psi_{j,0} = 0,$$

or, since $\psi_{N,0} = 1$

$$p_{i,N}(t) = -\sum_{j=1}^{N-1} p_{i,j}(t)\psi_{j,0}.$$

Since $\pi_j(t) = \sum_i \pi_i(0)p_{i,j}(t)$, this implies

$$\pi_N(t) = -\sum_{j=1}^{N-1} \pi_j(t)\psi_{j,0}, \ t > 0. \tag{8.34}$$

In other words, $\pi_N(t)$ can be expressed in terms of the $\pi_J(t)$, $j < N$. However, this equation only works for $t > 0$, but not for $t = 0$.

If zero is an eigenvalue, then it is possible to reduce the dimension of transition matrix. To show this, we use the transient equation $\pi(t + 1) = \pi(t)P$, which we write

$$\pi_j(t + 1) = \sum_{i=1}^{N-1} \pi_i(t)p_{i,j} + \pi_N(t)p_{N,j}. \tag{8.35}$$

We now eliminate $\pi_N(t)$ by using (8.34), and we also eliminate $p_{N,j}$ as follows: From $\phi_0 P = 0$, $\phi_0 = [\phi_{0,j}]$, with $\phi_{0,N}$ set to 1, we obtain

$$p_{N,j} = -\sum_{i=1}^{N-1} \phi_{0,i}p_{i,j}.$$

Substituting this and (8.34) in (8.35) yields

$$\pi_j(t + 1) = \sum_{i=1}^{N-1} \pi_i \left(p_{i,j} + \psi_{i,0} \sum_{k=1}^{N-1} \phi_{0,k}p_{k,j} \right).$$

Consequently, if

$$A = \left[p_{i,j} + \psi_{i,0} \sum_{k=1}^{N-1} \phi_{0,k} p_{k,j} \right],$$

then $\pi(t + 1) = \pi(t)A$. However, A may have negative elements, that is, its entries are no longer probabilities.

Instead of expressing $\pi_N(t)$ in terms of the transient probabilities of the other states, one could have used state j in place of state N and expressed $\pi_j(t)$ in terms of the other states. In fact, if P has rank d, then the geometric multiplicity of the eigenvalue $\gamma = 0$ is $N - d$, and there are $N - d$ different states, say $j_1, j_2, \ldots, j_{N-d}$ such that $\pi_{j_k}(t)$, $k = 1, 2, \ldots, d$ can be expressed in terms of $\pi_i(t)$, where i ranges over all states except for the states $j_1, j_2, \ldots, j_d$. This allows one to reduce the matrix to $N - d$ states. However, the new matrix may contain negative elements.

If the algebraic multiplicity of the eigenvalue $\gamma = 0$ exceeds the geometric one, we use

$$P = \Phi^{-1} \Lambda \Phi,$$

where Φ is determined such that Λ has the Jordan normal form. In particular, Λ has Jordan blocks with zeros on the diagonal. It is easy to verify that any Jordan block with a zero diagonal is *nilpotent*, that is, it becomes 0 for high enough powers: In fact, if J_0 is such a Jordan block, and if it has a dimension of ν, then $J_0^n = 0$ for $n \geq \nu$. Hence, if s is the dimension of the largest Jordan block, then P^s contains no Jordan block greater than 1, and the geometric multiplicity is equal to its algebraic one. From this point onward, if $\gamma = 0$ has a multiplicity of n, one can express n different state probabilities $\pi_i(t)$, $i = i_1, i_2, \ldots, i_n$ as a linear combination of the other $\pi_j(t)$. For $t < s$, this is not possible for all states. Hence, equations like (8.35) only hold for $t \geq s$.

There are examples where $\gamma = 0$ has at least one Jordan blocks greater than 1. In particular, the transition matrix of the remaining lifetimes introduced in Chapter 4, formulated as a phase-type distribution with absorption when the remaining lifetime is 0, forms a single Jordan block with the diagonal equal to 0.

8.2.11 Eigensolutions of Tridiagonal Transition Matrices

In this subsection, we consider transition matrices with generators or subgenerators of the following form, where $\lambda_i > 0$ and $\mu_i > 0$ for all i:

$$Q = \begin{bmatrix} -d_0 & \lambda_0 & 0 & \cdots\cdots & \cdots & \cdots & 0 \\ \mu_1 & -d_1 & \lambda_1 & 0 & \ddots & \ddots & 0 \\ 0 & \mu_2 & -d_2 & \ddots & 0 & \ddots & \vdots \\ \vdots & \ddots & \ddots & \ddots & \ddots & \ddots & \vdots \\ 0 & \ddots & \ddots & \ddots & 0 & \mu_{N-1} & -d_{N-1} & \lambda_{N-1} \\ 0 & \cdots & \cdots & \cdots\cdots & 0 & \mu_N & -d_N \end{bmatrix}.$$

If $d_0 = \lambda_0$ and $d_i = \lambda_i + \mu_i$, this is the transition matrix of a birth–death process. The waiting times LIFO and SIRO also have tridiagonal transition matrices. The main result is that in tridiagonal transition matrices, all eigenvalues are real. Moreover, in contrast to the general eigenvalue problem, the eigenvalues and eigenvectors are easy to find. If $\lambda_i = 0$ or $\mu_i = 0$ then the states form several communicating classes, and the eigenvalues of each class can be obtained separately. We therefore only need to consider the case where all μ_i and λ_i are greater than zero.

The classical method for finding the eigenvalues of tridiagonal matrices is to use determinants, which in the case of tridiagonal matrices are relatively easy to find. Here, we start directly with the problem at hand, which is to find a solution to the system of equations given by $0 = \phi(Q - \gamma I)$. For the eigenvector k, this equation expands to

$$0 = -(d_0 + \gamma)\phi_{k,0} + \mu_1\phi_{k,1} \tag{8.36}$$

$$0 = \lambda_{i-1}\phi_{k,i-1} - (d_i + \gamma)\phi_{k,i} + \mu_{i+1}\phi_{k,i+1}, \ 0 < i < N \tag{8.37}$$

$$0 = \lambda_{N-1}\phi_{k,N-1} - (d_N + \gamma)\phi_{k,N}. \tag{8.38}$$

In this way, we obtain the eigenvectors in addition to the eigenvalues. Since $\mu_i > 0$ for all i, equations (8.36) and (8.37) can be written as

$$\phi_{k,1} = \frac{d_0 + \gamma}{\mu_1}\phi_{k,0} \tag{8.39}$$

$$\phi_{k,i+1} = \frac{d_i + \gamma}{\mu_{i+1}}\phi_{k,i} - \frac{\lambda_{i-1}}{\mu_{i-1}}\phi_{k,i-1}, \ 0 < i < N. \tag{8.40}$$

To write the equation (8.38) in a similar way, we define

$$\phi_{k,N+1} = (d_N + \gamma)\phi_{k,N} - \lambda_{N-1}\phi_{k,N-1}. \tag{8.41}$$

The problem is to determine γ and $\phi_{k,i}$, $i = 0, 1, \ldots, N$, such that (8.38) is satisfied, or, equivalently, that $\phi_{k,N+1} = 0$. Note that $\phi_{k,0} \neq 0$, because otherwise, all $\phi_{k,i}$ become zero. Hence, we can set $\phi_{k,0} = 1$. Moreover, if for any i, $\phi_{k,i}$ and $\phi_{k,i+1}$ are both 0, all $\phi_{k,j} = 0$, $j = 0, 1, \ldots, N + 1$.

We now replace the $\phi_{k,i}$ in equations (8.39) to (8.41) by $\phi_i(\gamma)$. By definition, the vectors $\phi(\gamma) = [\phi_i(\gamma)]$ are eigenvectors only if $\phi_{N+1}(\gamma) = 0$. In other words, we need to find the zeros of $\phi_{N+1}(\gamma)$. To do this, we use a method due to Sturm [89]. In his paper, Sturm used the number of sign variations to find the number of real zeros of a polynomial, and we apply his ideas to our problem. A sign variation occurs if $\phi_i(\gamma)$ and $\phi_{i+1}(\gamma)$ have different signs, or, when $\phi_i(\gamma) = 0$, if ϕ_{i-1} and ϕ_{i+1} have different signs. We now relate the number of sign variations within a given interval to the number of real zeros within this interval. We have:

Theorem 8.7 *If $\mu_{i+1} > 0$ and $\lambda_i > 0$, $i = 0, 1, \ldots, N - 1$, then $\phi_{N+1}(\gamma)$ has $N + 1$ real zeros. Moreover, if $\nu(\gamma)$ is the number of sign variations from $-\infty$ to a, where a is not a zero of $\pi_{N+1}(\gamma)$, then the number of zeroes in the interval $\gamma > a$ is equal to $\nu(a)$.*

Proof Clearly, $v(\gamma)$ cannot change due to changes of γ unless at least one $\phi_i(\gamma)$, $0 \le i \le N + 1$ changes sign. Also note that equations (8.39) to (8.41) show that all $\phi_i(\gamma)$ are continuous and differentiable functions, which means that at least one $\phi_i(\gamma)$ must go through 0. Obviously, $\phi_0(\gamma)$, which is set to 1, cannot go through zero. We now prove that if any of the $\phi_i(\gamma)$, $1 \le i \le N$ goes through zero, $v(\gamma)$ will not change. The reason is that because equations (8.40) and (8.41), $\phi_{i-1}(\gamma)$ and $\phi_{i+1}(\gamma)$ have opposite signs when $\phi_i(\gamma) = 0$. This implies that there is exactly one sign variation between $i - 1$ and $i + 1$ both before and after $\phi_i(\gamma)$ goes through zero. It follows that the only way $v(\gamma)$ can change its value is when $\phi_{N+1}(\gamma)$ goes through zero, leading to a real zero of $\pi_{N+1}(\gamma)$. As $\phi_{N+1}(\gamma)$ goes through zero, $v(\gamma)$ changes by exactly 1, either up or down. Moreover, equations (8.39) to (8.41) show that $v(a) = N + 1$ as $a \to -\infty$, and that $v(a) = 0$ as $a \to \infty$. Consequently, $v(\gamma)$ must change $N + 1$ times, that is, $\phi_{N+1}(\gamma)$ must have at least $N + 1$ real zeros. On the other hand, there are only $N + 1$ eigenvalues, counting multiplicities, and consequently at most $N + 1$ zeros of $\phi_{N+1}(\gamma)$. It follows that $\phi_{N+1}(\gamma)$ has exactly $N + 1$ real zeros. In addition, $v(a)$ decreases or remains constant as γ increases. A little reflection shows that this implies that $v(a)$ yields the number of real zeros exceeding a. $\qquad\square$

In order to find all eigenvalues, note that according to equation (8.17), the eigenvalues must satisfy $-2 \max(d_i) \le \gamma \le 0$, with $\gamma = 0$ having no sign variations, and for $\gamma = -2 \max(d_i)$, there are $N + 1$ sign variation. One can now divide this interval into two subintervals, and search for eigenvalues in each subinterval. If there are no eigenvalues in a subinterval, that is, if $v(a) = v(b)$ for any interval $(a, b]$, then this subinterval contains no zero and can be discarded. If $v(a) \ne v(b)$, then one can half this interval again and continue, always discarding intervals with no zeros, until one has a single small subinterval for each eigenvalue [76]. At this point, one can use Newton's method to find the eigenvalues. In order to use Newton's method, one needs derivatives. Following [32], one can take the derivatives with respect to γ of equations (8.36) to (8.38) to obtain

$$0 = -1 + \mu_1 \phi_1'(\gamma)$$
$$0 = \lambda_{i-1}\phi_{i-1}'(\gamma) - (d_i + \gamma)\phi_i'(\gamma) - \phi_i(\gamma) + \mu_{i+1}\phi_{i+1}'(\gamma), \ 0 < i < N$$
$$\phi_{N+1}'(\gamma) = \lambda_{N-1}\phi_{N-1}'(\gamma) - (d_N + \gamma)\phi_N'(\gamma) - \phi_N(\gamma).$$

Solving for $\phi_i'(\gamma)$, $i = 1, 2, \ldots, N + 1$ yields

$$\phi_1'(\gamma) = \frac{1}{\mu_1}$$
$$\phi_{i+1}'(\gamma) = \frac{(d_i + \gamma)}{\mu_{i+1}} + \phi_i'(\gamma)\frac{1}{\mu_{i+1}}\phi_i(\gamma) - \frac{\lambda_{i-1}}{\mu_{i+1}}\phi_{i-1}'(\gamma), \ 0 < i < N$$
$$\phi_{N+1}'(\gamma) = \lambda_{N-1}\phi_{N-1}'(\gamma) - (d_N + \gamma)\phi_N'(\gamma) - \phi_N(\gamma).$$

This allows one to obtain all $\phi_i'(\gamma)$ recursively.

Once all eigenvalues γ_k and eigenvectors $\phi_k(\gamma)$, $k = 1, 2, \ldots, N$ are found, we have

$$\pi_j(t) = \sum_{k=1}^{N+1} c_k \phi_j(\gamma_k) e^{\gamma_k t},$$

where the c_k are the solutions of the equations

$$\pi_i(0) = \sum_{k=1}^{N+1} c_k \phi_i(\gamma_k), \ i = 1, 2, \ldots, N.$$

In conclusion, the eigenvalues and eigenvectors of tridiagonal matrices can be found without great effort, and the transient solutions can be determined in this fashion.

8.3 Conclusions

In the first section of this chapter, we discussed the structures of transition matrices, and their effects on the speed of convergence, whereas in the second section, we showed how eigenvalues determine the transient behavior of Markov chains. This gave us a good perspective to answer such questions as when fast convergence is expected, in when not. This topic turns out to be important for many applications. We mentioned already the application of our theory for the development of iterative methods to solve equilibrium equations. An area where it could also find useful applications is discrete-event simulation.

For ergodic Markov chains, we identified the following reasons why the convergence of $p_{i,j}(t)$ to π_j could be slow

1. Convergence is slow when all paths from i to j are long.
2. Convergence is slow when the transition matrix is almost decomposable, and i belongs to a set of states which is almost separate from the set of states to which j belongs.
3. The Markov chain is almost periodic. Note that even CTMCs can be almost periodic.

The slow convergence of $p_{i,j}(t)$ can lead to a slow convergence to $\pi_j(t) = \sum_i \pi_i(0)p_{i,j}(t)$, but it is possible that $\pi_j(t)$ converges surprisingly fast. In particular, if $\pi_{i_1}(0)$ and $\pi_{i_2}(0)$ are both positive, then it can happen that $p_{i_1,j}(t)$ is above its equilibrium value, while $p_{i_2,j}(t)$ is below, reducing thus the deviation of $\pi_j(t)$ from its equilibrium value. In particular, in almost decomposable matrices, convergence is improved if the initial non-zero $\pi_i(0)$ are from different blocks of states that are almost communicating classes. Similarly, in almost periodic Markov chains with r are almost periodic classes, convergence is improved if a single state from each of the r classes is selected, and for each i thus selected, $\pi_i(0) = 1/r$.

The second section deals with eigenvalues, which provide a different aspect of the speed of convergence of the $p_{i,j}(t)$ from their equilibrium. Consider first ergodic Markov chains, where the $p_{i,j}(t)$ converge to π_j. In an ergodic DTMC, the deviation from π_j are, as a rule, in the form $\sum_{i=2}^{N} c_k \gamma_k^t$, where the c_k may be positive, negative or complex. If γ_k is real and non-negative, then the resulting term decays geometrically, but if γ_k is negative or complex, then we get oscillation. In an ergodic CTMC, the deviations from the equilibrium are of the form $\sum_{k=2}^{N} c_k \exp(\gamma_k t)$. If γ_k is real, this means that the corresponding term decreases exponentially with t, and if γ_k is complex, we obtain sine waves. The behavior of $p_{i,j}(t)$ in both DTMCs and CTMCs is different when the algebraic multiplicity of an eigenvalue exceeds its geometric one. In this case, the exponential term in question must be multiplied by a polynomial.

As t gets large, the behavior of $p_{i,j}(t)$ is, as a rule, determined by the subdominant eigenvalue. In the case of a DTMC, the eigenvalue with the largest absolute value less than one is the subdominant eigenvalue, and in a CTMC, it is the eigenvalue with the largest real part. Consequently, if the subdominant eigenvalue is real, the equilibrium is eventually approached monotonically, whereas there are oscillations when the eigenvalue is complex. In this respect, it is of interest to determine which eigenvalues are real. As it turns out, in tridiagonal transition matrices, the eigenvalues are always real, and the same is true for bidiagonal matrices.

The subdominant eigenvalue depends on the transition matrix in its entirety. Consequently, if there is a single pair of states i and j such that $p_{i,j}(t)$ converges slowly toward its equilibrium, then the subdominant eigenvalue must be close to 1 if the Markov chain is discrete, or close to zero if it is continuous. This has the consequence that discrete-event systems containing states i and j where the shortest path between i and j is large, the subdominant eigenvalues are close to 1, respectively, 0. Whether or not this is relevant has to be decided on a case-by-case basis.

Also, if the subdominant eigenvalue is complex, a good choice of the initial conditions, or taking time averages can ameliorate the situation. In particular, in periodic DTMCs, several eigenvalues have an absolute value of 1, and no convergence occurs. However, the effect of the periodicity can be almost completely removed by giving each periodicity class an equal probability. A similar technique could be applied for Markov chains that are almost periodic. We believe that in this area, there is ample opportunity for further study.

Eigenvalues are also important for characterizing Coxian distributions. There, we found that all eigenvalues are real, including the subdominant eigenvalue. It follows that all Coxian distributions have an exponential tail, except in cases where the algebraic multiplicity of an eigenvalue exceeds the geometric one, in which case the tail is like the one of an Erlang distribution.

Problems

8.1 Why does any matrix where one column is a multiple of another column have an eigenvalue of zero?

8.2 Consider the following transition matrix of a DTMC

$$P = \begin{bmatrix} 0.4 & 0.6 & 0 & 0 \\ 0.3 & 0.7 - a & 0 & a \\ a & 0 & 0.5 - a & 0.5 \\ 0 & 0 & 0.4 & 0.6 \end{bmatrix}.$$

Set $a = 0.02$, and use MATLAB function "eig" to find the eigenvalues of this matrix. Then increase a 10 times in steps of 0.02. Comment on the results.

8.3 Show that transition matrices of DTMCs cannot have complex eigenvalues unless they have a size of at least three. Also give an example of a transition matrix of size three that has two complex eigenvalues. How is the situation for CTMCs?

8.4 In Coxian distributions, several diagonal entries λ_i may have identical values. When all subdiagonal entries μ_i are positive, then the algebraic multiplicity exceeds the geometric one. Is this also true when some μ_i vanish?

8.5 Find the eigenvalues for the SIRO waiting time distribution of an $M/M/1/3$ queue.

Chapter 9
Equilibrium Solutions of Markov Chains and Related Topics

In this chapter, we discuss how to find the equilibrium probabilities of ergodic, finite-state Markov chains. Mathematically, there is only a minor difference between the equilibrium equations of DTMCs and CTMCs. To show this, we write the equilibrium equations of DTMCs $\pi(P - I) = 0$, and when one only sees the entries of the matrix $P - I$, it is impossible to decide whether the entries are coming from a stochastic matrix or an infinitesimal generator. However, we will present probabilistic interpretations of our algorithms, and these are different for DTMCs and CTMCs.

One distinguishes between two classes of methods for solving $\pi = \pi P$ or $0 = \pi Q$ for π with the condition that the entries of π add up to 1. *Direct methods* find the solution in a finite number of steps, whereas *iterative methods* only approach the solution π without reaching it in a finite number of steps. Both methods are explained in details in this chapter.

The methods used for solving the equilibrium equations must be adapted to the matrices we are looking at. The transition matrices of discrete event systems are large, but sparse, and for this type of matrices, iterative methods tend to be faster. However, the theory of direct methods will be used to justify embedding methods (see Chapter 10) and matrix analytic methods (see Chapter 12). Also, most discrete event systems result in banded matrices, and direct methods can exploit bandedness.

In addition to solving the equilibrium equations, we also consider the behavior of Markov chains with absorbing states. In these systems, the expected time spent in a particular state has a finite limit, and this limit is of interest. This is also a long-range phenomenon, and it will be discussed in this chapter. In addition to this, the equations for finding the time integrals for transient states are very similar to the equilibrium equations of recurrent ergodic Markov chains. We also discuss some interesting relations between direct methods for solving equilibrium equations and finding the expected number of visits to the different states in Markov chains with substochastic matrices.

The outline of this chapter is as follows. We first discuss direct methods and their probabilistic interpretations. This brings us to the discussion of the limit of time averages, which will be discussed next. The last section presents some important iterative methods.

© The Author(s), under exclusive license to Springer Nature Switzerland AG 2022   187
W. Grassmann and J. Tavakoli, *Numerical Methods for Solving Discrete Event Systems*,
CMS/CAIMS Books in Mathematics 5, https://doi.org/10.1007/978-3-031-10082-6_9

9.1 Direct Methods

The equilibrium equations are systems of N linear equations, and the classical approach for solving linear equations are elimination methods, in particular Gaussian eliminations. Typically, one starts with equation 1 and eliminates the first variable, say x_1. Once this is done, one uses the second equation, with x_1 now eliminated, and eliminates x_2. In this fashion, one eliminates all x_i except for x_N. At this point, the back-substitution step begins, that is, one solves the last equation for x_N, the second last equation for x_{N-1}, and so on, until all x_i, $i = 1, 2, \ldots, N$ are determined. Here, we reverse the order of elimination: Our variables are the equilibrium probabilities π_i, $i = 1, 2, \ldots, N$, and we eliminate π_N first, followed by π_{N-1}, and so on until we have all π_i eliminated except for π_1. The reason is that, in queueing, long queues tend to be less likely than short queues. Consequently, the probabilities π_i decrease as i increases, and to reduce rounding, one should deal with small numbers first. Eliminating the π_i starting with $i = N$ is also necessary for the solution of infinite-state Markov chains.

A very interesting result is that the elimination procedure has a probabilistic interpretation, as we will show in detail. The resulting interpretation turns out to be crucial for some methods discussed in later chapters. This interpretation also means that we are dealing with probabilities, which are non-negative. In fact, as it turns out, we can solve the equilibrium equations without using subtractions, which ensures numerical stability.

Another difference between classical elimination and the solution of equilibrium equation is that the equilibrium equations are not independent. In fact, they are homogeneous, which means that if $\pi_1, \pi_2, \ldots, \pi_N$ is a solution, so is $d\pi_1, d\pi_2, \ldots, d\pi_N$. Here, the factor d can be found by the condition that the sum of all probabilities must be 1. We can thus find any solution of the equilibrium equation, say the solution found by setting $\pi_1 = 1$, and divide the π_i found in this way by their sum. Hence, if $\tilde{\pi}_1, \tilde{\pi}_2, \ldots, \tilde{\pi}_n$ are any solution of the equilibrium equations, we find the equilibrium probability of state i as $\tilde{\pi}_i / \sum_j \tilde{\pi}_j$. We call the $\tilde{\pi}_i$, $i = 1, 2, \ldots, N$ *equilibrium chances*. The conversion of the chances to probabilities will be referred to as *norming*. As it turns out, the equilibrium chances have an interesting probabilistic interpretation.

When solving linear equations, one often uses *pivoting*, that is, one exchanges rows or columns, either to reduce the rounding error, or to reduce the number of operations. For finding the equilibrium equations, pivoting is not recommended. Pivoting also destroys the probabilistic interpretation of the elimination method. An exception is a reordering of the states, which simultaneously changes the row and the column.

In this section, we first discuss a method related to Gaussian elimination, called *state elimination* and its interpretation for both DTMCs and CTMCs. Next, we provide an example to show how these methods are applied. We then discuss *block elimination*, as well as the *Crout method*, a variations of Gaussian elimination.

9.1.1 The State Elimination Method

In this section, we will present the state elimination algorithm for DTMCs, and show how to modify this algorithm to handle CTMCs.

9.1.1.1 State Elimination for DTMCs

For DTMCs with all states communicating, the equilibrium probabilities π_j, $j = 1, 2, \ldots, N$ can be found from the following equations:

$$\pi_j = \sum_{i=1}^{N} \pi_i p_{i,j}, \ j = 1, 2 \ldots, N. \tag{9.1}$$

We refer to the equations with π_j on the left as the *state j equation*. We solve the state N equation for π_N, yielding

$$\pi_N = \sum_{i=1}^{N-1} \pi_i \frac{p_{i,N}}{1 - p_{N,N}}.$$

After substituting all π_N in the other equilibrium equations, one has

$$\pi_j = \sum_{i=1}^{N-1} \pi_i \left(p_{i,j} + \frac{p_{i,N} p_{N,j}}{1 - p_{N,N}} \right), \ j < N.$$

This is a system with $N - 1$ equations for the $N - 1$ variables π_1 to π_{N-1}, and from this, we can eliminate π_{N-1}. This yields $N - 2$ equations with $N - 2$ variables, and in this fashion, one can continue. If $p_{i,j}^{(n)}$ is the coefficients of the system of equations once $\pi_{n+1}, \pi_{N+2}, \ldots, \pi_N$ have been eliminated, we obtain

$$\pi_n = \sum_{i=1}^{n-1} \pi_i \frac{p_{i,n}^{(n)}}{1 - p_{n,n}^{(n)}} \tag{9.2}$$

and

$$\pi_j = \sum_{i=1}^{n-1} \pi_i \left(p_{i,j}^{(n)} + \frac{p_{i,n}^{(n)} p_{n,j}^{(n)}}{1 - p_{n,n}^{(n)}} \right) = \sum_{i=1}^{n-1} \pi_i p_{i,j}^{(n-1)}, \ j < n.$$

Here, $p_{i,j}^{(N)} = p_{i,j}$ and

$$p_{i,j}^{(n-1)} = p_{i,j}^{(n)} + \frac{p_{i,n}^{(n)} p_{n,j}^{(n)}}{1 - p_{n,n}^{(n)}}, \ i, j < n \leq N. \tag{9.3}$$

As stated in Section 1.2, if π_n is a solution to the equilibrium equations, so is any multiple of π_n. Thus, the equilibrium equations can only be used to find π_n up to a factor, and to find this factor, the condition $\sum_{n=1}^{N} \pi_n = 1$ is used. We denote the solution of the equilibrium equations with $\pi_1 = 1$ by $\tilde{\pi}$. Given $\tilde{\pi}_1 = 1$, the remaining $\tilde{\pi}_n$ can be found from equation (9.2). The $\tilde{\pi}_n$ can be normed such that their sum is 1 by using

$$\pi_n = \frac{\tilde{\pi}_n}{\sum_{i=1}^{N} \tilde{\pi}_i}. \tag{9.4}$$

We now summarize the procedure. The $p_{i,j}^{(n)}$ are found from equation (9.3), which is applied until $n = 2$. Once all $p_{i,j}^{(n)}$ are known, (9.2) can be used to find the $\tilde{\pi}_i$, $i = 2, 3, \ldots, N$, starting with $\tilde{\pi}_1 = 1$. The π_n can then be found by (9.4).

Next, we count the number of flops needed to find the π_n. To save flops, we first calculate for each i and n

$$a_{i,n} = \frac{p_{i,n}^{(n)}}{1 - p_{n,n}^{(n)}}, \quad 1 \le i < n \tag{9.5}$$

and write

$$p_{i,j}^{(n-1)} = p_{i,j}^{(n)} + a_{i,n} p_{n,j}^{(n)}, \quad 1 \le i, j < n \le N.$$

For given i, j and n, this requires 2 flops, and and since there are $(n-1)^2$ different pairs i, j, this yields $(n-1)^2$ flops for each n. Since n varies from 2 to N, we get $\sum_{n=2}^{N} (n-1)^2$ flops, or, according to equation (6.1) of Chapter 6

$$2 \sum_{n=1}^{N-1} n^2 = \frac{2}{3}(N-1)^3 + O(N^2) \text{ flops.}$$

All operations remaining to find π_n only need $O(N^2)$ flops. To find the $a_{i,j}$ requires 1 flop for each n for a total $\frac{n(n-1)}{2}$ flops. To find $\tilde{\pi}_n$, we write equation (9.2) as

$$\tilde{\pi}_n = \sum_{i=1}^{n-1} a_{i,n} \tilde{\pi}_i, \quad 1 \le i < n,$$

and this yields $n(n-1)$ flops. The N flops for norming can be neglected. Hence, all together, we have $\frac{2}{3}N^3 + O(N^2)$ flops. As we will show later, the flop count can still be reduced when the transition matrix is banded.

In Chapter 6, we stated that subtractions can greatly increase previously committed rounding errors, and they should therefore be avoided. This can be done by using the so-called GTH method, named for its creators Grassmann, Taksar, and Heyman [38], [74], [86].

The GTH method is based on the following equation:

$$\sum_{j=1}^{n} p_{i,j}^{(n)} = 1. \tag{9.6}$$

Equation (9.6) holds because it holds for $n = N$, and once it holds for n, a simple calculation shows that it holds also for $n - 1$:

$$\sum_{j=1}^{n-1} p_{i,j}^{(n-1)} = \sum_{j=1}^{n-1} p_{i,j}^{(n)} + \sum_{j=1}^{n-1} \frac{p_{i,n}^{(n)} p_{n,j}^{(n)}}{1 - p_{n,n}^{(n)}} = \sum_{j=1}^{n-1} p_{i,j}^{(n)} + \frac{p_{i,n}^{(n)}}{1 - p_{n,n}^{(n)}} \sum_{j=1}^{n-1} p_{n,j}^{(n)}. \tag{9.7}$$

According to our assumption, $\sum_{j=1}^{n} p_{n,j}^{(n)} = 1$, which implies that $\sum_{j=1}^{n-1} p_{n,j}^{(n)} = 1 - p_{n,n}^{(n)}$, and (9.7) becomes $\sum_{j=1}^{n-1} p_{i,j}^{(n-1)} = \sum_{j=1}^{n} p_{i,j}^{(n)}$. Since $\sum_{j=1}^{N} p_{i,j}^{(N)} = 1$, (9.6) follows.

The following algorithm, called state elimination algorithm, summarizes the procedure given by equations (9.3) and (9.2), using also $a_{i,n}$ as defined in equation (9.5). Note that in iteration n, $\pi_{i,j}^{(n)}$, $i < n$ is stored in array location "pi(i,j)" and $a_{i,n}$ is stored in array location "pi(i,n)".

Algorithm 30. State Elimination Algorithm

```
for n = N :−1 : 1
    s = ∑_{j=1}^{n−1} p(n,j)
    for i = 1 : n − 1
        p(i,n) = p(i,n)/s
        for j = 1 : n − 1
            p(i,j) = p(i,j) + p(i,n)*p(n,j)
        end for
    end for
end for
pi(1) = 1
for n = 2 : N
    pi(n) = ∑_{i=1}^{n−1} pi(i) * p(i,n)
end for
for n = 1 : N
    pi(n) = pi(n)/∑_{i=1}^{N} pi(i)
end for
```

The algorithm also calculates $p_{1,1}^{(1)}$, which must be 1. Also, all $p_{i,j}^{(n)}$ are non-negative, and there are no subtractions, which ensures numerical stability as shown in [30] and [74].

Equation (9.3) has a probabilistic interpretation: In fact, the $p_{i,j}^{(n)}$ can be interpreted as the transition probabilities of a Markov chain reduced to the states $1, 2, \ldots, n$. In this Markov chain, $p_{i,j}^{(n)}$ is the probability of moving from i to j directly, whereas $p_{i,n}^{(n)}$ is the probability of moving from i to n, and $\frac{p_{n,j}^{(n)}}{1-p_{n,n}^{(n)}}$ is the probability to go from n to j given state n is left. We discuss this in Section 9.1.3.

9.1.1.2 State Elimination for CTMCs

For CTMCs, the equilibrium equations are

$$0 = \sum_{i=1}^{N} \pi_i q_{i,j}.$$

To solve these equations, one can use Algorithm 30 right above, except that the "p(i,j)" must be initialized with $q_{i,j}$ instead of $p_{i,j}$. We now show this in detail.

As in the DTMC, we first eliminate π_N, then π_{N-1}, and so on, and we obtain, similar to equation (9.2)

$$\pi_n = \sum_{i=1}^{n-1} \pi_i \frac{q_{i,n}^{(n)}}{-q_{n,n}^{(n)}}. \tag{9.8}$$

Instead of equation (9.3), one has

$$q_{i,j}^{(n-1)} = q_{i,j}^{(n)} + \frac{q_{i,n}^{(n)} q_{n,j}^{(n)}}{-q_{n,n}^{(n)}}. \tag{9.9}$$

Through methods similar to the ones used for the DTMC, one can prove that $-q_{n,n}^{(n)} = \sum_{j=1}^{n-1} q_{n,j}^{(n)}$, and

$$q_{i,j}^{(n-1)} = q_{i,j}^{(n)} + \frac{q_{i,n}^{(n)} q_{n,j}^{(n)}}{\sum_{n=1}^{n-1} q_{n,j}^{(n)}}. \tag{9.10}$$

This is exactly the formula used in Algorithm 30, which shows that the algorithm also works for CTMCs.

9.1.2 Banded Matrices

Direct methods have one major disadvantage: They make it difficult to exploit sparsity. Indeed, applying equation (9.3), $p_{i,j}^{(n-1)}$ may be different from 0 even when $p_{i,j}^{(n)} = 0$. Hence, though in every elimination step, the number of states is reduced, the number of non-zero entries in the transition matrix may increase.

As we pointed out, transition matrices generated by discrete event systems tend to be banded, and this feature can be exploited. By definition, a matrix is said to be (g, h) banded if $p_{i,j}$ can be non-zero only if $i - g \le j \le i + h$, where g and h are smaller than N, the size of the matrix. We say that the pair i, j is in the g, h band if:

$$i - g \le j \le i + h. \tag{9.11}$$

By subtracting $i + j$ from the relation above, and multiplying by -1, one obtains

$$j - h \leq i \leq j + g. \tag{9.12}$$

We now prove:

Theorem 9.1 *If the matrix* $P = [p_{i,j}]$ *is* (g, h) *banded, that is,* $p_{i,j} = 0$ *if* i, j *are outside the* g, h *band, so are the matrices* $P^{(n)} = [p_{i,j}^{(n)}]$.

Proof The proof is by complete induction. The theorem is true for $n = N$ because $p_{i,j}^{(N)} = p_{i,j}$, and we now show that if it is true for n, it is also true for $n - 1$. We have to show that $p_{i,j}^{(n-1)} = 0$ if i, j is outside the g, h band, provided $p_{k,\ell}^{(n)} = 0$ outside the g, h band for all k and ℓ. Clearly, $p_{i,j}^{(n-1)}$ is zero if $p_{i,j}^{(n)} = 0$, and either $p_{i,n}^{(n)} = 0$ or $p_{n,j}^{(n)} = 0$. By assumption, $p_{i,j}^{(n)} = 0$. Also, the pair i, j is outside the g, h band if either $j < i - g$ or $j > i + h$. Now $j < i - g$ and $i < n$ implies $j < n - g$, that is, $p_{i,n}^{(n)}$ is outside the band and therefore 0. If $j > i + h$, then $i < n$ implies $j > n + h$, and $p_{n,j}^{(n)}$ is outside the band. In either case, $p_{i,j}^{(n-1)} = 0$. □

The idea is now to restrict the range of i and j to the g, h band. This can be accomplished by modifying Algorithm 30. In fact, the lower limit of the range of i can be found from (9.12), and the lower limit of the range of j can be found from (9.11). In particular, in Algorithm 30, one has to replace the statement

for i = 1 : n − 1

by

for i = max(1, n − h) : n − 1

This is correct because p(i,n) is nonzero only if $i \geq n - h$. One also has to replace the statement

for j = 1 : n − 1

by

for j = max(1, n − g) : n − 1

In a similar fashion, one must replace $\sum_{j=1}^{n-1} p_{n,j}$ by $\sum_{j=\min(1,n-g)}^{n-1} p(n, j)$ and $\sum_{i=1}^{n-1} \pi_i p(i, n)$ by $\sum_{i=\max(1,n-h)}^{n-1} \pi_i p(i, n)$.

One can also save storage space by only storing the band. In this case, one needs an array "pb" with N rows and $h + g + 1$ columns, ranging from $-g$ to h, with "pb(i,k)" storing $p(i, i + k)$, $-g \leq k \leq h$. Hence, to calculate $p_{i,j}$, one has to use pb(i,j-i). This leads to the following algorithm:

Algorithm 31. Banded State Elimination Algorithm

for n = N : −1 : 2
   s = $\sum_{j=\max(1,n-g)}^{n-1}$ pb(n,j-n)
   for i = max(1, n − h) : n − 1
      pb(i,n-i) = pb(i,n-i)/s
      for j = max(1, n - g) : n − 1

pb(i,j-i) = pb(i,j-i) + pb(i,n-i)*pb(n,j-n)
  end for
 end for
next n
pi(1) = 1
for n = 2 : N
 $pi(n) = \sum_{i=\max(1,n-h)}^{n-1} pi(i) * pb(i,n-i)$
end for
for n = 1 : N
 $pi(n) = pi(n)/\sum_{i=1}^{N} pi(i)$
end for

If we ignore the case where n is less than $\max(g, h)$, we have $2hg$ flops to execute the innermost statement of the state elimination algorithm, leading to approximately $2Nhg$ flops in total.

To use this algorithm, we still need values for g and h. This is relatively easy if each event changes the state number by a constant amount, as is the case when the state space is Cartesian, and $f_k = X + a^{(k)}$, where $a^{(k)} = [a_i^{(k)}]$. In this case, one merely has to calculate $c_k = \sum_i a_i^{(k)} u_i$ by equation (3.1). The lowest c_k is then $-g$, and the highest c_k is h. When the matrix is given in a row, column, value format, then $-g$ is the minimum difference between column and row, and h is the maximum difference. Also, the row, column, value format must be converted into a banded transition matrix.

If matrices are banded, one is often tempted to use $p_{n,n-g}$ as the pivot element. Experience shows that unless $g = 1$, subtractive cancellation occurs, often with the result that some of the $p_{i,j}^{(n)}$ become negative. Hence, this method should not be used except in the case of small matrices.

9.1.3 Gaussian Elimination as Censoring

We mentioned that equation (9.3) has a probabilistic interpretation. This section explores this issue further. First, we define *censoring* as follows:

Definition 9.1 Given a Markov process $\{X(t), t \geq 0\}$, with $X(t) \in S$. If S_1 is a subset of S, then we say the states $k \in S_1$ are *censored* if the time spent in any of the states $k \in S_1$ is reduced to 0 and state k is in this way removed from the sample function.

For example, if a Markov chain has four states, say red, green, yellow, and blue, we can censor the states green and red, that is, any time spent in these states is reduced to 0 and then eliminated. For instance, if a sample function of the process starts with { blue, green, red, blue, yellow, blue, green, yellow, ... }, and red and green are eliminated, the censored process becomes { blue, blue, yellow, ... }. Note that this definition works for both DTMCs and CTMCs, provided the term "number of visits to state i" used in DTMCs is taken to be the time spent in state i. We now have the following theorem:

Theorem 9.2 *If, in a finite state Markov chain, some states are censored, the process is still Markovian. Also, in a DTMC with states $1, 2, \ldots, N$, once all states $k > n$ are censored, the transition matrix becomes $[p_{i,j}^{(n)}]$, where the $p_{i,j}^{(n)}$ are given by equation (9.3). Also, in a CTMC, the transition matrix becomes $[q_{i,j}^{(n)}]$, with the $q_{i,j}^{(n)}$ given by equation (9.10).*

Proof We first consider DTMCs, and we assume only state N is censored. This means that, for any $i, j \neq N$, a transition from i to j in the censored process occurs if, in the original process, we have either the sequence i, j, or the sequence i, N, j, or the sequence i, N, N, j, and so on. If the transition probabilities of the original process are $p_{i,j}$, the censored process has the transition probabilities $p_{i,j}^{(N-1)}$ given as

$$p_{i,j}^{(N-1)} = p_{i,j} + p_{i,n}p_{n,j} + p_{i,n}p_{n,n}p_{n,j} + p_{i,n}p_{n,n}^2 p_{n,j} + \cdots .$$

When using sums of geometric series, this yields

$$p_{i,j}^{(N-1)} = p_{i,j} + \frac{p_{i,N}p_{N,j}}{1 - p_{N,N}}.$$

Consider now censoring several states. These states can either be censored all at once or one can censor each separately. The result is obviously the same. If one censors the states $N, N-1, N-2, \ldots, n+1$, one after the other, one obtains

$$p_{i,j}^{(n-1)} = p_{i,j}^{(n)} + \frac{p_{i,n}^{(n)}p_{n,j}^{(n)}}{1 - p_{n,n}^{(n)}}.$$

This equation is identical to equation (9.3). Note that the $p_{i,j}^{(n-1)}$ are unique, that is, they depend only on $p_{i,j}$, and nothing else, which excludes any influence of the past history. This establishes that the process is Markovian. Consequently, the theorem is proven for the DTMC.

To prove the result for CTMCs, note that if only state N is to be censored, then the rate of going from i to j in the censored process is the rate of either going from i to j directly, or the rate of going from i to N, multiplied by the probability of going from N to j once leaving state N. By definition, the rate of going from i to N is $q_{i,N}$, and the probability of going from N to j is equal to the expected time spent in state N, which is $-1/q_{N,N}$, multiplied by the rate of going form N to j. It follows that the transition rate of the censored process is

$$q_{i,j}^{(N-1)} = q_{i,j} + \frac{q_{i,N}q_{N,j}}{-q_{N,N}}.$$

As in the case of the DTMC, we can now censor one state after the other, obtaining in this way equation (9.9). The remainder of the proof is now similar to the one of the DTMC.   □

By censoring states, one really uses embedding. Generally, we have

Definition 9.2 A process $\{U(n), n = 1, 2, 3, \ldots\}$ is said to be *embedded* in process $\{X(t), t > 0\}$ if $U(n) = X(t_n)$. The values of t_n can be influenced by $X(t), t \le t_n$, but not by $X(t), t > t_n$.

Note that $\{X(t), t > 0\}$ is not necessarily Markovian, but $\{U(n), n > 0\}$ typically is Markovian.

An example of an embedding process is used in the randomization method, discussed in Chapter 7. There, we had a CTMC $\{X(t, t > 0\}$, and a sequence t_n, which were generated by a Poisson process with rate q. Now, $U(n) = X(t_n)$, and the transition matrix of $U(n)$ was $P = Q/q + I$, where Q is the transition matrix of $\{X(t), t > 0\}$.

From the definition above, it follows that the process obtained by censoring is also an embedded process, and it is in fact a DTMC. In contrast to embedding used in randomization, censoring reduces the number of states. Specifically, if $X(t)$ is the original Markov chain, and $X^{(n)}$ is the Markov chain embedded at the times t_n, where $X(t) \le n$, then $X^{(n)}$ can only assume the values $1, 2, \ldots, n, n \le N$ whereas $X(t)$ can assume the values $1, 2, \ldots, N$.

By censoring, paths between non-eliminated states are merely shortened, but not eliminated, and neither are new paths added. For non-censored states i and j, it follows that if state i can reach state j in the original Markov chain, then i can reach j also in the censored Markov chain, and if i cannot reach j in the original Markov chain, it cannot reach j in the censored Markov chain either. Consequently, any communicating class in the censored Markov chain corresponds to a communicating class in the original Markov chain. In this context, we now have

Theorem 9.3 *If $p_{i,j}^{(n)}$ is the transition probability of a DTMC with state space 1, 2, $\ldots$, N and with states $k > n$ censored, then $\sum_{j=1}^{n-1} p_{n,j}^{(n)} = 0$ if and only if state n is the last non-eliminated state of a recurrent communicating class.*

Proof $\sum_{j=1}^{n-1} p_{n,j}^{(n)} = 0$ if and only if there is no path from n to any non-eliminated state, that is, if and only if state n is absorbing. Since we can associate a recurrent communicating class in the original Markov chain with each recurrent class C of the censored Markov chain, n must be the last non-censored state of C. On the other hand, if $\sum_{j=1}^{n-1} p_{n,j}^{(n)} > 0$, then there must be a path from n to at least one other state k, and state n is either not recurrent, or it is not the last non-censored state of a recurrent communicating class. $\square$

From this theorem, one concludes that Algorithm 30 will work if there is only one communicating class. Moreover, it can be modified to deal with several communicating classes: Whenever $\sum_{j=1}^{n-1} p_{i,j}^{(n)} = 0$, we have discovered a recurrent communicating class, and the equilibrium probabilities of this class can be calculated, independently of the other classes.

When solving systems of linear equations, one may encounter cases where the equations are almost dependent, and in these cases, rounding errors are magnified significantly. This means that even if $\sum_{j=1}^{n} p_{n,j}^{(n)} = 0$, the result obtained by floating point arithmetic may not exactly be zero. This is impossible when using the GTH method. If there is no path between n and j, no path is added, that is $p_{n,j}^{(n)} = 0$. Also, if there is a path, then $p_{n,j}^{(n)} > 0$. For a DTMC, this follows immediately from equation (9.3) which implies $p_{i,j}^{(n-1)} = 0$ if and only if $p_{i,j}^{(n)} = 0$ and $p_{i,n}^{(n)}p_{n,j}^{(n)} = 0$ for all i and j. None of these results are affected by rounding. If, however, one uses $1 - p_{n,n}^{(n)}$ instead of $\sum_{j=1}^{n-1} p_{n,j}^{(n)}$, then $p_{n,n}^{(n)}$ should be 1 when n is the last state of a recurrent communicating class. But, due to rounding, the calculated value of $p_{n,n}^{(n)}$ is typically not exactly 1, and $1 - p_{n,n}^{(n)}$ differs from 0.

Consider now CTMCs. If n is the last non-censored state of a communicating class, then $q_{n,n}^{(n)} = 0$, and this result is certainly obtained if one uses

$$q_{n,n}^{(n)} = -\sum_{j=0}^{n-1} q_{n,j}^{(n)}.$$

However, if standard eliminations procedures are used, one finds $q_{n,n}^{(n)}$ as

$$q_{n,n}^{(n)} = q_{n,n}^{(n+1)} + \frac{q_{n,n+1}^{(n+1)} q_{n+1,n}^{(n+1)}}{-q_{n+1,n+1}^{(n+1)}}.$$

Clearly, $q_{n+1,n+1}^{(n+1)} < 0$, which makes the fraction of the equation above positive. Since $q_{n,n}^{(n+1)}$ is negative, we have subtractive cancellation, which increases the rounding errors committed earlier. Consequently, if the calculated value of $q_{n,n}^{(n)}$ is small, the true value could very well be zero, and one is no longer sure whether or not n is an absorbing state.

The GTH formula is particularly useful for Markov chains that are almost decomposable. In DTMCs, $1 - p_{n,n}^{(n)}$ is close to zero if its transition matrix is almost decomposable, and then, $p_{n,n}^{(n)}$ is close to 1. In this case, the GTH algorithm obtains $\sum_{i=1}^{n-1} p_{n,j}$ with a small relative error, whereas the relative error of $1 - p_{n,n}^{(n)}$ increases without bound as $p_{n,n}^{(n)}$ approaches 1. A similar situation occurs for a CTMC.

One can easily modify Algorithm 30 such that it can handle several recurrent communicating classes, and calculate the individual equilibrium probabilities for each class. If $\sum_{j=1}^{n-1} p_{n,j}^{(n)} = 0$, respectively, $\sum_{j=1}^{n-1} q_{n,j}^{(n)} = 0$, then the communicating class C of state n is

$$C = \{m \mid p_{m,n}^{(n)} > 0\} \quad \text{for a DTMC}$$

$$C = \{m \mid q_{m,n}^{(n)} > 0\} \quad \text{for a CTMC}$$

At this point, one only considers the states from C and calculates their equilibrium probabilities.

When state n is the last non-censored state of a communicating class, all states with paths to n are transient, and they, too, can be eliminated. Consequently, one can eliminate all states i for which $p_{i,n}^{(n)} > 0$. Moreover, if k is one of these states, then all states satisfying $p_{i,k}^{(n)} > 0$ can also be removed, and so on. We leave it to the reader to work out the details.

9.1.4 A Two Server Queue

We now use the state elimination method for finding the equilibrium probabilities of a two-server queue with H_2 arrivals and E_2 service times for both servers. However, server 2 is slower than server 1, and when both servers are free, arrivals therefore choose server 1. For designing this system, management requires the following information:

1. There are only ten places for the queue, and when the queue has filled up, arrivals are lost. What is the probability that arrivals are lost?
2. What is the probability that server 1 is busy? What is the probability server 2 is busy?
3. What is the average number in the system?
4. How does the probability of customer loss change if arrival rate? increases by 5%

The parameters of this system are as follows. In arrival phase 1, which happens with probability $p = 0.8$, arrivals occur at a rate of $\lambda_1 = 1.5$, whereas in phase 2, $\lambda_2 = 3$. The service rate of server 1 is 1, whereas the service rate of server 2 is 0.8. It is for this reason that an arrival to the empty system is assigned to server 1.

The setup of this system is described in Section 5.3.2. To find the results, we ran this problem with the banded state elimination method. Here, $h = u_1 + u_2$, and $g = u_1 - 1$. We chose these values because according to Table 5.1, arrivals increase X_1 by 1, and possibly X_2 by 1, which increases the lexicographic code by $u_1 + u_2$. Also, the event "Next 1" combines with "Start 1" whenever $X_3 = 1$, and in this case, X_1 decreases by 1 and X_3 first decreases by 1, and then increases by $K_1 = 2$, that is, in the end, X_3 increases by 1, yielding a decrease of $u_1 - u_3$. A similar calculation for "Next 2" yields a decrease of $u_1 - u_4$, which is smaller. The upper bound for g becomes $u_1 - u_4 = u_1 - 1$. We ran the system with the input parameters given above. There were 98 states, and using the banded matrix algorithm, we found a probability of each state. From these probabilities, the information demanded by management has to be obtained. To do this, we first have to convert the state numbers into state vectors. As in Section 5.3.2, we do this by using the array "codes". Specifically, given a state number n, we find its code as "codes(n)". This code can then be decoded to find the state vector, and the information demanded by management can now be extracted easily.

Given the state vector, the expected number in the system, as well as the probability that the servers are idle, can be determined. Note that $X(1)$ is the number in the queue, excluding the elements served, $X(2)$ is the arrival phase, $X(3)$ the phases to go for server 1, and $X(4)$ the phases to go for server 2. If and only if $X(3)$ is zero, server 1 is idle, and a similar condition can be used to find if server 2 is idle. The number in the system is 0 if $X(1) = 0$, $X(3) = 0$ and $X(4) = 0$, it is 1 if either $X(3)$ or $X(4)$, but not both are 0. Otherwise, the number in the system is $X(1) + 2$. Note that the supplementary variables in this example are no longer purely supplementary: If $X(3)$ or $X(4)$ is zero, the corresponding server is idle, and this is visible, and, in this sense, indicates a physical state.

According to our program, server 1 is busy 90.2% of the time, and server 2 is busy 87.7% of the time. The average number in the system is 5.18. To find the probability of losing a customer, we have to divide the overall arrival rate by the rate at which customers are lost. The overall arrival rate is obtained by multiplying the probability of being in the different phases by the arrival rate of the respective phase. In our case, the overall arrival rate calculated that way turns out to be 1.8. Customers are lost when the buffer is full. At this point, the rate of loss depends on the arrival phase X_2. We use $r_{10,i}$ for the probability that the buffer is full, while the arrival state $X_2 = i$. The loss rate is then $\lambda_1 r_{10,1} + \lambda_2 r_{10,2}$, and this turns out to be 0.0676. It follows that the probability of losing a customer is $0.0676/1.8 = 0.0375$, or 3.75%. If the arrival rate increases by 5%, the loss rate increases to 0.0550.

9.1.5 Block-structured Matrices

Matrices are frequently block-structured. For DTMCs, this means that instead of transition probabilities $p_{i,j}$ we have entire matrices $P_{i,j}$. The transition matrix P is thus defined as follows:

$$
P = \begin{bmatrix}
P_{1,1} & P_{1,2} & \cdots & P_{1,N-1} & P_{1,N} \\
P_{2,1} & P_{2,2} & \cdots & P_{2,N-1} & P_{2,N} \\
\vdots & \vdots & \vdots\ \vdots & & \vdots \\
P_{N-1,1} & P_{N-1,2} & \cdots & P_{N-1,N-1} & P_{N-1,N} \\
P_{N,1} & P_{N,2} & \cdots & P_{N,N-1} & P_{N,N}
\end{bmatrix}.
$$

Matrices of CTMCs can have a similar block structure. However, we concentrate on DTMCs.

Block-structured matrices arise naturally when formulating discrete event systems. For instance, if one is mainly interested in one state variable, say X_1, the block $P_{i,j}$ may contain all probabilities where X_1 changes from i to j. In cases like this, the variable in question, whether it is X_1, or some other state variable, is called *level*. All other state variables are then combined and form the *sublevel*. Another word for sublevel is *phase*.

In this section, we apply ideas similar to the state elimination algorithm to find equilibrium probabilities of block-structured matrices. The resulting theory will be exploited in Chapter 12. Also, some languages, such as MATLAB and Octave, have convenient facilities for handling matrices, and this, too, can be exploited when dealing with block-structured matrices.

Let L be the level, and S be the sublevel. We want to find the equilibrium probabilities $\pi_{i,j} = P\{L = i, S = j\}$, where $\pi_i = [\pi_{i,j}, j = 1, 2, \ldots, n_i]$. The equilibrium probabilities can now be formulated as

$$\pi_j = \sum_{i=1}^{N} \pi_i P_{i,j}, \ j = 1, 2, \ldots, N.$$

Similar to the scalar case, we first set j to N and solve the resulting matrix equation for π_N to obtain

$$\pi_N = \sum_{i=1}^{N-1} \pi_i P_{i,N}(1 - P_{N,N})^{-1}.$$

This equation is used to eliminate π_N from all other matrix equations, which yields, after some obvious modifications:

$$\pi_j = \sum_{i=1}^{N-1} \pi_i \left(P_{i,j} + P_{i,N}(I - P_{N,N})^{-1} P_{N,j} \right), \ j = 1, 2, \ldots, N - 1.$$

If we set

$$P_{i,j}^{(N-1)} = P_{i,j} + P_{i,N}(I - P_{N,N})^{-1} P_{N,j},$$

then we obtain

$$\pi_j = \sum_{i=1}^{N-1} \pi_i P_{i,j}^{(N-1)}, \ j = 1, 2, \ldots, N - 1.$$

In analogy to the scalar case, we can now eliminate π_{N-1} from this expression, resulting in a system of matrix equations for the vector π_{N-2}, then we eliminate π_{N-2}, and so on. In general, we find

$$\pi_n = \sum_{i=1}^{n-1} \pi_i P_{i,n}^{(n)}(1 - P_{n,n}^{(n)})^{-1}, \tag{9.13}$$

where the $P_{i,j}^{(n)}$ are calculated recursively, using $P_{i,j}^{(N)} = P_{i,j}$ and

$$P_{i,j}^{(n-1)} = P_{i,j}^{(n)} + P_{i,n}^{(n)}(1 - P_{n,n}^{(n)})^{-1} P_{n,j}^{(n)}, \ n = N, N - 1, \ldots, 2. \tag{9.14}$$

Equation (9.14) is applied until one reaches $n = 2$, and at this point, one has to solve

$$\pi_1 = \pi_1 P_{1,1}^{(1)}.$$

Here, $P_{1,1}$ is the transition matrix once all levels except level 1 are eliminated, and the probabilities of level 1 can be obtained like in any other Markov chain.

If $Q = [Q_{i,j}]$ is a block-structured infinitesimal generator, then equations (9.13) and (9.14) become

$$\pi_n = \sum_{i=1}^{n-1} \pi_i Q_{i,n}^{(n)}(-Q_{n,n}^{(n)})^{-1},$$

$$Q_{i,j}^{(n-1)} = Q_{i,j}^{(n)} + Q_{i,n}^{(n)}(-Q_{n,n}^{(n)})^{-1}Q_{n,j}^{(n)}, \quad n = N, N-1, \ldots, 2.$$

Note that $(-Q_{n,n}^{(n)})^{-1}$ is positive as we will show later.

It does not matter whether one eliminates an entire level at once, or sublevel by sublevel. The result, and, if executed efficiently, even the actual operations are exactly the same. However, we will need the equation (9.13) later. Also, some languages, such as MATLAB, have very efficient methods to deal with matrices, making (9.14) more efficient.

As in the scalar case, the matrix $P^{(n)} = [P_{i,j}^{(n)}]$ is the transition matrix of the original chain with levels $m > n$ censored. This follows immediately from Theorem 9.2.

9.1.6 The Crout Method

In this section, we describe a slight variation of the state elimination method which is known as the Crout method [84, page 174]. We will use the Crout method to deal with changes in the transition matrix after the $p_{i,j}^{(n)}$ have been calculated.

To develop the Crout method for DTMCs, we use equation (9.3), which we repeat here in a slightly modified form

$$p_{i,j}^{(n)} = p_{i,j}^{(n+1)} + \frac{p_{i,n+1}^{(n+1)} p_{n+1,j}^{(n+1)}}{1 - p_{n+1,n+1}^{(n+1)}}.$$

This formula with n replaced by $n + 1$ can be used to substitute $p_{i,j}^{(n+1)}$ in the above formula, which yields

$$p_{i,j}^{(n)} = p_{i,j}^{(n+2)} + \frac{p_{i,n+2}^{(n+2)} p_{n+2,j}^{(n+2)}}{1 - p_{n+2,n+2}^{(n+2)}} + \frac{p_{i,n+1}^{(n+1)} p_{n-1,j}^{(n+1)}}{1 - p_{n+1,n+1}^{(n+1)}}.$$

Continuing this way, one finds, also noting that $p_{i,j}^{(N)} = p_{i,j}$:

$$p_{i,j}^{(n)} = p_{i,j} + \sum_{m=n+1}^{N} \frac{p_{i,m}^{(m)} p_{m,j}^{(m)}}{1 - p_{m,m}^{(m)}}, \quad n = N-1, N-2, \ldots, 1. \tag{9.15}$$

As one can see, we only need these $p_{i,j}^{(m)}$ in which $i < m$, $j = m$ or $i = m$, $j \leq m$. Now we define

$$a_{i,m} = \frac{p_{i,m}^{(m)}}{1 - p_{m,m}^{(m)}}, \quad i < m, \quad b_{m,j} = p_{m,j}^{(m)}, \quad j \leq m,$$

and then we can write

$$a_{i,n} = \frac{p_{i,n} + \sum_{m=n+1}^{N} a_{i,m} b_{m,n}}{1 - b_{n,n}}, \ i < n \tag{9.16}$$

$$b_{n,j} = p_{n,j} + \sum_{m=n+1}^{N} a_{n,m} b_{m,j}, \ j \le n. \tag{9.17}$$

The Crout method uses these two equations to find the $a_{i,n}$, and once this is completed, we have

$$\pi_n = \sum_{i=1}^{n-1} \pi_i a_{i,n}. \tag{9.18}$$

Using also the condition that $\sum_n \pi_n = 1$ allows one to find all π_n.

The formulas for CTMCs are similar, except the the $p_{i,j}$ are replaced by $q_{i,j}$, and the denominator in equation (9.16) is $-b_{n,n}$ instead of $1 - b_{n,n}$. Consequently

$$a_{i,n} = \frac{q_{i,n} + \sum_{m=n+1}^{N} a_{i,m} b_{m,n}}{-b_{n,n}}, \ i < n \tag{9.19}$$

$$b_{n,j} = q_{n,j} + \sum_{m=n+1}^{N} a_{n,m} b_{m,j}, \ j \le n. \tag{9.20}$$

Note that the operations of the Crout methods can be matched one to one with state elimination. Consequently, the number of flops is identical to the state reduction algorithm. The GTH method is also available, that is, the denominator $1 - b_{n,n}$ ($-b_{n,n}$ for CTMC) can be replaced by $\sum_{j=1}^{n-1} b_{n,j}$.

From equations (9.19) and (9.20), we can derive a UL factorization of Q, and an identical factorization is available for $P - I$, which is mathematically indistinguishable from an infinitesimal generator. First, we write equations (9.19) and (9.20) as follows, provided $a_{n,n}$ is set to -1:

$$-q_{i,n} = a_{i,n} b_{n,n} + \sum_{m=n+1}^{N} a_{i,m} b_{m,n} = \sum_{m=n}^{N} a_{i,m} b_{m,n}, \ i < n \tag{9.21}$$

$$-q_{n,j} = -b_{n,j} + \sum_{m=n+1}^{N} a_{n,m} b_{m,j} = \sum_{m=n}^{N} a_{n,m} b_{m,j}, \ j \le n. \tag{9.22}$$

Now, if we define

$$U = \begin{bmatrix} -1 & a_{1,2} & a_{1,3} & \dots & a_{1,N-1} & a_{1,N} \\ 0 & -1 & a_{2,3} & \dots & a_{2,N-1} & a_{2,N} \\ \vdots & \ddots & \ddots & & \vdots & \vdots \\ 0 & 0 & 0 & \dots 0 & & -1 \end{bmatrix}$$

$$L = \begin{bmatrix} b_{1,1} & 0 & 0 & \dots 0 & 0 \\ b_{2,1} & b_{2,2} & 0 & \dots 0 & 0 \\ \vdots & \ddots & \ddots & \dots \vdots & \vdots \\ b_{n,1} & b_{N,2} & b_{N,3} & \dots b_{N,N-1} & b_{N,N} \end{bmatrix}$$

then equations (9.21) and (9.22) can be written as

$$-Q = UL.$$

Consequently, we multiply the upper triangular matrix U with a lower triangular matrix L, obtaining thus a *UL factorization*. This corresponds to the LU factorization of linear algebra. The reason we have a UL rather than a LU factorization is that we eliminate the last variable first.

The Crout method is also available when using block-structured matrices. Specifically, equation (9.15) becomes

$$P_{i,j}^{(n)} = P_{i,j} + \sum_{m=n+1}^{N} P_{i,m}^{(m)}(I - P_{m,m}^{(m)})^{-1} P_{m,j}^{(m)}, \quad n = N-1, N-2, \dots, 1. \quad (9.23)$$

If

$$A_{i,m} = P_{i,m}^{(m)}(1 - P_{m,m}^{m})^{-1}, \quad B_{m,j} = P_{m,j}^{(m)},$$

then (9.23) becomes

$$A_{i,n} = \left(P_{i,n} + \sum_{m=n+1}^{N} A_{i,m} B_{m,n} \right)(1 - B_{m,m})^{-1}, \quad i < n \quad (9.24)$$

$$B_{n,j} = P_{n,j} + \sum_{m=n+1}^{N} A_{n,m} B_{m,j}, \quad j \le n. \quad (9.25)$$

Similar to equation (9.18), one has

$$\pi_n = \sum_{i=1}^{n-1} \pi_i A_{i,n}. \quad (9.26)$$

This completes our discussion of the Crout method.

9.2 The Expected Time Spent in a Transient State

For transient states, equilibrium probabilities do not exist. Instead, we can ask how much time the process spends in the different states as $t \to \infty$, and the latter is the topic of this section. For a DTMC, the expected time spent in the different states is equivalent to the expected number of visits to the states. For a CTMC, these expectations are different because the lengths of the visits vary.

Of course, for recurrent states, the expected time spent in the different states would converge to infinity and is therefore meaningless. On the other hand, the expected time between visits to a given state, say state 1, is meaningful for recurrent states as well. Indeed, the expected time in state j between visits to some base state, say state 1, is proportional to π_j. This shows that there is a close connection between equilibrium probabilities and the expected time spent in the different states.

The concept of expected time spent in the different transient states has a number of applications. It allows one to find the expected time to absorption, such as expected waiting times, or the expectation of a phase-type distribution with a given transition matrix. If a Markov chain has several recurrent communicating classes, the expected time in the different transient states of the process allows one to find the probabilities to be absorbed by the different recurrent classes.

The outline of this section is as follows. As it turns out, the expected times in the different states are given by the so-called fundamental matrix, which we discuss first. However, given the initial probabilities of the process, we can actually bypass the explicit calculation of the fundamental matrix, as will be shown. The fundamental matrix is also useful in providing a probabilistic interpretation of the main formulas used in block elimination. We also show that when $\tilde{\pi}_1 = 1$, the equilibrium chances $\tilde{\pi}_j$ can be interpreted as the expected time between two successive visits to state 1. Next, we derive formulas for the expectation and variance of the time to absorption, and we apply them to find the mean and variance of waiting times.

9.2.1 The Fundamental Matrix

If P is a substochastic matrix with all states communicating, then $M = (I - P)^{-1}$ is called the *fundamental matrix* of P [56]. Its entries $f_{i,j}$ can be shown to be the expected number of visits to state j, given the process starts in state i.

The fundamental matrix has its analog in CTMCs. If Q is the transition matrix, then $-Q^{-1}$ is its fundamental matrix. As in the discrete case, the entries $f_{i,j}$ of the fundamental matrix represent the expected times in state j, given the process starts in state i.

Below, we prove that the entries of the fundamental matrix are indeed the expected times spent in a transient state. We do this first for the DTMC, then for the CTMC. We also show how to use the fundamental matrix to obtain the probability of entering the different recurrent communicating classes of a Markov chain with several ergodic subchains.

9.2.1.1 Expected Number of Visits to a State in a DTMC

Consider a DTMC, and let $Z_{i,j}(t)$ be the actual number of visits in the interval $[0, t]$ to state j, given the process starts in state i. If $f_{i,j}(t) = E(Z_{i,j}(t))$, one has, as discussed in Chapter 7:

$$f_{i,j}(t) = E(Z_{i,j}(t)) = \sum_{\tau=0}^{t} p_{i,j}(\tau).$$

In matrix form, this becomes

$$[f_{i,j}(t)] = \sum_{\tau=0}^{t} P^{\tau}. \tag{9.27}$$

If P^{τ} converges to 0 as $\tau \to \infty$, then the sum on the right converges, and

$$\sum_{\tau=0}^{\infty} P^{\tau} = (I - P)^{-1}. \tag{9.28}$$

This immediately follows by solving the following identity for $\sum_{\tau=0}^{\infty} P^{\tau}$.

$$(I - P) \sum_{\tau=0}^{\infty} P^{\tau} = I.$$

If $f_{i,j} = \lim_{t \to \infty} f_{i,j}(t)$, then (9.27) implies

$$[f_{i,j}] = (I - P)^{-1}.$$

This proves that the entries of the fundamental matrix $(I - P)^{-1}$ are the expected number of visits to state j, given the process starts in state i [56].

Books in linear algebra stress that where possible, matrix inversion should be avoided, and this can indeed be done, provided we know for all i that the probability of starting in state i is $\pi_i(0)$, $i = 1, 2, \ldots, N$. If $\pi(0) = [\pi_1(0), \pi_2(0), \ldots, \pi_N(0)]$, then f_j, $j = 1, 2, \ldots, N$, the expected number of visits in state j before absorption becomes

$$f = [f_1, f_2, \ldots, f_N] = \pi(0)(I - P)^{-1}. \tag{9.29}$$

Consequently, we have to solve the equation $\pi(0) = f(I - P)$ for f. We first write this equation as

$$[1, f] = [1, f] \begin{bmatrix} 0 & \pi(0) \\ s & P \end{bmatrix}, \tag{9.30}$$

where s is determined such that the sum across the row is 1, which makes

$$P^+ = \begin{bmatrix} 0 & \pi(0) \\ s & P \end{bmatrix}$$

a stochastic matrix. To show that (9.30) yields the same f as $\pi(0) = f(I - P)$, expand the vector–matrix multiplication to obtain $[fs, \pi(0) + fP]$. Equation (9.30) is now almost identical to the equation for equilibrium probabilities, and it can be solved by Algorithm 30, except that there is no norming.

Frequently, one selects a subset of states, say S_1, and one is interested in the number of visits to the different states of $j \in S_1$ until S_1 is left forever. In this case, one forms the transition matrix P_1, which contains the transition probabilities $p_{i,j}$, $i, j \in S_1$. The $f_{i,j}$ corresponding to P_1 are the expected number of visits to state $j \in S_1$ before leaving S_1. Expressed mathematically $[f_{i,j}] = (I - P_1)^{-1}$.

Of interest is also the state $j \notin S_1$ that is entered when S_1 is left. If $p_{i,j}$, $i \in S_1, j \notin S_1]$ is the probability of ending in state j outside S_1, and $R = [p_{i,j}, i \in S_1, j \notin S_1]$, then the probability vector of ending in the different states outside S_1 is given by the matrix

$$(I - P_1)^{-1}R. \tag{9.31}$$

Equation (9.31) allows one to find the long-run probabilities of Markov chains with several communicating classes, some of which are recurrent. For simplicity, let there be three communicating classes, with state spaces S_1, S_2 and S_3. States in S_1 are transient, and states in S_2 and S_3 are recurrent. The transition matrix, written in block form, is given as

$$P = \begin{bmatrix} P_{1,1} & P_{1,2} & P_{1,3} \\ 0 & P_{2,2} & 0 \\ 0 & 0 & P_{3,3} \end{bmatrix}.$$

Here, the $P_{i,j}$ reflect the transitions from set S_i to S_j. Given the process starts in S_1, the probability to end in S_i, $i = 2, 3$ is according to equation (9.31):

$$(I - P_{1,1})^{-1}P_{1,i}\mathbf{e}.$$

Here, $\mathbf{e}$ is the column vector with all its entries equal to 1. If π_i, $i = 2, 3$ are the vectors of the equilibrium probabilities of the states in S_i, then given that the process starts in a state of S_1, the long-run probabilities will be

$$(I - P_{1,1})^{-1}(P_{1,2}\mathbf{e}\pi_2 + P_{1,3}\mathbf{e}\pi_3).$$

This result can easily be extended to more than two recurrent classes.

9.2.1.2 The Expected Time Spent in a Transient State in a CTMC

If Q is the transition matrix of a CTMC, and $p_{i,j}(t)$ is the probability of being in state j at t, given the process starts in state i , then as shown in equation (7.2) of Chapter 7, the expectation of $Z_{i,j}(t)$, the time the process spends in state j, given the system starts in state i are given as

$$f_{i,j}(t) = \int_0^t p_{i,j}(\tau)d\tau.$$

Now, $[p_{i,j}(\tau)] = \exp(Qt)$. If Q is a proper subgenerator with one recurrent communicating class, then Q^{-1} exists. In this case, we find

$$[f_{i,j}(t)] = \int_0^t \exp(Q\tau)d\tau = Q^{-1}(\exp(Qt) - I).$$

Since $\exp(Qt)$ converges to 0 as $t \to \infty$, we have

$$[f_{i,j}] = \lim_{t\to\infty} f_{i,j}(t) = -Q^{-1}.$$

Hence, to find the expected time to be in state j, given the process starts in state i one has to invert $-Q$.

If the probability of starting in state i is $\pi_i(0)$, and $\pi(0) = [\pi_1, \pi_2, \ldots, \pi_N]$, then the expected time in state j becomes

$$f = \pi(0)(-Q)^{-1}. \tag{9.32}$$

In this expression, we use $f = [f_1, f_2, \ldots, f_N]$ for the expected times spent in the different states. In this case, the matrix inversion can be avoided by bringing (9.32) into a form that is similar to the equilibrium equations for a CTMC:

$$fQ + \pi(0) = 0, \tag{9.33}$$

or

$$0 = f^+ Q^+.$$

Here $f^+ = [1, f_1, f, \ldots, f_N]$ and

$$Q^+ = \begin{bmatrix} -1 & \pi(0) \\ u & Q \end{bmatrix},$$

with the column vector u determined such that the sum across the rows is 0. This makes Q^+ an infinitesimal generator, and the equilibrium equations corresponding to Q can be solved by a slight modification of Algorithm 30. Consequently, the f_i can be interpreted as a solution of the equilibrium equations of the CTMC corresponding to Q^+.

As in the discrete-time case, we apply the above theory to find the expected time spent in the different states of a subset S_1 before going to some state not in S_1. If $R = [q_{i,j}, i \in S_1, j \notin S_1]$, the continuous-time analog of equation (9.31) becomes

$$(-Q_1)^{-1} R.$$

As before, this formula allows one to find the probabilities of entering the different recurrent communicating classes, and it, therefore, allows one to find the long-term probabilities to be in the different states.

9.2.1.3 Fundamental Matrix, Block Elimination, and Equilibrium Chances

In this section, we explore the relationship of the fundamental matrix with block elimination and with equilibrium chances. When discussing block elimination, we actually encountered the fundamental matrix in equations (9.13) and (9.14). In particular, the i, jth element of $(I - P_{n,n}^{(n)})^{-1}$ can now be interpreted as the expected number of visits to sublevel j of level n before returning to a level below n, given that level n was entered in sublevel i. It follows that $P_{i,n}^{(n)}(I - P_{n,n}^{(n)})^{-1}$ is the expected number of visits to the different sublevels of level n before returning to a level below n, given the level n was entered from level $i < n$. For CTMCs, a similar argument indicates that $(-Q_{n,n})^{-1}$ is the expected time spent in the different sublevels of level n, and $Q_{i,n}(-Q_{n,n}^{(n)})^{-1}$ therefore is the expected time spent in the different sublevels of level n, given level n was entered from level $i < n$.

In Algorithm 30, we first found equilibrium chances. As it turns out, in DTMCs, the chances $\tilde{\pi}_i$ can be interpreted as the expected number of visits to state i between successive visits of state 1, and a similar interpretation is available for CTMCs. This will now be demonstrated.

For the DTMC, equation (1.4) in Chapter 1 was given to find the equilibrium chances $\tilde{\pi}$

$$\tilde{\pi}^- = u(I - P^-)^{-1}.$$

Here, $\tilde{\pi}^-$ is $\tilde{\pi}$, with the first element deleted, and P^- is the transition matrix P with the first row and column deleted. Also, $u = [p_{1,j}, \ j = 2, 3, \ldots, N]$. Obviously, $(I - P^-)^{-1}$ is the fundamental matrix, and its i, j entries, let us call them $r_{i,j}$, provide the expected number of visits to state j, given the process starts in state i. Now, $u(I - P^-)^{-1}$ yields

$$\tilde{\pi}^- = \left[\sum_{i=2}^{N} q_{1,i} r_{i,j}, \ j = 2, 3, \ldots, N \right],$$

that is,

$$\tilde{\pi}_j = \sum_{i=2}^{N} q_{1,i} r_{i,j}.$$

Consequently, $\tilde{\pi}_j$ is the expected number of visits to state j between two visits to state 1. Stated differently, $\tilde{\pi}_j = f_j$, where f_j is the expected number of visits to state j, given the initial probabilities are $p_{1,j}$. This shows that $\tilde{\pi}_j$ is indeed the expected number of visits to state j between two visits of state 1.

If T_1 is the time from the moment state 1 is entered, until the moment it is entered again, then $E(T_1) = 1 + \sum_{j=1}^{N} f_j$. The proportion of time during this interval the system is in state i is now

$$\pi_i = f_i/E(T_1). \tag{9.34}$$

Also, $\pi_1 = 1/E(T_1)$ implies $E(T_1) = 1/\pi_1$.

What is true for state 1 is also true for any other state k, because by reordering the states, one can designate state k as the first state. Consequently, $\pi_i = 1/E(T_i)$, where T_i is the number of visits to states other than i before returning to state i again.

In CTMCs with states $1, 2, \ldots, N$ and with the transition matrix Q, we find when $\tilde{\pi}_1 = 1$

$$\tilde{\pi}^- = u(-Q^-)^{-1}. \tag{9.35}$$

Here, $u = [q_{1,1}, q_{1,2}, \ldots, q_{1,N}]$, and Q^- is Q, with the first row and the first column deleted. We denote the i, j entry of $(-Q^-)^{-1}$ as $r_{i,j}$, where $r_{i,j}$ is the expected time spent in state j, given the system starts in state i. To find the expected time spent in state j between two visits of state 1, we note that the initial probabilities are now $q_{i,j}/q_1$, where q_1 is the rate of leaving state 1, or $q_1 = -q_{1,1}$. Hence, we divide both sides of (9.35) by q_1 to obtain

$$\tilde{\pi}^-/q_1 = (u/q_1)(-Q^-)^{-1}.$$

Using arguments similar to the ones used for the DTMC, we conclude that $\tilde{\pi}_i/q_1$ is the expected time spent in state i. To find this time directly, one can set $\tilde{\pi}_1 = 1/q_1$, and calculate the $\tilde{\pi}_i$ recursively as before, and with this initialization, $\tilde{\pi}_i$ is the expected time spent in state i between two visits of state 1. If $E(T_1)$ is the expected time from the moment state 1 is entered to the moment state 1 is entered again, one finds, if $\tilde{\pi}_1 = 1/q_1$:

$$E(T_1) = \sum_{i=1}^{N} \tilde{\pi}_i,$$

and

$$\pi_i = \frac{\tilde{\pi}_i}{E(T_1)}.$$

The expected time spent in state i is even available in infinite-state Markov chains that diverge, meaning that no equilibrium probabilities exist. In this situation, the interpretation of $\tilde{\pi}_i$ as the expected time spent in state i is still available, however.

9.2.2 Moments of the Time to Absorption

In this section, we show how to find the mean and the variance of T, the time until absorption. In the DTMC, this can be achieved by using z transforms, whereas, in the CTMC, we need moment generating functions (see Section 1.3.4). Since f_i is the expected time in state i, and the expectation of a sum is equal to the sum of expectations, we have

$$E(T) = \sum_{j=1}^{N} f_j.$$

If e is the vector $[1, 1, \ldots, 1]^T$, then this expression can also be written as

$$E(T) = f\mathbf{e}. \tag{9.36}$$

It only remains to find the variance by using generating functions. However, as a check, we will also find the expectation. We do this first for the DTMC, then for the CTMC.

9.2.2.1 Finding Moments for the DTMC

To calculate expectation and variance of the time to absorption for the DTMC with the substochastic matrix $P = [p_{i,j}]$, we use the z transform $G(z)$. According to Section 1.3.4, one has

$$G(z) = \sum_{i=0}^{\infty} p_i z^i.$$
$$E(T) = G'(1)$$
$$\text{Var}(T) = E(T^2) - E^2(T) = G''(1) + E(T) - E^2(T). \tag{9.37}$$

To apply these formulas, we have to find p_t, the probability that absorption occurs at time t. Clearly, absorption has not yet occurred if the system is in one of the transient states, that is

$$P\{\text{not absorbed at } t\} = \pi(0)P^t\mathbf{e}, \ t > 0.$$

For $t = 0$ the probability of not being absorbed is $\pi(0)$. Clearly:

$$p_t = P\{\text{not absorbed at } t - 1\} - P\{\text{not absorbed at } t\} = \pi(0)P^{t-1}(I - P)\mathbf{e}, \ t > 0.$$

Note that P^{t-1} and $I - P$ commute, that is

$$P^{t-1}(I - P) = P^{t-1} - P^t = (I - P)P^{t-1}.$$

The generating function is therefore

$$G(z) = 1 - \pi(0)\mathbf{e} + \sum_{t=1}^{\infty} \pi(0)(I - P)P^{t-1}z^t\mathbf{e}$$
$$= 1 - \pi(0)\mathbf{e} + z\pi(0)(I - P)(I - Pz)^{-1}\mathbf{e}.$$

To find the derivatives of this expression, note that

$$\frac{d}{dz}(I - Pz)^{-1} = P(I - Pz)^{-2}$$
$$\frac{d}{dz}z(I - Pz)^{-1} = zP(I - Pz)^{-2} + (I - Pz)^{-1}$$
$$= (I - Pz)^{-2}(Pz + I - Pz)$$
$$= (I - Pz)^{-2}.$$

Consequently,

$$G'(z) = \pi(0)(I - P)(I - Pz)^{-2}\mathbf{e}$$
$$G''(z) = 2\pi(0)(I - P)(I - Pz)^{-3}P\mathbf{e}.$$

We set $z = 1$ to obtain

$$G'(1) = \pi(0)(I - P)^{-1}\mathbf{e}$$
$$G''(1) = 2\pi(0)(I - P)^{-2}P\mathbf{e}.$$

It follows that:

$$E(T) = \pi(0)(I - P)^{-1}\mathbf{e} \tag{9.38}$$
$$Var(T) = 2\pi(0)(I - P)^{-2}P\mathbf{e} + E(T) - E^2(T). \tag{9.39}$$

Note that the right-hand side of (9.38) is, in effect, $\sum_{i=1}^{N} f_i$, confirming (9.36).

We now give an efficient method to calculate $G''(1)$. We first calculate

$$f^{(2)} = \pi(0)(I - P)^{-2}.$$

We have, since $f = \pi(0)(I - P)^{-1}$:

$$\pi(0)(I - P)^{-2} = \left(\pi(0)(I - P)^{-1}\right)(I - P)^{-1} = f(I - P)^{-1}.$$

Consequently

$$f^{(2)} = f(I - P)^{-1}.$$

This equation can be written as

$$f^{(2)} = f + f^{(2)}P,$$

or

$$f^{(2)+} = f^{(2)+}P^+,$$

where $f^{(2)+} = [0, f]$ and

$$P^+ = \begin{bmatrix} 0 & f \\ s & P \end{bmatrix}.$$

Except for the first row, this is identical to (9.30). If the $p_{i,j}^{(n)}$ are defined by equation (9.3), then except for the first row, the $a_{i,n} = p_{in}^{(n)}/(1 - p_{n,n}^{(n)})$ and $b_{n,j} = p_{n,j}^{(n)}$ are already available and do not need to be recalculated. To find the $a_{0,j}$, we use the Crout formula given by equation (9.5), which becomes

$$a_{0,n} = \frac{1}{1 - b_{n,n}} \left(f_n + \sum_{m=n+1}^{N} a_{0,m} b_{m,n} \right).$$

Next, we find for $n = 1, 2, \ldots, N$

$$f_n^{(2)} = a_{0,n} + \sum_{i=1}^{n-1} f_i a_{i,n}.$$

Var(T) is now given by

$$\text{Var}(T) = 2f^{(2)}P\mathbf{e} + \text{E}(T) - \text{E}^2(T),$$

and we are finished. Note that the vector $P\mathbf{e}$ is merely the sum across the rows of the substochastic matrix P.

9.2.2.2 Finding Moments for the CTMC

In CTMCs, we first determine the complementary cumulative distribution of T, the time to absorption as

$$1 - F_Z(t) = \pi(0) \exp(Qt)\mathbf{e}.$$

This yields the density

$$f_T(t) = \pi(0) \exp(Qt)Q\mathbf{e}. \tag{9.40}$$

We now obtain the moment generating function of this expression as discussed in Section 1.3.4.

$$M(s) = \int_0^\infty e^{-st} f_T(t)dt.$$

Once $M(s)$ is known, we find $\text{E}(T) = M'(0)$, $\text{E}(T^2) = M''(0)$.

For $f_T(t)$ given in (9.40), the moment generating function becomes

$$M(s) = \int_0^\infty e^{-st} f_T(t)dt = \int_0^\infty e^{-st} \pi(0) \exp(Qt)Q\mathbf{e}dt$$

$$= \pi(0) \int_0^\infty \exp(t(Q - sI))dtQ\mathbf{e}.$$

If $\exp(t(Q - sI))$ converges to 0 as $t \to \infty$, we obtain

$$M(s) = \pi(0)(sI - Q)^{-1}Q\mathbf{e}.$$

The first two derivatives of this expression are

$$M'(s) = -\pi(0)(sI - Q)^{-2}Q\mathbf{e}$$
$$M''(s) = 2\pi(0)(sI - Q)^{-3}Q\mathbf{e}.$$

We set $s = 0$ in $M'(s)$ to obtain $\text{E}(T)$ as

$$\text{E}(T) = \pi(0)(-Q)^{-1}\mathbf{e}, \quad \text{E}(T^2) = 2\pi(0)Q^{-2}\mathbf{e}. \tag{9.41}$$

The variance is now

$$\text{Var}(T) = 2\pi(0)Q^{-2}\mathbf{e} - \text{E}^2(T).$$

To calculate $\text{Var}(T)$, we note again that $\pi(0)(-Q)^{-1} = f$. If $f^{(2)} = \pi(0)Q^{-2} = f(-Q)^{-1}$, we have

$$f = -f^{(2)}Q.$$

We solve this equation for $f^{(2)}$ and obtain $\text{Var}(T)$ as

$$\text{Var}(T) = 2f^{(2)}\mathbf{e} - E^2(T).$$

This completes our discussion of how to find the variance of the time to absorption.

9.2.3 Finding Expectation and Variance for $M/M/1$ Waiting Times

The method just discussed shall now be used to find the expectation and variance for the FIFO and SIRO waiting time distributions of the $M/M/1/N$ queue. To simplify our discussion, note that we can always use the expected service time as our time unit, which means that the service rate μ becomes 1. The arrival rate λ is then equal to the traffic intensity ρ.

Note that in equilibrium, Little's law is valid. For the $M/M/1/N$ queue, r_i, the probability that there are i in the system while it is in equilibrium becomes:

$$r_i = \frac{1}{d}\rho^i, \ d = \sum_{i=0}^{N} \rho^i.$$

The expected number in the queue is now

$$L_q = \frac{1}{d} \sum_{i=1}^{N-1} i\rho^{i+1}.$$

Because arrivals are lost when the buffer is full, the effective arrival rate is $\lambda_e = \frac{\lambda}{d} \sum_{i=0}^{N-1} \rho^i$. Consequently, for any queueing discipline, we have, with $\lambda = \rho$

$$E(T) = \frac{L_q}{\lambda_e} = \frac{\sum_{i=1}^{N-1} i\rho^i}{\sum_{i=0}^{N-1} \rho^i}. \tag{9.42}$$

We only consider those arrivals that actually join, which implies that the probability that an arrival that joins finds at most $N - 1$ elements in the system.

9.2.3.1 Expectation and Variance of the FIFO Waiting Time

We have found earlier the FIFO waiting time distribution in Section 7.6.1, and we now find its mean and its variance. The transition matrix is the same as the one used there, and the initial probabilities are p_i, where p_i is the probability that the arrival finds i elements in the system. Note that the transition matrix for the FIFO waiting time is very simple. We have N states, starting with state 0, and ending with state $N-1$. Except for state 0, all diagonal entries are -1, all subdiagonal entries are 1, and all other entries are 0. We now find the $a_{i,n}$ and $b_{n,j}$ as they are given by equations (9.19) and (9.20). It is easy to see that for such a matrix, $b_{i,i-1} = 1$ and $b_{i,i} = -1$. Also, $a_{i,n} = 0$ for $i > 0$. To find the $a_{0,j}$, one can use the Crout method. The $a_{0,j}$ are the initial probabilities, which are given by the p_n, the probabilities of having n elements in the system at the time immediately before an arrival. Since $b_{n-1,n} = 1$, the Crout method given by equation (9.19) yields

$$a_{0,N} = p_N, \quad a_{0,n-1} = p_{n-1} + a_{0,n}.$$

Here, n goes from $N-1$ to 2. Since for $i > 0$, $a_{i,j} = 0$, we can now find f_n as follows

$$f_{N-1} = p_{N-1}, \quad f_n = a_{0,n} = p_n + f_{n+1}.$$

Effectively, this implies

$$f_n = \sum_{i=n}^{N-1} p_i.$$

Now

$$E(T) = \sum_{n=1}^{N-1} f_n = \sum_{n=1}^{N-1} \sum_{i=n}^{N-1} p_i = \sum_{i=1}^{N-1} i p_i.$$

This makes sense because if $\mu = 1$, the expected waiting time is equal to the number the arrival meats when joining the line. In equilibrium, $p_i = \frac{\rho^i}{\sum_{j=0}^{N-1} \rho^j}$, and we obtain

$$E(T) = \sum_{i=1}^{N-1} i \frac{\rho^i}{\sum_{j=0}^{N-1} \rho^j},$$

which confirms Little's law.

To find $f^{(2)}$, we use the same $a_{i,n}$ and $b_{n,j}$ as before, except that row 0 is now f_n rather than p_n. Consequently,

$$f_{N-1}^{(2)} = f_{N-1}, \quad f_n^{(2)} = f_n + f_{n+1}^{(2)}.$$

When using $f_n = \sum_{i=n}^{N-1} p_i$, this yields

$$f_n^{(2)} = \sum_{j=n}^{N-1} f_j = \sum_{j=n}^{N-1} \sum_{i=j}^{N-1} p_i = \sum_{i=n}^{N-1} i p_i.$$

The variance becomes now

$$\text{Var}(T) = 2 \sum_{n=1}^{N-1} \sum_{i=n}^{N-1} i p_i - E^2(T) = \sum_{i=1}^{N-1} i(i+1) p_i - E^2(T).$$

This formula can also be found directly. Given an arrival meets i in system, and $\mu = 1$, then the expectation $E(T \mid i) = i$, and the variance $Var(T \mid i) = i$ also. According to the formulas about the variance of sums and mixtures (equations (5.2) and (5.10) of Section 5.1), we now have

$$Var(T) = E(Var(T \mid i)) + Var(E(T \mid i))$$

$$= \sum_{i=1}^{N-1} p_i Var(T \mid i) + \left(\sum_{i=1}^{N-1} p_i i^2 - E^2(T) \right)$$

$$= \sum_{i=1}^{N-1} i(i+1)p_i - E^2(T).$$

There is also another way to find the f_i. If there are i elements in the system immediately preceding the arrival, then the states i, state $i-1, \ldots$, and state 1 are all visited exactly once. Now state i is also visited if there were $i+1, i+2, \ldots, i+N-1$ elements in the system at the point of arrival. Consequently, $f_i = p_i + p_{i+1} + \ldots, p_{N-1}$.

9.2.3.2 Mean and Variance of the SIRO Waiting Time

For the SIRO waiting time in the $M/M/1/N$ queue, we use numerical methods. Hence, assume that $N = 6$, which implies that if an arrival joins, there are at most five elements in the system, otherwise the arrival would have been lost. Also, $\mu = 1$, and $\lambda = \rho = 0.8$. The transition matrix Q is as follows

$$Q = \begin{bmatrix} -1.8 & 0.8 & 0 & 0 & 0 \\ 0.5 & -1.8 & 0.8 & 0 & 0 \\ 0 & 0.6667 & -1.8 & 0.8 & 0 \\ 0 & 0 & 0.75 & -1.8 & 0.8 \\ 0 & 0 & 0 & 0.8 & -1 \end{bmatrix}.$$

The matrices $U = [a_{i,n}]$, (with $a_{i,i} = -1$) and $L = [b_{n,j}]$ are now

$$U = \begin{bmatrix} -1 & 0.5779 & 0 & 0 & 0 \\ 0 & -1 & 0.6237 & 0 & 0 \\ 0 & 0 & -1 & 0.6897 & 0 \\ 0 & 0 & 0 & -1 & 0.8 \\ 0 & 0 & 0 & 0 & -1 \end{bmatrix},$$

$$L = \begin{bmatrix} -1.5110 & 0 & 0 & 0 & 0 \\ 0.5 & -1.3842 & 0 & 0 & 0 \\ 0 & 0.6667 & -1.2828 & 0 & 0 \\ 0 & 0 & 0.75 & -1.16 & 0 \\ 0 & 0 & 0 & 0.8 & -1 \end{bmatrix}.$$

With this, we can now calculate all $a_{0,i}$, using

$$a_{0,N-1} = p_{N-1}, \quad a_{0,n} = \frac{1}{-b_{n,n}}(p_n + a_{0,n+1}b_{n+1,n}).$$

Once this is done for $n = N - 1, N - 2, \ldots, 1$, we have

$$f_0 = 1, \quad f_n = a_{0,n} + a_{n-1,n}f_{n-1}.$$

The calculation of the $f_i^{(2)}$ is similar, except that the p_i must be replaced by f_i. The results are as follows:

$$[a_{0,n}] = [0, 0.2168, 0.2216, 0.2000, 0.1570, 0.0888]$$
$$f = [0.2168, 0.3470, 0.4163, 0.4441, 0.4441].$$

This yields $E(T) = 1.868$. For the variance, we obtain

$$f^{(2)} = [0.3424, 0.7989, 1.2257, 1.5344, 1.6716].$$

Consequently, we have

$$\mathrm{Var}(T) = 2\sum_{n=1}^{5} f_n^{(2)} - E^2(T) = 7.655.$$

As a check, we also calculated $E(T)$ by Little's theorem, and we also obtained $E(T) = 1.868$.

9.3 Iterative Methods

In the previous section, we solved the equilibrium equations by direct methods, as they were implemented in Algorithm 30. This algorithm needed $\Theta(N^3)$ flops to find the equilibrium probabilities of models with N states. This is a huge number when N is high, and in discrete event systems, N tends to be huge because the number of states increases exponentially with the dimensions of the problems. Of course, in our case, the transition matrices are typically banded, which implies that one needs only $\Theta(Ngh)$ flops when the band extends from g below to h above the diagonal. However, the bandwidth also increases exponentially with the number of state variables. To verify this, note that g and h depend on u_1 as defined by equation (3.1) of Section 3.1, and this equation shows that u_1 increases exponentially with the problem dimension. Hence, even when exploiting bandedness, the amount of time required to solve the equilibrium equations by direct methods may be unacceptable.

As indicated earlier, the transition matrices of discrete event systems tend to be sparse because the number of events is small when compared to the number of states. As it turns out, the so-called iterative methods can exploit the sparsity of the transition matrix. Many iterative methods for solving the equilibrium equations of Markov chains are related to transient solutions, and indeed, transient solutions approach the equilibrium solutions without ever reaching them. We will discuss two iterative methods based on this idea: First, we consider finding the solution given by the equilibrium equation $\pi Q = 0$ by calculating the transient solutions of the Markov process with transition matrix $P = Q/q + T$. Alternatively, one can use the so-called *jump matrix*, which will be described later in this section. Another class of methods starts directly with a trial solution of the equilibrium equations, which is then iteratively improved. These methods have been used extensively to solve linear equations in many different disciplines. We discuss here the method of Jacobi and the method of Gauss–Seidel.

All iterative methods discussed here exploit sparsity, and this can make a huge difference in time complexity. In fact, for Poisson event systems, the complexity of one iteration in the iterative methods we consider here are all $\Theta(Ne)$, where e is the number of events. Of course, we have to consider the number of iterations, but according to our experience, the number of iterations necessary to find the equilibrium solution at an acceptable precision does not depend on the dimension of the problem. In short, if N is large, and e small, then the iterative method can be orders of magnitude faster than direct methods.

Iterative methods have their own problems. In particular, a decision must be made about many iterations to do until the result can be accepted. Some of the iterative methods converge faster, and they therefore require fewer iterations, whereas others are slower. Of the methods discussed here, the method of Gauss–Seidel tends to be the fastest, but it is also the most difficult to program. Whatever method is used, each iteration should bring us closer to the desired result. If very precise results are required, then more iterations may be necessary.

In discrete-event systems, one is usually not interested in the individual equilibrium probabilities, but in measures such as the expectations and marginal distributions. For this reason, we are using rewards as defined in Section 7.1. For our purpose, we define the expected reward here as

$$\bar{R} = \sum_{i=1}^{N} \pi_i r_i,$$

where r_i is the reward per time unit the process is in state i. As mentioned in Section 6.4.1, in stochastic systems, the data errors are seldom better than $\pm 1\%$, which means that an error of $\pm 1\%$ of the rewards is acceptable.

Even when there is a clear-cut precision requirement, it is not easy to determine after an iteration if this requirement is met, or if more iterations are required. This issue will also be addressed in this section, and to accomplish this, we use eigenvalues. All iterative methods require a starting vector to begin the iterations, and we want to find a good starting vector, an issue which will also be discussed.

In this section, we assume that the equilibrium equations are all in the form $\pi Q = 0$. This also covers the DTMC with the transition matrix P if we set $Q = P - I$.

Aside from the methods we discuss here, there are many other methods in use. We mention, in particular, projection methods, as they are discussed in [86]. But before using these more complex methods, we suggest to use the methods discussed here. Only if their performance is unacceptable, one should proceed to those more complex methods.

9.3.1 Equilibrium Probabilities Found as Limits of Transient Probabilities

One way to determine the equilibrium probabilities of Markov chains is to use the fact that the transient probabilities $\pi(t)$ converge to the equilibrium probabilities π. There are two methods to be discussed here: the uniformization–matrix method, and the jump-chain method. In both cases, we have vector–matrix multiplications of the form $\pi^{n+1} = \pi^n V$, where V is called *iteration matrix*. When the transition matrix is stored in a row, column, value format, Algorithm 25 can be used to perform vector–matrix multiplications. Also, there is computer software, such as MATLAB, which implements vector–matrix multiplications efficiently, particularly for sparse matrices.

9.3.1.1 The Uniformization–matrix Method

In the uniformization–matrix method, we use the fact that the DTMC with transition matrix $P = Q/q + I$, $q \geq -q_{j,j}$, has the same equilibrium probabilities as a Markov chain with the rate matrix Q, that is, $\pi Q = 0$ and $\pi = \pi P$ have the same solution vector π, as is easily verified. The iterative algorithm thus consists of choosing an initial vector π^0 and calculating

$$\pi^{n+1} = \pi^n P, \ n \geq 0$$

until π^{n+1} is close enough to the equilibrium vector π. The iteration matrix in this case is P.

The convergence of this method depends on q. To estimate the speed of convergence as a function of q, we use the fact that $1/q$ is the expected time between steps in the DTMC with transition matrix P. Hence, if q doubles, then the time to reach equilibrium should be cut in half, which means q should be as small as possible. However, this may lead to a Markov chain that is periodic or almost periodic, and periodic Markov chains never reach equilibrium. There is thus an optimal q, which may be greater $\max_i(-q_{i,i})$. To demonstrate this, consider the following transition matrix:

$$Q = \begin{bmatrix} -\lambda & \lambda \\ \lambda & -\lambda \end{bmatrix}.$$

In this case, one could set $q = \lambda$, but this leads to the Markov chain

$$P = \begin{bmatrix} 0 & 1 \\ 1 & 0 \end{bmatrix},$$

which is periodic with a period of 2, and the iteration $\pi^{n+1} = \pi^n P$ never reach equilibrium. As it turns out, using $q = 2\lambda$ is best. This value yields the transition matrix

$$P = \begin{bmatrix} 0.5 & 0.5 \\ 0.5 & 0.5 \end{bmatrix},$$

and the uniformization–matrix method converges in one iteration.

9.3.1.2 The Jump-chain Method

We can reduce the expected time spent in the different states by embedding the process at the times when the state changes. This brings us to the *jump-matrix*, that is, the matrix embedded at the moment the Markov chain changes its state. The transition probabilities of the jump matrix are given as, if we set $q_i = -q_{i,i}$:

$$p_{i,j} = \begin{cases} q_{i,j}/q_i & i \neq j \\ 0 & i = j. \end{cases}$$

Hence, if $P = [p_{i,j}]$, then to use the jump-chain method, we select an initial value $\hat{\pi}^0$, and calculate recursively

$$\hat{\pi}_j^{n+1} = \hat{\pi}^n P. \tag{9.43}$$

Note that $\hat{\pi}$ does not converge to the equilibrium vector π given by $\pi Q = 0$. However, $\hat{\pi}_i = \lim_{n \to \infty} \hat{\pi}_i^n$ is related to π_i as follows:

$$\pi_i = \frac{\hat{\pi}_i/q_i}{\sum_j \hat{\pi}_j/q_j}. \tag{9.44}$$

Essentially, this relation says that in order to obtain π, the probabilities $\hat{\pi}_i$ have to be weighted by the expected time the system spends in state i, which is $1/q_i$. Equation (9.44) is typically proven through the theory of semi-Markov processes, which will be discussed in Chapter 10. Here, we provide a purely arithmetic proof. If D is the diagonal matrix with the diagonal entries q_i, then the iteration matrix P can be written as

$$P = D^{-1}Q + I. \tag{9.45}$$

Now, the equilibrium distribution $\hat{\pi}$ satisfies $\hat{\pi} = \hat{\pi}P$, and from this, we conclude

$$0 = \hat{\pi}(P - I) = \hat{\pi}(D^{-1}Q) = (\hat{\pi}D^{-1})Q.$$

This is equivalent to $\pi Q = 0$ if and only if $\hat{\pi}D^{-1} = \pi$. Consequently, if we set $\pi = \hat{\pi}D^{-1}$, then $\pi Q = 0$. To complete the proof of (9.44), we only need to norm the π_i such that their sum is 1.

The jump-chain method rather frequently leads to transition matrices P that are periodic and π^n does not converge. Methods to overcome this difficulty will be discussed later.

9.3.2 Methods based on Successive Improvements

In the methods of successive improvements for solving the equation $0 = \pi Q$, one is given an initial approximation $\pi^0 = [\pi_1^0, \pi_2^0, \ldots, \pi_N^0]$, which is then improved to obtained π^1, followed by a further improvement to obtain π^2, and so on, until one believes that π^n is close enough to the equilibrium vector π. The best known successive improvement methods are the method of Jacobi and the method of Gauss–Seidel. Both are discussed below.

9.3.2.1 The Method of Jacobi

To derive the method of Jacobi, we write the equilibrium equations as follows:

$$0 = \sum_{i=1}^{j-1} \pi_i q_{i,j} - \pi_j q_j + \sum_{i=j+1}^{N} \pi_i q_{i,j}, \quad j = 1, 2, \ldots, N. \tag{9.46}$$

We solve this equation for π_j, which yields

$$\pi_j = \sum_{i=1}^{j-1} \pi_i q_{i,j}/q_j + \sum_{i=j+1}^{N} \pi_i q_{i,j}/q_j. \tag{9.47}$$

If we have an approximation π_i^0 for π_i, $i = 1, 2, \ldots, N$, then this equation can be used to find an improved solution: On the right side of equation (9.47), we replace the π_i by an approximation π_i^n, to find improved approximations π_j^{n+1}, that is,

$$\pi_j^{n+1} = \sum_{i=1}^{j-1} \pi_i^n q_{i,j}/q_j + \sum_{i=j+1}^{N} \pi_i^n q_{i,j}/q_j. \tag{9.48}$$

This is the method of Jacobi. We start with an initial solution, say $\pi_1 = 1$, $\pi_i^0 = 0$ for $i = 2, 3, \ldots, N$. Next, we use equation (9.48) with $j = 1, 2, \ldots, N$ to find an improved approximation π_j^1, $j = 1, \ldots, N$. These values can be used to find the next approximation π_j^2 by (9.48). This continues until one is close enough to the equilibrium solution.

Equation (9.48) can be expressed in matrix form as follows:

$$\pi^{n+1} = \pi^n V,$$

where the iteration matrix $V = QD^{-1} + I$ and D is the diagonal matrix with entries q_i.

The method of Jacobi is very similar to the jump matrix method given by (9.43), except that $\hat{\pi}_j$ replaces π_j, and $q_{i,j}$ is divided by q_j rather than q_i. Hence, whereas the iteration matrix is $D^{-1}Q + I$ in the jump-chain method, it is $QD^{-1} + I$ in the method of Jacobi. As it turns out, these two matrices have the same eigenvalues. This follows from the fact that if A and B are two square matrices of the same size, AB and BA have the same eigenvalues. For the proof, see [48, Theorem 1.3.20]. Hence, QD^{-1} and $D^{-1}Q$ have the same eigenvalues, and adding the identity matrix only changes the eigenvalues by 1. Consequently, if the jump chain $\hat{P} = D^{-1}Q + I$ is periodic, so is $V = QD^{-1} + I$ in the method of Jacobi.

9.3.2.2 The Method of Gauss–Seidel

Like the method of Jacobi, the method of Gauss–Seidel starts with equation (9.47) to improve approximation n. However, it makes use of the fact that when finding a new value of π_j, one uses the most recently calculated values of π_i. This leads to the following expression for the new value for π_j:

$$\pi_j^{n+1} = \sum_{i=1}^{j-1} \pi_i^{n+1} q_{i,j}/q_j + \sum_{i=j+1}^{N} \pi_i^n q_{i,j}/q_j. \tag{9.49}$$

Thus, one starts with an initial solution, say $\pi_2^0 = 1$, and $\pi_i^0 = 0, i = 1, 3, 4, 5, \ldots, N$, and calculates π_1^1 according to equation (9.49). Next, π_2^1 is calculated, using now the new value π_1^1, but π_i^0 for $i > 1$. After that, π_3^1 is obtained, using π_1^1 and π_2^1 as calculated. This continues, and when calculating π_j^1, one uses π_i^1 for $i < j$, and π_i^0 for $i > j$. The step from n to $n + 1$ works the same way. This approach uses the fact that at the moment π_j^{n+1} is calculated, the most recent approximation for $\pi_i, i < j$ is π_i^{n+1} rather than π_i^n. Notice that we cannot initialize π_1^0 to 1 because this would result $\pi_1^1 = 0$, and as a consequence, $\pi_j^1 = 0$ for all j.

Obviously, π^{n+1} as calculated by equation (9.49) depends on how the states are ordered. If the states are in a different order, say with j decreasing, a different result is obtained. For a good performance, one should number the states such that as many $q_{i,j}, i < j$ are non-zero. It also helps if for $i < j$, one has large $q_{i,j}$. On the other hand, the number of non-zero $q_{i,j}, i > j$ should be small.

If the transition matrix is stored in a row, column, value format, then the Gauss–Seidel method requires that the non-zero entries of the transition matrix are sorted according to the column number before any vector–matrix multiplication can be used, such as the one given in Algorithm 25. The algorithm must also be slightly modified. To sort according to the column number, one can use a so-called *counting sort* [14]. The time complexity of this sort is proportional to the number of non-zero entries in the transition matrix. In this sort, one counts the number of non-zero entries in each column, which then allows one to put them all into the correct slot. For the details of this algorithm, see [14] or the article "Counting Sort" in Wikipedia.

One disadvantage of the Gauss–Seidel method is that we cannot use the standard vector/matrix multiplications provided by some languages, such as MatLab. Moreover, Algorithm 25 must also be modified such that as soon as a column is processed, the new vector, "b(j)" in Algorithm 25 is changed to "x(j)", which stands for π_j^{n+1}. These changes are rather trivial, however.

9.3.3 Convergence Issues

Obviously, π^n in iterative methods should converge as fast as possible toward their equilibrium vector π. To investigate the speed of convergence, we bring all methods into the form

$$\pi^{n+1} = \pi^n V,$$

where V is the iteration matrix. In fact, except for the method of Gauss–Seidel, all methods discussed so far are already in this form, with $V = P$. To deal with the method of Gauss–Seidel, we define the three matrices $U = [q_{i,j}, \ 1 \le i < j \le N]$, $D = \operatorname{diag}(q_j)$, $L = [q_{i,j}, \ 1 \le j < i \le N]$. This allows us to write equation (9.46) as:

$$0 = \pi U - \pi D + \pi L.$$

We replace πU by $\pi^{n+1} U$, πD by $\pi^{n+1} D$ and πL by $\pi^n L$ and solve for π^{n+1} to get:

$$\pi^{n+1} = \pi^n L (D - U)^{-1}.$$

If we set $V = L(D - U)^{-1}$, then we have

$$\pi^{n+1} = \pi^n V,$$

which is the desired vector–matrix multiplication.

Note that the matrix $U - D$ is a subgenerator, and $(D - U)^{-1}$ is its fundamental matrix. Its entries $(U - D)_{i,j}$ are the times spent in state j before entering a state below i, provided the process starts in i. This matrix is a dense upper triangular matrix, which makes its use unattractive for numerical purposes. Here, we use it only for the purpose of estimating how fast the iterations converge.

We now look at the convergence of $\pi^{n+1} = \pi^n V$, where the iteration matrix V depends on the method used. For the methods under consideration, we have

1. $V = Q/q + I$, with $q \ge q_i$ for all i. This is the uniformization–matrix method.
2. $V = D^{-1} Q + I$. This is the jump-chain method.
3. $V = L(D - U)^{-1}$. This is the method of Gauss–Seidel.

In all cases, the iteration matrix is non-negative. This is clear for the uniformization method, provided $q \ge q_i$ for all i, and for the jump-chain method. For the method of Gauss–Seidel, it follows from the fact that $(D - U)^{-1}$ is the fundamental matrix.

When V has no negative elements, the recursion $\pi^{n+1} = \pi^n V$ has no subtractions which makes it very resistant to rounding errors.

The convergence of methods based on vector–matrix multiplications is best analyzed by using eigenvalues. They will give us the foundation to obtain results about stopping criteria, as well as how to choose the initial value π^0. In many cases, one has to handle a matrix V that is periodic. We discuss how to avoid periodicity, and if it cannot be avoided, how to proceed.

9.3.3.1 Eigenvalues and Convergence

Eigenvalues were discussed in Section 8.2 in connection with stochastic matrices, but they can be used here as well. In particular, we can apply Theorem 8.2, which indicates that given an initial vector π^0, we have, provided V has a sufficient number of eigenvectors

$$\pi^n = \sum_{k=1}^N c_k \phi_k \gamma_k^n,$$

where the γ_k are the eigenvalues of the matrix V, ϕ_k the corresponding eigenvectors, and the c_k are chosen such that the initial approximation π^0 is equal to

$$\pi^0 = \sum_{k=1}^N c_k \phi_k.$$

Since both the uniformization matrix and the jump matrix are stochastic matrices, they have an eigenvalue that is equal to 1, and this is also their spectral radius. The eigenvector for the eigenvalue of 1 is π. This is also true if $V = L(U - D)^{-1}$. Indeed, V was determined in such a way that as $\pi^{n+1} = \pi^n = \pi$, the equilibrium equations hold, and this implies that $\gamma = 1$ is an eigenvalue, and its eigenvector is $\phi = \pi$. It can also be shown that the spectral radius of $L(U - D)^{-1}$ is 1 [86, page 129]. Note that V is a non-negative matrix, and therefore satisfies the theorem of Perron–Frobenius in its extended form. Also, when the iteration matrix V is given by $L(U - D)^{-1}$, we can define paths like in stochastic matrices, which means that we can form communicating classes and periodicity classes, and the notions of recurrent and transient states still apply.

Let us number the eigenvalues of V such that $|\gamma_i| \geq |\gamma_{i+1}|$, $i = 1, 2, \ldots, N - 1$. Thus, $\gamma_1 = 1$ is the eigenvalue with the largest absolute value, and γ_2 is the eigenvalue with the second largest absolute value, or the *subdominant* eigenvalue. Since $\phi_1 = \pi$, and since we have enough degrees of freedom to set $c_1 = 1$, we can write, provided there are enough eigenvectors

$$\pi^n = \pi + \sum_{k=2}^N c_k \phi_k \gamma_k^n. \tag{9.50}$$

In this equation, the sum $\sum_{k=2}^{N} c_k \phi_k \gamma_k^n$ represents the deviation between the present approximation for π and its true value. If $c_2 \neq 0$, then for large n, this deviation is dominated by γ_2. However, even if c_2 is initially 0, rounding errors may move c_2 away from 0, and eventually, the term with γ_2 still dominates. Of course, since the matrices we are dealing with have no negative entries and there are therefore no major errors due to rounding.

Equation (9.50) can be generalized to hold for any reward. Clearly, one has

$$\bar{R}_n = \sum_{i=1}^{N} r_i \pi_i + \sum_{k=1}^{N} \sum_{i=1}^{N} r_i c_k \phi_{k,i} \gamma_k^n = \bar{R} + \sum_{k=1}^{N} d_k \phi_{k,i} \gamma_k^n,$$

where r_i is the reward for visiting state i and

$$d_k = c_k \sum_{i=1}^{N} r_i \phi_{k,i}.$$

This shows that deviations of the rewards from their true value are also dominated by the subdominant eigenvalue if n is large enough.

The iterations converge to π if, except for $\gamma_1 = 1$, all eigenvalues have an absolute value that is less than 1. There are only two cases where the modulus of any eigenvalue can reach 1:

1. The iteration matrix is periodic.
2. The iteration matrix V is decomposable.

The case where the iteration matrix is periodic will be covered in Section 9.3.4 below, and we introduce methods of removing the periodicity. If the iteration matrix is decomposable, then each component should be dealt with on its own. Indeed, each component has its own set of eigenvalues. If one starts in state i, that is, if $\pi_i^0 = 1$ and $\pi_j^0 = 0$, $j \neq i$, and if state i belongs to a recurrent class, then the process finds the equilibrium probabilities of the recurrent class to which i belongs.

There are cases where V is almost periodic, and in this case, the subdominant eigenvalue γ_2 has a modulus close to 1. The same is true if the process is almost completely decomposable. Another case where we observed that γ_2 can approach 1 is when the range of a state variable is large, and it has a slowly decaying tail. For example, the $M/M/c/N$ queue with a high traffic intensity has a slowly decaying tail, and as N increases, the subdominant eigenvalues increase with N. In fact, if $\rho \to 1$ and $N \to \infty$, γ_2 approaches 1 as can be seen from the formulas for γ_2 provided for the $M/M/c/N$ queue in [40]. In all these cases, convergence is slow. In the case of periodic Markov chains, this problem of slow convergence can partially be overcome as shown in Section 9.3.4.

9.3.3.2 Starting and Stopping the Iterations

One obviously wants to choose an initial vector such that the number of iterations required to obtain results at the prescribed precision is minimized, and one wants to stop iterations as soon as this objective is reached.

Intuitively, one wants to start with an initial vector that is as close to π as possible, but since π is not known, other considerations are needed. Also, the rate of convergence depends on c_k, especially c_2, the weight of the subdominant eigenvalue, and this weight is also influenced by the initial conditions. Indeed, as we show below, if V is periodic, then by choosing π^0 such that the sum of the probabilities has the same value for each periodicity class, then the periodicity of the sequence $\{\pi^n, n \geq 0\}$ disappears. In effect, the c_k for the γ_k with a modulus of 1 becomes 0. Similar results also apply to matrices V that are almost periodic.

If the iteration matrix is almost decomposable, then the time needed for moving between the almost decomposable parts of the matrix is substantial, delaying thus the convergence. To counteract this, it is best to aim for initial vectors π^0 that assign to each of the classes in question a probability mass that is as close as possible to their equilibrium value. One also has to ask how trustworthy the equilibrium vectors are in this case. Typically, in almost decomposable transition matrices, small changes in the transition rates between the almost decomposable transition matrices lead to major changes in the equilibrium vector. If the transition rates are based on observations and therefore subject to data errors, then the accuracy of the results is questionable.

In cases that are neither almost periodic nor almost decomposable, the effect of the initial vector π^0 tends to be small, and to look for good starting solutions may not be worth the effort.

If the desired objective is to find a reward, one may be tempted to use an initial vector that yields a reward that is close to its true value. For instance, if one wants to find the expected number in line in a $GI/G/1$ queue, one could set all π_i^0 to zero except for the probabilities of states where the line length is close to the estimated line length. However, this method is not foolproof as shown in [33].

When initializing the Gauss–Seidel iterations, care must be taken that π^1 does not become 0. This happens when $\pi_1^0 = 1, \pi_i^0 = 0, i > 1$, because then, in iteration 1, we have

$$\pi_1^1 = \sum_{i=2}^{N} \pi_i^0 q_{i,1}/q_1 = 0.$$

It is thus necessary that some of the $\pi_i^0 q_{i,1}$ are not zero, and this can be ensured by setting at least one π_i^0 with $q_{i,1} > 0$ to a non-zero value. If the transitions are arranged by columns, a necessity when using Gauss–Seidel, this is relatively easy to accomplish. We decided to select m states with $q_{i,1} > 0$ and set $\pi_i^0 = 1/m$ for all these states. Even with this method, it is theoretically possible that for $n > 0, \pi^n = 0$.

Next, we have to decide when to stop the iterations as the results have the required error bounds. In other words, we need some stopping criterion. Let us start with the simplest possibility: Stop as soon as for all i, $|\pi_i^{n+1} - \pi_i^n| < \epsilon$, where $\epsilon > 0$ is given.

However, if convergence is slow, this gives no indication that $|\pi_i^n - \pi_i| < \epsilon$. To demonstrate this, suppose that the subdominant eigenvalue is real and that n is large enough such that $\pi_i^n \approx \pi_i + c_2\phi_{2,i}\gamma_2^n$. Using this approximation, one finds

$$\pi_i^n - \pi_i^{n+1} = c_2\phi_{2,i}\gamma_2^n(1 - \gamma_2).$$

Since the difference between $\pi_i^{n+1} - \pi_i$ is $c_2\phi_{2,i}\gamma_2^{n+1}$, the ratio of the true error to the improvement obtained from one additional iteration is

$$\frac{\pi_i^{n+1} - \pi_i}{\pi_i^n - \pi_i^{n+1}} = \frac{c_2\phi_{2,i}\gamma_2^{n+1}}{c_2\phi_{2,i}\gamma_2^n(1 - \gamma_2)} = \frac{\gamma_2}{1 - \gamma_2}.$$

It follows that as $\gamma_2 \to 1$, this ratio goes to infinity, that is, the stopping criterion in question can be extremely poor. Even if γ_2 is as low as 0.9, this ration is 9, that is, $\pi_i - \pi_i^{n+1}$ is 9 times higher than $\pi_i^{n+1} - \pi_i^n$. On the other hand, this stopping criterion would work well if the dominant eigenvalue is negative and close to 1. In this case, the error is first positive, then negative, which implies that $\pi_i^{n+1} - \pi_i^n$ exaggerates the error.

Theoretically, in any sequence of the form $d_n = a + b\gamma_2^n$, one only needs three values in succession to find γ_2 as

$$\gamma_2 = \frac{d_n - d_{n+1}}{d_{n-1} - d_n}.$$

Applying this formula may, however, be problematic because it neglects the contributions of γ_i, $i > 2$. Moreover, the formula is invalid when γ_2 is complex.

Of course, γ_2 could be complex, that is

$$\gamma_2 = \rho(\cos\alpha + i\sin\alpha).$$

It follows

$$\gamma_2^n = \rho^n(\cos n\alpha + i\sin n\alpha).$$

Note that in this case, there is also a γ_3, which is the conjugate complex value of γ_2. We only follow γ_2^n here, since, for γ_3, one only needs to take the conjugate complex value of the error term.

The angle α can be approximated arbitrary closely by $2\pi\frac{\kappa}{\nu}$, where κ and ν are non-negative integers. The important fact is now that if n is a multiple of ν, then $\cos(n\alpha) = 1$ and $\sin(n\alpha) = 0$. Consequently, for any n that is a multiple of ν, we have

$$\pi_i^n = \pi_i + c_2\phi_{2,i}\gamma_2^n = \pi_i + c_2\phi_{2,i}\rho^n(\cos(n\alpha) + i\sin(n\alpha))$$
$$= \pi_i + c_2\phi_{2,i}\rho^n.$$

Thus, one would have to sample at a distance of ν in order to estimate ρ^ν. To estimate ν, one could, in principle, take the distance between two maxima, or take the distance from a maximum to a minimum, and multiply this distance by 2. This method is particularly appropriate if the iteration matrix is periodic or almost periodic.

If v is known, then it is advantageous to take the average over v successive values of π_i^n. The reason is that $\cos(n\alpha)$, $n = m + 1, m + 2\ldots, m + v$ can either be positive or negative, and the positive values partially cancel the negative values. In fact, one has, since $\gamma_2^v = \rho^v$:

$$\sum_{n=0}^{v-1} \gamma_2^n = \frac{1 - \rho^v}{1 - \gamma_2}.$$

and if ρ is close to 1, then this sum is very small. In fact, for periodic iteration matrices, $\rho = 1$, and the sum is zero. In this way, the effect of γ_2, and its conjugate complex companion, is eliminated.

There is a great variety of stopping criteria, in particular, we use the following:

1. Let L be a reward, for instance, the expected line length in a queue and let L_n be this number as calculated based on π^r. In this case, one can stop iterating as soon as $|L_{n-1} - L_n| < \epsilon$.
2. One can stop iterating as $|\pi_i^n - \pi_i^{n-1}| < \epsilon$ for all i.
3. One can stop iterating as soon as $\sum_{i=1}^{N} |\pi_i^n - \pi_i^{n-1}| < \epsilon$.
4. One can stop iterating as soon as $\frac{1}{N} \sum_{i=1}^{N} |\pi_i^n - \pi_i^{n-1}| < \epsilon$.

In addition, in all these cases, one can use relative errors as opposed to absolute errors. For instance, one can require that $\sum_{i=1}^{N} \frac{|\pi_i^n - \pi_i^{n-1}|}{|\pi_i^n|} < \epsilon$.

When comparing the convergence behavior of systems of different sizes, it is important that the stopping criterion is reasonably independent of N, the number of states. For instance, as N increases, then $\sum_{i=1}^{N} |\pi_i^n - \pi_i^{n-1}|$ increases with N. It is therefore better to divide this number by N. Also, the probability that $\sum_{i=1}^{N} |\pi_i^n - \pi_i^{n-1}| < \epsilon$ for all i is more stringent for larger values of N. This should be kept in mind when comparing the speed of iteration for models with different N.

In addition to the proposed stopping criteria, one may consider alternatives. For instance, in a one-server queue, one may require that $|\lambda(1 - \pi_N^n) - \mu(1 - \pi_0^n)| < \epsilon$, meaning that the expected number of arrivals to the system must equal the expected number of departures. A similar rule can be used in tandem queues.

9.3.4 Periodic Iteration Matrices

In Section 8.1.3, we indicated that state i is periodic with a period of r if any path starting in i and returning to i is a multiple of $r > 1$. We also mentioned that all states in the same communicating class have the same periodicity. In each communicating class, the states can be partitioned into *periodicity classes*. For now, we assume there is only one communicating class, and state 1 is in communicating class 1. Then, all states that can be reached in mr steps, where m is an integer, also belong to communicating class 1. All states that can be reached in one step from any state of class 1 belong to class 2, and any state that can be reached in one step from any state in class 2 belongs to class 3. This continues until we obtain r periodicity classes. Periodicity classes can be formed based on any matrix $V = [v_{i,j}]$. Matrices with a periodicity r are called r-*cyclic matrices*, a cycle being a path that starts and ends in the same state. Actually, one frequently uses the letter p instead of r for the length of the period, which means that in literature, the term "p-cyclic matrix" is used for a matrix with period $r > 0$.

If we number the states such that all states of periodicity class k precede the states of periodicity class $k + 1$, $k < r$, then the matrix in question, whether it is a transition matrix or an iteration matrix, can be written as

$$V = \begin{bmatrix} 0 & V_1 & 0 & \dots & 0 \\ 0 & 0 & V_2 & \ddots & \vdots \\ \vdots & \ddots & \ddots & \dots & 0 \\ 0 & \dots & \dots & 0 & V_{r-1} \\ V_r & 0 & \dots & \dots & 0 \end{bmatrix}. \tag{9.51}$$

The V_k are matrices, and they must be either periodic or their only non-zero entries of V_k are of the form $v_{i,j}$, where j is in periodicity class k and i is in the immediately preceding class. In other words, i is in class $k - 1$ or, if $k = 1$, in class r. The requirement that $v_{i,j} = 0$ if i and j are in the same periodicity class excludes matrices of the following form:

$$V = \begin{bmatrix} 0 & V_1 \\ V_2 & 0 \end{bmatrix},$$

with $V_1 = [1]$ and

$$V_2 = \begin{bmatrix} 0.5 & 0.5 \\ 0 & 1 \end{bmatrix}.$$

In the method of Gauss–Seidel, $V = L(D - U)^{-1}$ has rows of zeros. Indeed, since the first row of L is zero, the first row of V is also 0. There may be other rows of L that are zero. In this case, the corresponding states can be ignored when forming periodicity classes.

9.3.4.1 Models Leading to Periodic Matrices

We already showed that the uniformization matrix can be periodic. As an example where the jump matrix is r-cyclic, consider, the $M/M/1/N$ queue, the one-server queue with finite state space N discussed in Section 2.4.2. If λ is the arrival rate, and μ is the service rate, the jump matrix becomes

$$\begin{bmatrix} 0 & 1 & 0 & \cdots \cdots \cdots \cdots & 0 \\ \frac{\mu}{\lambda+\mu} & 0 & \frac{\lambda}{\lambda+\mu} & 0 & \cdots \cdots \cdots & 0 \\ 0 & \ddots & \ddots & \ddots & \ddots & \ddots & 0 \\ \vdots & \ddots & \ddots & \ddots & \ddots & \ddots & 0 \\ 0 & \cdots \cdots \cdots & 0 & \frac{\mu}{\lambda+\mu} & 0 & \frac{\lambda}{\lambda+\mu} \\ 0 & \cdots \cdots \cdots \cdots & 0 & 1 & 0 \end{bmatrix}.$$

This jump chain is obviously periodic with a periodicity of 2, and periodicity class 1 consists of even-numbered states and the odd-numbered states form periodicity class 2.

Another example where the jump chain is periodic is the tandem queue with d stations, say 1, 2, $\ldots$, d. The event types are "arrival", "from k to $k+1$", $k = 1, 2, \ldots, d-1$, and "departure", which occurs from station d. To prove periodicity, we consider two points in time, say t_1 and t_2, and we find the conditions that must be satisfied such that $X(t_1) = X(t_2)$, which implies that we return to the same state. Here $X(t) = [X_1(t), X_2(t), \ldots, X_d(t)]$. To formulate the conditions to ensure $X(t_1) = X(t_2)$, let n_1 be the number of arrivals between t_1 and t_2, let n_k, $k = 2, 3, \ldots, d$ be the number of the moves from line $k-1$ to line k between t_1 and t_2, and let n_{d+1} be the number of departures. Since we are looking at the jump chain, there is a state change in each time unit, that is, $t_2 - t_1 = n_1 + n_2 + \ldots + n_{d+1}$. Note that $X_1(t_1) = X_1(t_2)$ if and only if the number of arrivals in $(t_1, t_2]$ is equal to the number of moves from line 1 to line 2, that is, if and only if $n_1 = n_2$. Similarly, $X_i(t_1) = X_i(t_2)$ if and only if $n_i = n_{i+1}$, $i = 2, 3, \ldots d - 1$, and this also holds for $X_d(t_1) = X_d(t_2)$. It follows that all n_i, $i = 1, 2, \ldots, d + 1$ must have the same value, and their sum must therefore be a multiple of $d + 1$. Since $t_2 - t_1 = n_1 + n_2 + \cdots + n_{d+1} = (d + 1)n_1$, $X(t_1)$ cannot be equal to $X(t_2)$ unless $t_2 - t_1$ is divisible by $d + 1$. We conclude that the process is periodic with a period of $d + 1$.

It is also possible that the Gauss–Seidel iteration matrix is periodic. As an example, consider a CTMC with the following transition matrix:

$$Q = \begin{bmatrix} -\lambda & 0 & \lambda \\ \lambda & -\lambda & 0 \\ 0 & \lambda & -\lambda \end{bmatrix}. \tag{9.52}$$

The method Gauss–Seidel now yields:

$$\pi_1^{n+1} = \pi_2^n$$
$$\pi_2^{n+1} = \pi_3^n$$
$$\pi_3^{n+1} = \pi_1^{n+1}.$$

We now apply these equations, starting with $\pi_2^0 = 1$, $\pi_1^0 = \pi_3^0 = 0$. One obtains

i	π_i^0	π_i^1	π_i^2	π_i^3	π_i^4 ...
1	0	1	0	1	0 ...
2	1	0	1	0	1 ...
3	0	1	0	1	0 ...

This is clearly a periodic sequence. Indeed

$$V = L(D - U)^{-1} = \begin{bmatrix} 0 & 0 & 0 \\ \lambda & 0 & 0 \\ 0 & \lambda & 0 \end{bmatrix} \begin{bmatrix} \lambda & 0 & -\lambda \\ 0 & \lambda & 0 \\ 0 & 0 & \lambda \end{bmatrix}^{-1} .$$

Since

$$\begin{bmatrix} \lambda & 0 & -\lambda \\ 0 & \lambda & 0 \\ 0 & 0 & \lambda \end{bmatrix}^{-1} = \begin{bmatrix} \lambda^{-1} & 0 & \lambda^{-1} \\ 0 & \lambda^{-1} & 0 \\ 0 & 0 & \lambda^{-1} \end{bmatrix} ,$$

we obtain

$$V = \begin{bmatrix} 0 & 0 & 0 \\ 1 & 0 & 1 \\ 0 & 1 & 0 \end{bmatrix} .$$

This matrix is periodic with one periodicity class consisting of states 1 and 3, and the other one of state 2.

This example also shows that $\sum_{i=1}^{N} \pi_i^n$ is no longer equal to 1, and this means that the probabilities need to be normed when using Gauss–Seidel. Indeed, in this example, $\pi_1^n + \pi_2^n + \pi_3^n$ is 1 for n even, and 2 for n odd.

The periodicity in this example can easily be removed by renumbering. If we exchange states 1 and 3, the transition matrix given in equation (9.52) becomes

$$Q = \begin{bmatrix} -\lambda & \lambda & 0 \\ 0 & -\lambda & \lambda \\ \lambda & 0 & -\lambda \end{bmatrix} . \tag{9.53}$$

Applying Gauss–Seidel now yields

$$\pi_1^{n+1} = \pi_3^n$$
$$\pi_2^{n+1} = \pi_1^{n+1}$$
$$\pi_3^{n+1} = \pi_2^{n+1} .$$

If we set $\pi_3^0 = 1$, $\pi_1^0 = \pi_2^0 = 0$, then these equations yield $\pi_1^1 = \pi_2^1 = \pi_3^1 = 1$, and these values do not change any more. With this setup. equilibrium was reached in one step. Indeed, we find

$$V = \begin{bmatrix} 0 & 0 & 0 \\ 0 & 0 & 0 \\ 1 & 1 & 1 \end{bmatrix} .$$

Note that in our example, we have a Markov chain with a periodicity 3. In the numbering scheme given by equation (9.52), we numbered the states against the flow, whereas, in the numbering given by equation (9.53), we numbered the equations with the flow. In fact, one has:

Theorem 9.4 *If the states of a DTMC with transition matrix P are numbered such that the states of periodicity class i precede the states of periodicity class i + 1, then the sequence obtained by using Gauss–Seidel is aperiodic.*

Proof Let π_i^n denote the vector containing the probabilities to be in a state of periodicity class i at iteration n, and let P be of the form given by (9.51), with P_i substituted for V_i. Gauss–Seidel now yields

$$\pi_1^{n+1} = \pi_r^n P_r$$
$$\pi_2^{n+1} = \pi_1^{n+1} P_1 = \pi_r^n P_r P_1$$
$$\pi_3^{n+1} = \pi_2^{n+1} P_2 = \pi_r^n P_r P_1 P_2$$
$$\vdots = \vdots$$
$$\pi_r^{n+1} = \pi_{r-1}^{n+1} P_{r-1} = \pi_r^n P_r P_1 P_2 \dots P_{r-1}. \tag{9.54}$$

Written in matrix notation, this yields the matrix V as follows:

$$\pi^n V = \pi^n \begin{bmatrix} 0 & 0 & \dots & \dots & 0 \\ 0 & 0 & \dots & \dots & 0 \\ \vdots & \vdots & \vdots & \vdots & \vdots \\ 0 & 0 & \dots & \dots & 0 \\ P_r & P_r P_1 & P_r P_1 P_2 & \dots & P_r P_1 P_2 \dots P_{r-1} \end{bmatrix}.$$

To establish the presence or absence of periodicity, we simply apply equation (9.54), which provides us with π_r^n for all n without calculating any probability vector π_i^n, $i < r$. We have, in fact, eliminated all periodicity classes except class r. Consequently, the only way the iteration matrix V could be periodic is that the matrix product $P_r P_1 P_2 \dots P_{r-1}$ is periodic, say with a period of r_1. In this case, the periodicity class r is visited only every rth time, but if i is a state in the rth periodicity class, it will take at least r_1 visits to periodicity class r before i is visited again. The periodicity of the original Markov chain is thus $r r_1$ and not r, contrary to our assumption. □

Though periodicity in Gauss–Seidel can be eliminated by numbering the states in a certain way, the required numbering is often unnatural. In fact, in Section 9.3.5, we show that the lexicographical order can lead to a periodic iteration matrix when using the method of Gauss–Seidel.

9.3.4.2 Efficient Ways to Deal with Periodicity

Using the jump matrix for iterating is easier to program than Gauss–Seidel. Moreover, it allows one to use the efficient vector–matrix multiplications available in some computer languages. As shown above, the jump matrix may be periodic or almost periodic, but there are methods to overcome this difficulty.

For our discussion, we assume that the matrix has a period of r. We define S_i to be the set of states that belong to periodicity class i, $i = 1, 2, \ldots, r$. We now show that even when π^n is periodic, and never reaches an equilibrium, π^{n+mr} does when n is fixed and m varies. In effect, we embed the Markov chain at points r time units apart, and to do this, we calculate P^r, where

$$
P = \begin{bmatrix}
0 & P_1 & 0 & \ldots & 0 \\
0 & 0 & P_2 & \ddots & \vdots \\
\vdots & \ddots & \ddots & \ldots & 0 \\
0 & \ldots & \ldots & 0 & P_{r-1} \\
P_r & 0 & \ldots & \ldots & 0
\end{bmatrix}.
$$

We now have

$$
P^r = \begin{bmatrix}
P_1 P_2 \ldots P_r & 0 & \ldots & \ddots & 0 \\
0 & P_2 P3 \ldots P_r P_1 & \ddots & \vdots & 0 \\
\vdots & & \ddots & \ddots & \vdots \\
0 & \ldots & 0 & P_{r-1} P_r P_1 \ldots P_{r-2} & 0 \\
0 & 0 & \ldots & 0 & P_r P_1 P_2 \ldots P_{r-1}
\end{bmatrix}.
$$

Clearly, P^r is decomposable, and there is one communicating class for each periodicity class, with the transition matrix for class k given as

$$
\bar{P}_k = \prod_{v=k}^{r} P_v \prod_{v=1}^{k-1} P_v.
$$

We thus have r different processes, and as shown in the proof of Theorem 9.4 they are all aperiodic, and therefore each converges toward its particular equilibrium, as is true for any decomposable matrix. Any linear combination of these equilibria leads to an overall equilibrium, that is, the overall equilibrium is not unique.

Reaching an equilibrium implies that $\pi^n = \pi^{n+r}$, and as a consequence

$$
\sum_{v=n}^{n+r-1} \pi^v = \sum_{v=n}^{n+r-1} \pi^v - \pi^n + \pi^{n+r} = \sum_{v=n+1}^{r} \pi^v.
$$

In other words, if $u_n = \sum_{v=n}^{n+r-1} \pi^v$, then as soon as $u_n = u_{n+1} = u$, the vectors u_n have reached an equilibrium. Moreover, u satisfies

$$
u_{n+1} = \sum_{v=1}^{r-1} \pi^{v+1} = \sum_{v=1}^{r-1} \pi^v P = u_n P.
$$

If $u_n = u$ for all n, we have $u = uP$. If we compare this with the equilibrium equation for π, $\pi = \pi P$, we conclude that the vector u differs from π only by a factor.

Note that to have $\pi^n = \pi^{n+1}$, we must have for all k and ℓ

$$\sum_{i \in S_k} \pi_i = \sum_{i \in S_\ell} \pi_i.$$

The reason is that as the process goes from n to $n + 1$, all sample functions enter the next periodicity class, yet the sum of probabilities in class k at time n is the same as the sum of all probabilities of the new class, which is $k + 1$ for $k < r$ and 1 for $k = r$. If these sums are different for the different classes, the sums of probabilities in the different classes change, and we are not in equilibrium, contrary to our assumption.

Based on the results above, we have a variety of methods at our disposal for determining the equilibrium probabilities of periodic processes. Here, we suggest the following method. Select a starting periodicity class, say class k, and give the π_i^0 a non-zero value only if $i \in S_k$. Next, calculate for $n = 1, 2, \ldots$:

$$\pi^{n+1} = \pi^n P_{\mathrm{mod}(k+n-1)+1,r}, \; n > 0. \tag{9.55}$$

Here, we follow the standard definition of $\mathrm{mod}(m, r)$ as $m - \lfloor \frac{m}{r} \rfloor r$. We use $\mathrm{mod}(k + n - 1) + 1$ to insure this expression is between 1 and r. For instance, when $n = 0$, then $\mathrm{mod}(k + n - 1, r) + 1 = k$, and the above equation specializes to $\pi^1 = \pi^0 P_k$. This still holds when $k = r$.

The advantage of the method suggested by equation (9.55) is that the resulting matrices have only a fraction of the dimension of P and that we consistently use a vector–matrix multiplication. When $\pi^{r(m-1)}$ is close enough to π^{rm}, say at $m = m_e$, we continue for another $r - 1$ iterations, because, in this way, one presumably reaches an equilibrium for the other periodicity classes within the prescribed tolerance. In fact, the π^{rm_e+n} are obtained by multiplying π^{rm_e} by the appropriate matrices P_ℓ, and if π^{rm_e} has its equilibrium value, so do all π^{rm_e+n}. Thus, as soon as π^{rm_e} is close enough to its equilibrium value, we form

$$\pi = u = \sum_{n=m_e}^{m_e+r-1} \pi^n.$$

Consequently, our objective has been reached.

We doubt that initializing states of different classes can reduce the computational effort significantly. Indeed, if one has the equilibrium probabilities for a single class, one can readily calculate the equilibrium probabilities for all other classes in $r - 1$ steps, and the effort for doing this is essentially the same as for a single iteration when the π_i^0 are given initial values for all i. In both cases, one has to multiply a vector with P_k, where k varies from 1 to r. If for some reason, all π_i^0 should be initialized, it pays to ensure that for all k and ℓ,

$$\sum_{i \in S_k} \pi_i^0 = \sum_{i \in S_\ell} \pi_i^0.$$

In this case, a similar relation holds for all π_i^n, including for large n when an equilibrium is reached within a prescribed precision. In fact, with this method, π_i^n converges toward an equilibrium in the sense that π_i^n approaches π_i^{n+1}, that is, one does not even have to compare π^n with π^{n+r} to check for equilibrium.

In some cases, one is not sure whether or not P is periodic. To verify this, we set $\pi_i^0 = 1$ for exactly one i. and run for a limited number of iterations. The patterns of the π_i^n that are zero then allow one to discover if the transition matrix is periodic.

Jump chains and uniformization matrices may not be strictly periodic, but they may be almost periodic. Even in these cases, one can apply some of the results pertaining to periodic iteration matrices. In particular, if r is known, taking the sum of r successive values is helpful as shown in Section 9.3.3.2. It should also be helpful to initialize the π_i^0 such that one obtains the same sum of all probabilities in each periodicity class.

9.3.5 Examples

We now present a number of examples. With these examples, we want to explore how to avoid periodicity by using proper initialization, how to handle state variables that have tails that decrease slowly, and how the execution times increase as the size of the problem increases.

We use essentially two examples: the $M/M/1/N$ queue and the tandem queue, and we solve these two models with both the jump-chain method and the Gauss–Seidel method. We use VBA as it comes with Microsoft Excel$^{©}$. This may not be the fastest system, but as we found in earlier studies, the actual execution times reflect the theoretical time complexity much better than when programming in Octave.

First, consider the $M/M/1/5$ queue with $\lambda = 0.8$ and $\mu = 1$. We ran this system, first with the jump-chain method, starting with the initial value of $\pi^0 = [0, 1, 0, 0, 0]$. The result for the first 8 iterations was as follows

n	0	1	2	3	4	5	6	7	8
π_0^n	0	0.556	0	0.446	0	0.392	0	0.358	0
π_1^n	1	0	0.802	0	0.705	0	0.645	0	0.606
π_2^n	0	0.444	0	0.466	0	0.456	0	0.446	0.
π_3^n	0	0	0.198	0	0.256	0	0.287	0	0.307
π_4^n	0	0	0	0.088	0	0.153	0	0.196	0
π_5^n	0	0	0	0	0.039	0	0.068	0	0.087

One clearly sees that the system is periodic with $r = 2$ because π_i^n is zero for every other n. We also see that when only considering every other n, the system seems to be converging. Indeed, after 26 iterations, the absolute difference between π_i^{24} and π_i^{26} for even i, π_i^{23} and π_i^{25} for odd i, is less than 0.001. To bring these differences below 0.0001, one needs 36 iterations.

It may be instructive to consider the case where we initialize both π_0^0 and π_1^0 to 1. As one can see, the results are similar. Indeed, where there was no zero in our first initialization, the results are exactly the same. Clearly, if we had initialized with $\pi_0^0 = \pi_1^0 = 0.5$, the process would have converged to the equilibrium probabilities of the jump chains. In fact, since the numbers are essentially the same, the rate of convergence is also the same.

n	0	1	2	3	4	5	6	7	8
π_0^n	1	0.556	0.556	0.446	0.446	0.392	0.392	0.358	0.358
π_1^n	1	0.000	0.802	0.802	0.705	0.705	0.645	0.645	0.606
π_2^n	0	0.444	0.444	0.466	0.466	0.456	0.456	0.446	0.446.
π_3^n	0	0.000	0.198	0.198	0.256	0.256	0.287	0.287	0.307
π_4^n	0	0.000	0.000	0.088	0.088	0.153	0.153	0.196	0.196
π_5^n	0	0.000	0.000	0.000	0.039	0.039	0.068	0.068	0.087.

For the sake of completeness, we present the true values with the values obtained after 26 iterations. we have

	π_0	π_1	π_2	π_3	π_4	π_5
26 iteration	0.149	0.269	0.214	0.171	0.136	0.0604
True value	0.149	0.268	0.214	0.171	0.137	0.0609.

In the uniformization–matrix method of the $M/M/1/5$ queue, we have an almost periodic iteration matrix. In this case, the results are as follows:

n	0	1	2	3	4	5	6	7	8
π_0^n	0	0.556	0.309	0.446	0.324	0.392	0.316	0.358	0.306
π_1^n	1	0.000	0.494	0.137	0.381	0.178	0.328	0.193	0.297
π_2^n	0	0.444	0.000	0.329	0.061	0.278	0.094	0.248	0.116
π_3^n	0	0.000	0.198	0.000	0.195	0.027	0.184	0.054	0.173
π_4^n	0	0.000	0.000	0.088	0.000	0.108	0.022	0.113	0.043
π_5^n	0	0.000	0.000	0.000	0.039	0.017	0.056	0.034	0.065.

Here, one sees that π_i^n rises and falls with n. For instance, π_0^7 is greater than π_0^6, but less than π_0^8. In this case, it pays to use the average of two successive values to improve convergence.

We now use the method of Gauss–Seidel with $\pi_1^0 = 0$, $\pi_i^0 = 0$, $i \neq 1$ and find for the first 8 iterations:

n	0	1	2	3	4	5	6	7	8
π_0^0	0	1.250	0.694	0.557	0.490	0.448	0.421	0.404	0.392
π_0^1	1	0.556	0.446	0.392	0.358	0.337	0.323	0.314	0.308
π_0^2	0	0.247	0.259	0.253	0.248	0.244	0.242	0.241	0.240.
π_0^3	0	0.110	0.142	0.160	0.171	0.178	0.182	0.185	0.187
π_0^4	0	0.049	0.085	0.109	0.124	0.134	0.141	0.145	0.147
π_0^5	0	0.039	0.068	0.087	0.099	0.107	0.112	0.116	0.118

Clearly, one must not set $\pi_0^0 = 1$ because it is overwritten right away, and if all other $\pi_i^0 = 0$, then all π_i^n vanish. When using $\pi_1 = 1$, $\pi_i = 0$, $i \neq 1$, periodicity disappears. Gauss–Seidel converges faster than the methods used before: The maximum difference between π_i^{n-1} and π_i^n is already below 0.001 after 14 iterations, and it is below 0.0001 after iteration 19.

Next, we consider the tandem queue. In our first run, we look at only two stations. Buffer sizes were chosen to be 3 and 2. The arrival rate is 0.75, and both service rates are 1. We only used the method of Gauss–Seidel. The states were in lexicographic order. We set $\pi_1^0 = 1$, and $\pi_i^0 = 0$, $i \neq 1$. Here, state 1 is the state with $X_1 = 0$, $X_2 = 1$. Unfortunately, the iteration matrix is 2-cyclic, as can be seen from the following table:

n	0	1	2	3	4	5	6	7	8
$[0, 0]$	0	1.333	0	0.435	0	0.297	0	0.245	0
$[0, 1]$	1	0	0.327	0	0.223	0	0.184	0	0.163
$[0, 2]$	0	0	0	0.102	0	0.097	0	0.090	0
$[1, 0]$	0	0.571	0	0.288	0	0.224	0	0.195	0
$[1, 1]$	0	0	0.178	0	0.170	0	0.158	0	0.154
$[1, 2]$	0	0	0	0.110	0	0.121	0	0.127	0

From this table, one can also see the states $[0, 0]$, $[0, 2]$, $[1, 0]$ and $[1, 2]$ are in the same periodicity class. The states $[0, 1]$ and $[1, 1]$ are part of the other periodicity class.

To remove the periodicity, we renumber the states such that X_1 changes least, and X_2 most. In other words, the coding formula is changed to $c = X_1 u_1 + X_2 u_2$, where $u_1 = 1$ and $u_2 = N_1 + 1 = 4$. To find the starting state, we use the equilibrium equation for state $0 = [0, 0]$, which is $\pi_0 q_0 = \pi_4 q_{4,0}$, where state 4 represents the state vector $[0, 1]$. We therefore set $\pi_4^0 = 1$. The first six rows of the solution for n up to 8 are now

n	0	1	2	3	4	5	6	7	8
$[0, 0]$	0.000	1.333	0.435	0.297	0.245	0.218	0.205	0.199	0.196
$[1, 0]$	0.000	0.571	0.288	0.224	0.195	0.181	0.174	0.171	0.170
$[2, 0]$	0.000	0.245	0.190	0.176	0.172	0.171	0.170	0.169	0.169
$[3, 0]$	0.000	0.184	0.185	0.211	0.224	0.230	0.233	0.234	0.235
$[0, 1]$	1.000	0.327	0.223	0.184	0.163	0.154	0.149	0.147	0.146
$[1, 1]$	0.000	0.178	0.170	0.158	0.154	0.151	0.150	0.150	0.150

One sees that the periodicity has been removed. Indeed, the π_i^n is mostly decreasing in n. Convergence is fast: After 10 iterations, $|\pi_i^n - \pi_i^{n-1}|$ is already less than 0.001 for all i, and it is less than 0.0001 after 14 iterations.

Convergence is poor if one of the state variables has a wide range and a slowly decreasing tail. To demonstrate that, we chose the $M/M/1/N$ queue with $\lambda = 0.95$, $\mu = 1$ and $N = 50$. We used the jump-chain method with the starting vector $\pi_0^0 = 1$, $\pi_i^0 = 0$, $i > 0$. We obtained the average of π_i^{n-1} and π_i^n every 100 iterations, and calculated the expected line length, using the formula

$$L_n = \sum_{i=0}^{50} i \frac{1}{2} (\pi_i^{n-1} + \pi_i^n) q_i$$

with $q_0 = \lambda = 0.95$, $q_i = \lambda + \mu$, $i = 1$ to 49, and $q_{50} = \mu = 1$. We compared L_n after n iterations with the expected number in the $M/M/1/50$ queue with its true value, which is $L = 14.978$, and found L_n/L. Here are the results:

iterations	100	200	500	1000	2000
L	14.978	14.978	14.978	14.978	14.978
L_n	6.466	8.617	11.932	14.028	14.893
L_n/L	0.432	0.575	0.797	0.937	0.994

Clearly, after 1000 iterations, one still has only reached 93.7% of the true value. Only 2000 iterations, with an accuracy of 99.4%, give reasonable results.

If $N = 100$, the approach to the equilibrium is still worse. In this case, we get

Iterations	100	200	500	1000	2000	4000
L	18.429	8.617	18.429	18.429	18.429	18.429
L_n	6.466	14.978	12.002	14.664	16.903	18.150
L_n/L	0.351	0.575	0.651	0.796	0.917	0.985

This shows that if N is increased from 50 to 100, convergence suffers significantly: Whereas we get 99.4% of the correct value when $N = 50$ when $n = 2000$, one reaches only 91.7% of the true value when $N = 100$ and $n = 2000$.

However, for this model, one has to ask if the arrival rate can be found with a reasonable accuracy. In fact, for $N = 50$, an increase of the arrival rate from 0.95 to 0.96, a 1% increase, changes L from 14.978 to 22.337, an increase of almost 20%. Also, the question arises whether or not an equilibrium expectation is really appropriate when it is reached only after 4000 events. If the events are arrivals or departures of human customers, this is unrealistic. It may be realistic in computer applications. Also, the queueing behavior may be quite different when the line approaches 50 or even 100.

The method of Gauss–Seidel converges much faster when applied to our model with $N = 50$. There, we get the following results:

Iterations	100	200	500	1000	2000
L	14.978	14.978	14.978	14.978	14.978
L_n	8.934	11.33	14.074	14.888	14.977
L_n/L	0.596	0.757	0.940	0.994	1.000

As one can see, here, Gauss–Seidel is roughly twice as fast as the jump-chain method. Of course, the difference between methods is reduced when languages are used with optimized facilities for vector–matrix multiplications.

To demonstrate how this model works for larger problems, we considered a tandem queue with 5 stations, where each station has a buffer of 9. We also did a run with the same buffer sizes, but only with 4 stations. The computer used was a ThinkPad computer from the year 2008, and we used Microsoft Excel© 2010, meaning we used relatively old equipment. Today's computers are faster. The model with 5 stations has $10^5 = 100,000$ states, and the model with 4 stations has 10,000 states. In both models, the arrival rate was 0.75, and the service rates were all 1. We used u as our stopping criterion, where

$$u = \frac{1}{N} \sum_{i=1}^{N} |\pi_i^{n-1} - \pi_i^n|.$$

We checked every 10 iterations to check if $u < 0.0001$ and stopped as soon as it was. The results for the 5-station model were as follows. It took 2 seconds, and 20 iterations until u fell below 0.0001. After 10 iterations, u was 1.12E-06, and after 20, it was 3.97E-7. In the case of the 4-station model, u was below 0.0001 after 20 iterations, and the execution time was 0.093 seconds. In this model, after 10 iterations, we found $u = 9.34E-6$, and after 20, $u = 3.16E-06$. What is remarkable is that u hardly changes when the number of stations changes. This is counter-intuitive because it takes longer to move through 5 rather than 4 stations. This suggests that in this model, the number of iterations does not seem to depend on the dimensionality of the problem. In the 5-station model, we also changed the arrival rate. The results are as follows:

arrival rate	0.7	0.75	0.8
u after 20 iteration	3.33E-07	3.92E-07	4.48E-07

As one can see, u increases with the arrival rate, but this increase may be less than one would expect.

Note that in the case of the tandem queue, the number of iterations is relatively low, compared to the number of iterations required for the $M/M/1/100$ queue. This seems to indicate that the range of the state variables has a stronger influence on the number of iterations than the dimension of the model, especially if the system has slowly decaying tails.

9.4 Conclusions

In this chapter, we discussed both direct and iterative methods to find the equilibrium solutions for Markov chains. As it turns out, for problems with high dimensions, iterative methods are the only viable method. To demonstrate this, consider the 5-station tandem queue, with buffer sizes of 9 for each station. This model has $N = 10^5 = 100,000$ states, and we got our result within 20 iterations. One finds that the number of flops for each iteration is $2Ne$, where $e = 6$ is the number of events. Hence, we need $12N = 1,200,000$ flops per iteration. With 20 iterations, the value we get for our example is 24 million flops or 2.4E07 flops. Consider now solving the same problem by state elimination, which is a direct method. In this case, we need $2hgN$ flops to find the equilibrium probabilities. Here, h must be at least 10^4 and g must be at least 9×10^3, which means we have $1.8 \times 10^8 N$ flops, or 1.8×10^{13} flops. Thus, to solve this problem by state elimination, we need roughly a million times more flops than by using Gauss–Seidel, and instead of the two seconds needed to find the equilibrium probabilities, we would need 2,000,000 seconds, which makes roughly 24 days. The same type of calculations can be done with any other model: If the problem is large enough, iterative methods are preferable. If even the iterative methods lead to unacceptable execution times, simulation must be used.

On the other hand, state elimination is ideal for small problems. For example, consider the 3-station tandem queue, each with a buffer size of 9. This model has $N = 1000$ states, with $h = 100$ and $g = 90$, and one needs $2Nhg = 9000000$ flops to find the equilibrium probabilities. This requires less than a second on a computer, and looking for faster methods may not be worth the effort.

An issue in iterative methods, that does not exist in direct methods, is how many iterations are required until the results are within acceptable error bounds. In this case, one has to decide which stopping criterion to use, and how the stopping criterion is affected by N, the number of states.

Iterative methods converge poorly if the iteration matrix has a subdominant eigenvalue that is 1 or close to 1. This can occur when the original Markov chain is periodic or almost periodic, or when the original Markov chain is decomposable or almost decomposable. Periodic Markov chains cause no problems in direct methods. In fact, it can potentially be used to improve the algorithm. Periodicity in iterative methods means that there is no convergence at all, but this problem can be overcome. One can, for instance, take the average of r successive values, where r is the length of the period. A similar method can be used when the iteration matrix is almost periodic. If the method of Gauss–Seidel is used, one can avoid periodicity by an appropriate numbering of the states.

If the Markov chain is decomposable, then equilibrium probabilities should be found for each individual component. This is true for both iterative and direct methods. There are major problems in dealing with almost decomposable Markov chains. In this case, the rates governing the transitions between the almost decomposable components are small, and even minor absolute changes in these rates will cause significant changes in the equilibrium probabilities. If direct methods are used, it is therefore essential to minimize rounding. This is easily achieved by using the GTH method. If iterative methods are used, one will need a high number of iterations, and in this situation, it is difficult to find appropriate stopping criteria. Therefore, one should consider methods that are designed to deal with almost decomposable Markov chains as they are described in [86].

In iterative methods, we provided evidence that the convergence is not affected by the number of state variables used. The convergence deteriorates, however, if some state variables have slowly decreasing tails. In such cases, execution times increase with the upper limit of the state variable with the slowly decreasing tail.

So far, we only have dealt with problems where the number of states is finite. Many queueing models deal with models where there is no upper bound on the length of the line, and this requires methods for solving infinite state problems. In fact, whereas direct methods can be extended to deal with infinite state models, we are not aware of any way to extend iterative methods to deal with infinite state problems.

Problems

9.1 In the textbooks on linear algebra, one usually eliminates the variable with the lowest subscript first, which would be π_1 in our case. Implement Algorithm 30, for this case.

9.2 In an $M/E_2/1$ queue, derive a formula for the expected waiting time and its variance under the FIFO queueing discipline.

9.3 In an $M/E_2/1/5$ queue with an arrival rate of 0.9 and a service rate of 1, calculate the expected waiting time and its variance under the SIRO queueing discipline.

9.4 Write a program to solve the three-way stop problem described in Section 2.6.2 by using the Gauss–Seidel method. Discuss whether or not you could encounter periodicity, how to recognize it, and how to avoid it.

9.5 Find the equilibrium solution of a queueing system as follows: There are two exponential servers with service rates $\mu_1 = 1$ and $\mu_2 = 0.8$. Arrivals are Poisson, with rate 1.5. All arrivals join the shorter line, and if the two lines are of equal length, they join line 1. Each line has a length of at most 8.

1. Use the state elimination method to solve this problem.
2. Use Gauss–Seidel to solve this problem.

9.6 In equation (9.49), we started calculating the π_j^{n+1} in ascending order of j.

1. Change equation (9.49) such that the π_j^{n+1} are calculated in descending order of j.
2. Design an algorithm for calculating π_i^{n+1} in ascending order when n is even, and in descending order when n is odd. Discuss the advantages and disadvantages of this method as opposed to the straightforward application of the Gauss–Seidel method.

Chapter 10
Reducing the Supplementary State Space Through Embedding

The state space of Markov chains increases exponentially with the number of state variables, and embedding helps reduce the need for supplementary state variables. For a definition of embedding, see Chapter 9, Definition 9.2. Generally, for embedding, we need a countable sequence $t_1, t_2, t_3, \ldots$ of times, and if $\{X(t), t > 0\}$ is a given process, the embedded process is $\{X(t_n), \ n > 0\}$. We will call the times t_n *embedding points*. The following are examples of embedded processes

1. The jump chain is an embedded process, with state changes used as embedding points.
2. The randomization method uses embedding, with the embedding points generated by state changes, combined with a Poisson process.

To use embedding to reduce the supplementary state space, we have to embed at the times when an event occurs, which allows us to remove the supplementary state variable connected with that event. This method is well known in queueing theory. In particular, in a $GI/M/1$ queue, the points immediately preceding an arrival, or the points immediately following an arrival are used as embedding points. At the points immediately before an arrival, we know that the remaining lifetime is 0, and if the time to the next arrival has been used as a supplementary variable in the original process, it no longer needs to be recorded in the embedded process, which in turn reduces the state space. Similarly, if the age of the arrival is used as a supplementary state variable, it no longer needs to be recorded if the points immediately following an arrival are used as embedding points. In an $M/G/1$ queue, one can use the points of departure as embedding points, and the corresponding supplementary state variable no longer needs to be recorded.

In a discrete event system, the process at the times where the physical state changes is an embedded process. The embedding points are the points where any of the events occur. In this case, the supplementary variable of the event that changed the physical state no longer needs to be recorded. This reduces the state space.

When using the remaining lifetime as a supplementary variable, then the only states that can occur before an event are states with at least one $Y_k = 0$. Clearly, if all Y_i are integers that range from 0 to $M_i - 1$, then there are $M_1 M_2 \ldots M_e$ vectors of the

W. Grassmann and J. Tavakoli, *Numerical Methods for Solving Discrete Event Systems*, CMS/CAIMS Books in Mathematics 5, https://doi.org/10.1007/978-3-031-10082-6_10

form $[Y_1, Y_2, \ldots, Y_e]$ in the original chain. From this number, we have to deduct all vectors with no $Y_i = 0$, and their number is $(M_1 - 1)(M_2 - 1) \cdots (M_e - 1)$. It follows that the number of supplementary states is

$$M_1 M_2 \cdots M_e - (M_1 - 1)(M_2 - 1) \cdots (M_e - 1). \tag{10.1}$$

If all M_i are equal, say they are all equal to M, then the number of states that can precede an event is approximately eM^{e-1}. To show this, we write

$$M^e - (M - 1)^e = M^e - \left(M^e + \sum_{n=1}^{e} \binom{e}{n} M^{e-n} (-1)^n \right) \approx eM^{e-1},$$

and the result follows. Hence, one can reduce the number of states almost by an order of magnitude.

Consider now the age approach. If $Y_k(t)$ is the age, then only states with at least one $Y_k(t) = 1$ can follow an event. Counting these states will lead to the same number of states as when remaining lifetimes are used, which is given in expression (10.1).

There is one crucial difference between picking only one event type, say arrivals, and using these as embedding points, versus using the occurrences of any event type as embedding points. In the first case, $X(t)$ may change between event points, and in the second case, it does not. Both of these cases will be discussed in this chapter.

We will call the equilibrium probabilities $P\{X(t_n) = i\}$ the embedding point probabilities, and the probabilities of $P\{X(t) = i\}$ the random-time probabilities. Note that these probabilities tend to be different, that is, the distributions of $X(t_n)$ and $X(t)$ differ. However, there are methods for finding the random-time probabilities from the embedding point probabilities. We only consider systems where the distribution of $X(t)$ has reached its equilibrium.

If $X(t)$ is a discrete-time process, $X(t_n)$ can no longer reach all states. The states $X(t_n)$ cannot reach must be eliminated. Since the elimination of state probabilities amounts to censoring states, embedding is related to censoring.

Embedding is related to the so-called *semi-Markov process*, which is a process where $X(t_n)$ is Markovian, and the times τ_n between t_n and t_{n+1} depend on $X(t_n)$, but are otherwise independent.

As before, t is the time immediately before the event, whereas t^+ was the time immediately after the event. In the case of embedding, one has to decide whether to use t_n or t_n^+ as the embedding point. In renewal event systems, with $Y(t)$ used as the remaining lifetimes at t, and event Ev_k happens at t, $Y_k(t)$ is zero, whereas $Y_k(t^+)$ is given a non-zero value. If Y_k is an age, however, $Y_k(t)$ had a non-zero value, whereas $Y_k(t^+) = 0$, which implied $Y_k(t + 1) = 1$.

We also tried embedding with physical state changes as embedding points when the supplementary variables are of phase-type, but there seemed to be no real savings. The reason is that the states one could possibly eliminate have many targets, and this makes elimination not only more complicated, it also makes the transition matrices denser. We therefore do not consider eliminating phase-type supplementary variables.

The outline of this chapter is as follows. First, we discuss semi-Markov chains and show how embedding relates to state elimination. Next, we discuss the case where the embedding points include all points where the physical state changes, and finally, we discuss the case where a certain event is selected, and the occurrences of this event are the embedding points.

10.1 The Semi-Markov Process (SMP)

A semi-Markov process, abbreviated SMP, is a generalization of a Markov chain in that the times between transitions are not 1, as in DTMCs, but random variables. Specifically, the process $X(t)$ is a SMP if the following conditions are satisfied

1. $X(t)$ can only change at times $t_1, t_2, t_3, \ldots$. The transition probabilities $P\{X(t_n) = j \mid X(t_n) = i\}$ will be denoted by $p_{i,j}$.
2. The process $\{X(t_n), \ n > 0\}$ is a Markov process.
3. At each time t_n, a random variable τ is drawn, and $t_{n+1} = t_n + \tau$. The distribution of τ depends of $X(t_n^+)$, but not on the past history of the process.

There is also a substantial literature on the process $\{X(t_n), t_n, \ n > 0\}$, which is called *Markov renewal process*. We will not discuss it here but refer to [13].

If $E_i(\tau)$ is the expectation of τ, given $X(t_n^+) = i$, one has

$$P\{X(t) = i\} = \frac{P\{X(t_n) = i\}E_i(\tau)}{\sum_j P\{X(t_n = j)\}E_j(\tau)}. \tag{10.2}$$

If we define $P_i = P\{X(t) = i\}$ and $\pi_i = P\{X(t_n) = i\}$, then this equation becomes

$$P_i = \frac{\pi_i E_i(\tau)}{\sum_j \pi_j E_j(\tau)}.$$

Also note that in equilibrium, $P\{X(t_n) = i\} = P\{X(t_{n+1}) = i\}$, and since $P\{X(t_{n+1} = i)\} = P\{X(t_n^+) = i\}$, we have

$$P\{X(t_n) = i\} = P\{X(t_n^+) = i\}.$$

We only prove equation (10.2) for the case that τ is a discrete random variable. Since continuous random variables can be approximated arbitrarily close by discrete random variables, one can, using the appropriate ϵ/δ logic, generalize the proof to hold for continuous τ_n, but we will not do this here. When τ_n is discrete, we can use the two dimensional DTMC $\{X(t), Y(t), \ t = 1, 2, \ldots\}$, where $X(t)$ is the state of the SMP at time t, and $Y(t)$ is the age. We define

$$\pi_{i,j} = P\{X = i, Y = j\}.$$

Consequently

$$P_i = \sum_{j=1}^{\infty} \pi_{i,j}, \quad \pi_i = \pi_{i,1}.$$

The second equation follows from the fact that each change of X can be matched with an entry to a state where the age is 1.

According to equation (9.34) of chapter 9, in DTMCs, the equilibrium probabilities are proportional to the expected number of visits to state i between two successive visits of base state. Hence, if $d_{i,j}$ is the expected number of visits to state i, j in the process $((X(t), Y(t)), \; t > 0)$, then one has

$$\pi_{i,j} = P\{X = i, Y = j\} = \frac{d_{i,j}}{\sum_{k,m} d_{k,m}}. \tag{10.3}$$

To visit state (i, j), state $(i, 1)$ must be visited first, and after state $(i, 1)$ has been visited, state (i, j) is visited only if $\tau \geq j$. It follows that $d_{i,j} = d_{i,1} P\{\tau \geq j\}$. Consequently

$$\sum_{j=1}^{\infty} d_{i,j} = d_{i,1} \sum_{j=1}^{\infty} P\{\tau \geq j\} = d_{1,j} E_i(\tau).$$

Since according to equation (10.3), the $d_{i,j}$ are proportional to the $\pi_{i,j}$, equation (10.2) now follows easily.

To show how our result is related to state reduction, we now give two additional proofs for equation (10.2), one based on age as the supplementary variable, the other one on remaining lifetimes. For this purpose, we define

$$p_k^{(i)} = P\{\tau = k \mid X(t_n^+) = i\}, \; k \geq 1.$$

10.1.1 Using Age as the Supplementary Variable

We now prove equation (10.2) by Gaussian elimination. We assume again that τ is a discrete random variable. For a continuous version of the proof that follows, see [47, page 216ff]. We define $\pi_{i,j}$ as the probability that $X = i$ and the age $Y = j$. We now find the equilibrium equations for $\pi_{i,j}$. The probability P_i that $X = i$ can then be found as

$$P_i = \sum_j \pi_{i,j}.$$

To formulate the equilibrium equations, we define $r_k^{(i)}$ as the probability of a renewal if $Y(t) = k$, $X(t) = i$. We have

$$r_k^{(i)} = \frac{p_k^{(i)}}{\sum_{n=k}^{\infty} p_n^{(i)}}, \quad 1 - r_k^{(i)} = \frac{\sum_{n=k+1}^{\infty} p_n^{(i)}}{\sum_{n=k}^{\infty} p_n^{(i)}}. \tag{10.4}$$

When there is no renewal, $X(t)$ does not change, while $Y(t)$ increases by 1 and if $Y(t) = k$, then this happens with a probability of $1 - r_k$, that is

$$\pi_{i,k+1} = \pi_{i,k}(1 - r_k^{(i)}), \ k = 1, 2, \ldots \tag{10.5}$$

On the other hand, when there is a renewal, then there is a transition from k to 1, while X changes from i to j with probability $p_{i,j}$, $i, j = 1, 2, \ldots, N$, where n is the number of states. Consequently,

$$\pi_{j,1} = \sum_{i=1}^{N} \sum_{k=1}^{\infty} \pi_{i,k} r_k^{(i)} p_{i,j}. \tag{10.6}$$

We now eliminate all $\pi_{i,k}$ with $k > 1$. Applying (10.5) repeatedly leads to

$$\pi_{j,k} = \pi_{j,1} \prod_{n=1}^{k-1}(1 - r_n^{(j)}), \ j = 1, 2, \ldots, N.$$

By using (10.4), with $\sum_{n=1}^{\infty} p_n^{(j)} = 1$, one concludes

$$\pi_{j,k} = \pi_{j,1} \prod_{n=1}^{k-1}(1 - r_n^{(j)}) = \pi_{j,1} \sum_{n=k}^{\infty} p_n^{(j)}. \tag{10.7}$$

Substituting this expression into (10.6) yields

$$\pi_{j,1} = \sum_{i=1}^{N} \sum_{k=1}^{\infty} \pi_{i,1} \sum_{n=k}^{\infty} p_n^{(i)} r_k^{(i)} p_{i,j}.$$

Now, since $r_k^{(i)} = \dfrac{p_k^{(i)}}{\sum_{n=k}^{\infty} p_n^{(i)}}$, we have

$$\sum_{n=k}^{\infty} p_n^{(i)} r_k^{(i)} = p_k^{(i)}.$$

Consequently:

$$\pi_i = \pi_{j,1} = \sum_{i=1}^{N} \pi_{i,1} p_{i,j} \sum_{k=1}^{\infty} p_k^{(i)} = \sum_{i=1}^{N} \pi_{i,1} p_{i,j}.$$

To find P_i, we take the sum of all $\pi_{i,k}$ over k to obtain from equation (10.7):

$$P_i = \sum_{k=1}^{\infty} \pi_{i,k} = \pi_{i,1} \sum_{k=1}^{\infty} \sum_{n=k}^{\infty} p_n^{(i)}.$$

We now use

$$\sum_{k=0}^{\infty} \sum_{n=k+1}^{\infty} p_n^{(i)} = \sum_{n=1}^{\infty} n p_n^{(i)} = E_i(\tau),$$

to find

$$P_i = \pi_i E_i(\tau).$$

The equilibrium equations determine the $\pi_{i,k}$ only up to a factor, and the same is true for π_i. To make sure that the sum of all probabilities is 1, we have divide the right side by $\sum_j \pi_{j,1} E_j(\tau)$, in which case we obtain equation (10.2).

10.1.2 Using the Remaining Lifetime as Supplementary Variable

We can also use remaining lifetimes to prove equation (10.2). To do this, we define $\pi_{i,j}$ as the probability that $X = i$ and the remaining lifetime $Y = j$. Note that now, each occurrence when $Y = 0$ can be matched with a state change, that is, $\pi_i = \pi_{i,0}$. We assume that the random variable τ cannot exceed the constant M_i when $X = i$. To simplify the notation, we set $p_\tau^{(i)}$ to 0 for $\tau > M_i$.

The equilibrium equations are now

$$\pi_{j,k} = \pi_{j,k+1} + \sum_{i=1}^{N} \pi_{i,0} p_{i,j} p_{k+1}^{(j)}, \quad 0 \le k \le M_j - 2 \qquad (10.8)$$

$$\pi_{j,M_j-1} = \sum_{i=1}^{N} \pi_{i,0} p_{i,j} p_{M_j}^{(j)}. \qquad (10.9)$$

To prove equation (10.8), note that $Y(t + 1) = k$ can be reached either when $Y(t) = k + 1$, or when $Y(t) = 0$. In the case $Y(t) = k + 1$, $Y(t)$ decreases by 1, yielding $Y(t + 1) = k$. In the case where $Y(t) = 0$, a new $Y(t^+)$ is selected, its value has to be $k + 1$ in order to have $Y(t) = k$. In addition to that, $X(t)$ changes from i to j. Equation (10.9) is just (10.8) with $\pi_{j,M_j} = 0$ missing because $Y(t) < M_j$.

Substituting k in (10.8) by $M_j - 2$ and, in addition to that, $\pi_{j,k+1}$ by (10.9), one finds

$$\pi_{j,M_j-2} = \sum_{i=1}^{N} \pi_{i,0} p_{i,j} p_{M_j}^{(j)} + \sum_{i=1}^{N} \pi_{i,0} p_{i,j} p_{M_j-1}^{(j)} = \sum_{i=1}^{N} \pi_{i,0} p_{i,j} \sum_{m=M_j-1}^{M_j} p_m^{(j)}.$$

Continuing this way, one obtains

$$\pi_{j,k} = \sum_{i=1}^{N} \pi_{i,0} p_{i,j} \sum_{m=k+1}^{M_j} p_m^{(j)}. \qquad (10.10)$$

Note that $\sum_{i=1}^{N} \pi_{i,0} p_{i,j}$ is the probability that right after an embedding point, but before determining the remaining lifetimes, the system is in state j. Since π_j is this probability, we find

$$\pi_{j,k} = \pi_j \sum_{m=k+1}^{M_j} p_m^{(j)}.$$

It follows that

$$P_j = \sum_{k=0}^{M_j-1} \pi_{j,k} = \pi_j \sum_{k=0}^{M_j-1} \sum_{m=k+1}^{M_j} p_m^{(j)} = \pi_j E_j(\tau).$$

After insuring that $\sum_{j=1}^{N} P\{X(t) = j\} = 1$, this yields (10.2) as claimed after changing i to j.

If there is no upper bound M_i for τ when $X = i$, then one can take the sum of both sides of (10.8) over k ranging from 0 to infinity, and obtain (10.10) as before. Hence, the result does not depend on the finiteness of M_i.

10.2 Embedding at Changes of the Physical State

In this section, we discuss the Markov chain embedded at the points where the physical state changes. The section has three subsections. The first two subsections deal with the supplementary variables, which can either be remaining lifetimes or ages. In either case, we build use SMPs. The third subsection presents numerical results.

10.2.1 Creating the Supplementary State Space

When remaining lifetimes are used as supplementary variables, the physical state can only change when at least one remaining lifetime is zero, that is, there is at least one k with $Y_k = 0$. These states must be enumerated. To do this, note that the states fall into 2^e groups, depending on which ones of the Y_k are zero, and in each group, the non-zero Y_k form a Cartesian product, with the non-zero Y_k ranging from 1 to $M_k - 1$. The states in each group can be enumerated readily. The enumeration is practically identical when ages are used as supplementary variables.

When using the reachability method as described in Section 3.6, one has to find all states that can be reached from a given state. We only discuss this for the case where the remaining lifetimes are used. If only one $Y_k = 0$ for only one k, then one first has to find $T^{(k)}$ the length of the inter-event time for the event Ev_k, and this time can have any value from 1 to M_k. We thus have M_k different possibilities, and for each possibility, we have to find the target and the corresponding probability. The target of the physical state is obvious: it changes from X to $f_k(X)$. To find the new supplementary states, the states at the next embedding point, we proceed as follows. Let $Y^+ = [Y_1^+, Y_2^+, \ldots, Y_e^+]$ be the vector defined by setting $Y_m^+ = Y_m$, $m \neq k$, and

$Y_k^+ = T^{(k)}$. The next embedding point is reached as soon as the first Y_m reaches 0. This happens d time units from now, where $d = \min_m Y_m^+$. The new supplementary state is then $[Y_m^+ - d, m = 1, 2, \ldots, e]$. In this new vector, at least one Y_ℓ is zero. This calculation has to be done for all possible values of $T^{(k)}$.

If several Y_k are zero, one has to apply all event functions for the events with $Y_k = 0$, and this can be done as shown in Chapter 4. The possible targets are now given by the Cartesian product of the sets $\{1, 2, \ldots, M_k\}$, where the range of k is given by the Y_k that are zero, and for each point in the Cartesian product, Y_m^+, $m = 1, 2, \ldots, e$ must be determined. Once this is done, one finds, as before, $d = \min_m Y_m^+$ and the new supplementary state is given by $[Y_m^+ - d, m = 1, 2, \ldots, e]$. The procedure must be modified when some events are disabled.

10.2.2 The Physical States of Embedded Markov Chains can Form Semi-Markov Processes

For the purpose of this section, we distinguish between two types of states: stable states, that is, those states where the physical state does not change, and change states, that is, states where it does. If we use the points where the physical state changes as embedding points, we essentially eliminate all stable states. If remaining life times are used, these are the states where no Y_k is zero.

Let $\{X(t_n), Y(t_n), n = 1, 2, \ldots\}$ be the Markov chain embedded at the change points. We now define a SMP $\{V(t), t > 0\}$ as follows. If $t_n \leq t < t_{n+1}$, then $V(t) = (X(t_n), Y(t_n))$. This implies that $V(t)$ is a change state. Between t_n and t_{n+1}, $V(t)$ does not correctly represent the supplementary states because it omits all stable states. This is not relevant since we do not need the supplementary states. What is important is that $V(t)$ represents the physical state correctly.

Since the process $\{V(t_n), n > 0\}$ is a SMP, equation (10.2) applies, and to use this equation, we need the transition probabilities $p_{i,j} = P\{V(t_{n+1}) = j \mid V(t_n) = i\}$ and $E_j(\tau)$, which we abbreviate by β_j. Details of this will now be shown, first using remaining lifetimes as supplementary variables, then ages.

10.2.2.1 Remaining Lifetimes Approach

When remaining lifetimes are used, the transition probabilities $p_{i,j}$ of the embedded process can be found as follows. First, consider how $X(t_{n+1})$ is related to $X(t_n)$. By definition, $X(t)$ changes at time t_n to $X(t_n^+)$, but it then remains unchanged until the next event occurs, that is, $X(t_{n+1}) = X(t_n^+)$. The method to find $X(t_n^+)$ is the identical to the one discussed in Chapter 4. The same is true for $Y(t_n)$ and $Y(t_n^+)$. Note that given $Y(t_n)$, there are many possible vectors $Y(t_n^+)$, and the probability of each has to be obtained. How this is done is also discussed in Chapter 4. Given $Y(t_n^+) = [Y_1^+, Y_2^+, \ldots, Y_e^+]$, we can find $Y(t_{n+1})$ as follows. The next embedding point occurs when the first remaining life is zero, that is, we first have to find the minimum of the Y_k^+. Let d be this minimum, hence $d = \tau$. Between t_n and t_{n+1}, all Y_k^+ decrease by τ, that is, $Y(t_{n+1}) = [Y_1^+ - \tau, Y_2^+ - \tau, \ldots, Y_e^+ - \tau]$, and we have found $Y(t_{n+1})$. The vector $Y(t_{n+1})$ can be converted into a state number, which provides the column for a non-zero transition probability to be found. How to accomplish this is best explained by using a simple example.

Suppose we have a system with three events, say an $GI/G/2$ queue with the events arrival, dep1 and dep2. The event dep1 occurs when server 1 has finished service, and dep2 occurs when server 2 has finished service. Let Y_1 be the remaining inter-arrival time Y_2 the remaining service time of server 1, and let Y_3 be the remaining service time of server 2. Also, suppose the times between arrivals are between 1 and 4, with a probability of p_i that the inter-arrival time is i, $i = 1, 2, 3, 4$.

We consider now a specific state and find its targets. To have an example, we use $Y(t_n) = [0, 2, 4]$, that is, at time t_n, an arrival is imminent, while the remaining service times are 2 and 4. The arrival will increase X by 1, that is, $X(t_{n+1}) = X(t_n) + 1$. At this time, a new inter-arrival time must also be selected. If the inter-arrival time is 1, which happens with probability p_1, then $Y^+ = [1, 2, 4]$. The minimum of the entries of Y^+ is 1, and $\tau = 1$ with probability p_1. Also,

$$Y(t_{n+1}) = [1 - \tau, 2 - \tau, 4 - \tau] = [0, 1, 3].$$

More generally, if the inter-arrival time is i with probability p_i, then $Y^+ = [i, 2, 4]$, and for these values, $Y(t_{n+1})$ and the corresponding probabilities can be found in the table below: In this table, column $Y(t_{n+1})$ gives the supplementary state at time t_{n+1},

Table 10.1 Targets and their probabilities for a $GI/G/2$ queue

$Y(t_n)$	$Y(t_n^+)$	$Y(t_{n+1})$	probability	τ
[0,2,4]	[1,2,4]	[0,1,3]	p_1	1
[0,2,4]	[2,2,4]	[0,0,2]	p_2	2.
[0,2,4]	[3,2,4]	[1,0,2]	p_3	2
[0,2,4]	[4,2,4]	[2,0,2]	p_4	2

which together with $X(t_n)$ determines the state number, and hence the column in the transition matrix. The probability to be entered into the transtion matrix is given under the heading "probability". Note that in all cases, the supplementary state at t_{n+1} contains at least one 0. If i is the state number of the present state $[X, 0, 2, 4]$, then $E_i(\tau) = p_1 + 2p_2 + 2p_3 + 2p_4$.

To summarize: The transition matrix of $V(t_n)$ to $V(t_{n+1})$ can be created as follows. For the physical variables, one obtains $X(t_n^+)$, and for $Y(t_n)$, one obtains the different $Y(t_n^+) = [Y_1^+, Y_2^+, \ldots, Y_e^+]$, which is given by the joint distribution of the $T^{(k)}$, where k assumes all values for which $Y_k = 0$. This is described in Chapter 4. Now, $X(t_{n+1}) = X^+(t_n)$, and the values of the different $Y^+(t_{n+1})$ are given by $[Y_1^+ - \tau, Y_2^+ - \tau, \ldots, Y_e^+ - \tau]$, where τ is the minimum of Y_k^+, $k = 1, 2, \ldots, e$. With this, the $p_{i,j}$ of the embedded Markov chain are determined. To find $\beta_j = E_j(\tau)$ when $V(t_n) = j$, one multiplies the values of τ obtained for the different targets with the probabilities of the target and adds the products. In the example of Table 10.1, this yields $p_1 + 2(p_2 + p_3 + p_4)$.

Once all $p_{i,j}$ are found, the equilibrium probabilities $\pi_i = P\{V(t_n) = i\}$ can then be obtained. We now have all we need in order to apply equation (10.2), to find the equilibrium probabilities. The distribution of the $X(t_n)$ is now obtained by taking, for any $X = i$, the sum over all possible values of supplementary states of the process $\{V(t), t > 0\}$.

$$P\{X = i\} = \sum_m P\{X = i, Y = m\}. \tag{10.11}$$

10.2.2.2 The Age Approach

To find the embedding points when using ages as supplementary variables, we again have to find the transition probabilities $p_{i,j}$ of the embedded Markov chain, and we have to determine β_i, the expected times between embedding points. The discussion is greatly simplified by using sets. We define $\mathcal{H}(t)$ as the set of subscripts of the events that occur at time $t_n + t$. For instance, if at time $t_n + t$, the enabled events are Ev_1, Ev_2 and Ev_3, $\mathcal{H}(t)$ can possibly be realized by any of these 8 sets:

$$\{\}, \{1\}, \{2\}, \{3\}, \{1,2\}, \{1,3\}, \{2,3\}, \{1,2,3\}.$$

Here, $\{\}$ is the empty set, also denoted by $\emptyset$, and $\{1, 2, 3\}$ is the set of all enabled events, which we denote by Ω. To find the entries of the transition matrix, we have to find the probabilities for these sets at time t, under the condition that no event has occurred at τ, $0 \leq \tau < t$.

We also define $T^{(k)}$ to be the inter-event times for Ev_k. Let

$$p_i^{(k)} = P\{T^{(k)} = i\}$$

$$S^{(k)}(i) = P\{T^{(k)} \geq i\} = \sum_{j=i}^{\infty} p_j^{(k)}.$$

To determine the transition probabilities of the embedded Markov chain, consider first event Ev_k on its own. If $Y_k(t_n) = i_k$, then certainly $T^{(k)} \geq i_k$. For the probability that Ev_k happens at time $t_n + t$, but not before, we have

$$P\{T^{(k)} = i_k + t \mid T^{(k)} = i_k\} = \frac{P\{T^{(k)} = i_k + t\}}{P\{T^{(k)} \geq i_k\}} = \frac{p_{i_k+t}^{(k)}}{S^{(k)}(i_k)}. \tag{10.12}$$

Similarly, the probability that Ev_k happens after $t_n + t$ equals

$$P\{T^{(k)} > i_k + t \mid T^{(k)} = i_k\} = \frac{S^{(k)}(i_k + t + 1)}{S^{(k)}(i_k)}. \tag{10.13}$$

The probability that no event has occurred before $t_n + t$, and $\mathcal{H}(t)$ is realized at time $t_n + t$ can now be found as the product of the probabilities given by (10.12) for the events in $\mathcal{H}$, and multiplying the result by the product of the probabilities given by (10.13) for the events not in $\mathcal{H}$:

$$P\{\text{no event before } t_n + t\,, \mathcal{H}(t)\} = \prod_{k \in \mathcal{H}(t)} \frac{P_{i_k+t}^{(k)}}{S^{(k)}(i_k)} \prod_{k \notin \mathcal{H}(t)} \frac{S^{(k)}(i_k + t + 1)}{S^{(k)}(i_k)}. \quad (10.14)$$

If $\mathcal{H}(t)$ is not empty, then X changes at time $t_n + t^+$, and $t_{n+1} = t_n + t + 1$. Also

$$Y_k(t_{n+1}) = \begin{cases} Y_k(t_n) + t + 1 & \text{if } k \notin \mathcal{H}(t) \\ 1 & \text{if } k \in \mathcal{H}(t). \end{cases}$$

If $f_{\mathcal{H}(t)}(X)$ is the composition of the functions indexed by $\mathcal{H}(t)$, then

$$X(t_{n+1}) = f_{\mathcal{H}(t)}(X(t_n)).$$

With this, the targets of the embedded Markov chain, as well as their probabilities, are fully determined, and the equilibrium probabilities of $X(t_n) \frown Y(t_n)$ can be calculated.

In order to find the equilibrium distribution of the original process, we need $\beta_i = E_i(\tau)$. Clearly, $T > t$ if no event happens in the interval $[t_n, t_n + t]$, and the probability of this is equal to

$$\prod_{k \in \Omega} \frac{S^{(k)}(i_k + t + 1)}{S^{(k)}(i_k)}.$$

Consequently, if the state vector $Y(t_n)$ is equal to $i = [i_k, k \in \Omega]$, one has

$$E(T) = \sum_{t=0}^{\infty} \prod_{k \in \Omega} \frac{S^{(k)}(i_k + t + 1)}{S^{(k)}(i_k)}. \quad (10.15)$$

After applying equation (10.2) to find the random-time probabilities for the states that can occur at t_n, we obtain the marginal distributions for the physical state vector X, as stated by equation (10.11).

If $T^{(k)}$ is bounded by M_k, then $S^{(k)}(i) = 0$ for $i > M_k$. This will limit the lengths of the intervals between event points. Therefore, the product given by (10.14) is zero as soon as any $S^{(k)}(i_k + t) = 0$. It follows that all products with $i_k + t > M_k$ vanish, and for $t \geq \min_{k \in \Omega}(M_k - i_k)$, one therefore does not need to evaluate this product. Also, the upper bound in equation (10.15) can be set to $\min_{k \in \Omega}(M_k - i_k)$.

10.2.3 Numerical Experiments

We wrote a program to find the equilibrium solutions for a tandem queue with two stations, as well as for the $GI/G/2$ queue with homogeneous servers. Several inter-arrival times and service times were used. The largest problem involved a tandem queue with inter-arrival times uniformly distributed between 1 and 10, and both service times uniformly distributed between 1 and 8. The first line was restricted to 10, and the second one to 4. This problem has 10,945 states and 133,672 non-zero entries in the transition matrix. A simple calculation shows that only 0.112% of the $(10945)^2$ entries of the transition matrix are non-zero, that is, the transition matrix is very sparse. On the ThinkPad with an INTEL COREi5vPro CPU, it took 0.75 seconds to find the transition matrix, a time that is negligible. We then found the equilibrium solutions by using the method of Gauss–Seidel, which took only 1.66 seconds, again a negligible amount of time. The largest $GI/G/2$ problem that was solved involved uniformly distributed inter-arrival between 1 and 5 and uniformly distributed service times between 1 and 10, with a buffer size of 20. This problem has 3696 states and 48335 transitions. It took 0.2734 seconds to find the transition matrix, and 1.156 seconds to find the equilibrium distribution.

The details of our implementation of Gauss–Seidel were as follows: we gave all probabilities the same initial value, and then we iterated. After each iteration, the probabilities were scaled such that their sum was 1. Every 10 iterations, we checked for convergence. Since we were not interested in the Y-values, we based the convergence tests strictly on the distribution of the X-values. We continued iterating until none of the probabilities in question deviated by more than 0.0001 from the one 10 iterations earlier. The tandem queue example required 160 iterations, and the GI/G/2 example required 300 iterations.

10.3 Embedding at Specific Event Types

In this section, we will cover embedded Markov chains where a single event Ev_k is selected, and the points used for embedding are the times when Ev_k occurs. We will refer to Ev_k as *embedding event*. For instance, one can embed the process at the times when a departure occurs. As embedding points, one can either use t_n, the time immediately before the embedding event occurs, or t_n^+, the time immediately after that. If $\{X(t), t > 0\}$ is the original process, then the embedded process is $\{X(t_n), n > 0\}$ if we choose the times immediately before the embedding events, and it is $\{X(t_n^+), n > 0\}$ if we choose the times immediately after the embedding events. It turns out to be convenient to use t_n^+ rather than t_n, that is, we consider the process $\{X(t_n^+), n = 1, 2, \ldots\}$. If k is the embedding event, then $X(t_n^+) = f_k(X(t_n))$. For our future discussion, we define $g(X) = f_k(X)$. Note that the state space of $X(t_n^+)$ may be different from the one of $X(t)$, which we assume to be the set of integers from 0 to N. For instance, if $g(X) = \max(0, X - 1)$, the state space of $X(t_n^+)$ includes the integers from 0 to $N - 1$ instead of the integers from 0 to N.

To find the transition probabilities $p_{i,j}$ of the process $X(t_n^+)$, we consider the process $\{Z(s), 0 \leq s \leq t_{n+1} - t_n^+\}$, the process between embedding points. Hence, let $Z(s)$ be the state of the system s time units after t_n, but before t_{n+1}. We call the process *inter-embedding process*. By definition, $Z(0) = X(t_n^+)$. Moreover, notice that the inter-embedding process cannot contain embedding events. For instance, if, in a $GI/G/1$ queue, the embedding events are the arrivals, $\{Z(s), 0 \leq s \leq t_{n+1} - t_n^+\}$ cannot contain arrivals. The embedding process is rather crucial for our discussions.

We restrict our attention to the case where the process creating embedding points are renewal processes, with $F(t)$ being the cumulative distribution of the times between renewals. The case where the renewal process can be disabled requires special treatment, and we therefore will first discuss the case where the renewal process cannot be disabled.

The inter-embedding process is assumed to be a CTMC. However, other setups have been used in literature. For instance, Yang and Chaudhry [96] used embedding to analyze the discrete-time $GI/G/1$ queue, where the inter-embedding process is a DTMC.

In this section, we will first present the main formulas used, and we then present an example. This is followed by the discussion of the $M/G/1/N$ queue and the $GI/M/1/N$ queue.

10.3.1 The Main Formulas

We assume there is a discrete event system $\{X(t), t > 0\}$ as follows. All event types in this process are Markovian, usually Poisson, except that one event type is generated by a renewal process. This event is used as an embedding event. It follows that the inter-embedding process $\{Z(s), 0 \leq s \leq t_{n+1} - t_n\}$ is a CTMC. Since the process $\{X(t), t > 0\}$ is not Markovian with a countable state space, its direct analysis is difficult. It is easier to first find the equilibrium probabilities $\pi_i = P\{X(t_n^+) = i\}$ and use them to find the random-time probabilities $P_i = P\{X(t) = i\}$.

To determine π_i, we need the transition probabilities $p_{i,j} = P\{X(t_{n+1}^+) = j \mid X(t_n^+) = i\}$. To obtain these, we consider the transient probabilities of $Z(s)$ and define

$$\hat{p}_{i,j}(s) = P\{Z(s) = j \mid Z(0) = i\}.$$

Note that

$$P\{X(t_{n+1}) = j \mid X(t_n^+) = i \text{ and } t_{n+1} - t_n = t\} = \hat{p}_{i,j}(t).$$

We uncondition to arrive at

$$\hat{p}_{i,j} = P\{X(t_{n+1} = j \mid X(t_n^+) = i\} = \int_0^\infty \hat{p}_{i,j}(t)dF(t).$$

If $X(t_{n+1}) = k$, then $X(t_{n+1}^+) = g(k)$, which implies

$$p_{i,j} = \sum_{k:g(k)=j} \hat{p}_{i,k}. \tag{10.16}$$

With this, the transition probabilities $p_{i,j}$ of the embedded process are determined.

As stated above, $\{Z(s),\ s > 0\}$ is typically a CTMC, and the randomization method can be used to find $\hat{p}_{i,j}(t)$. Let Q be the transition matrix guiding $Z(s)$, and let $\bar{P} = [\bar{p}_{i,j}] = Q/q + I$. We calculate $P^n = [\bar{p}_{i,j}^n]$ to obtain

$$\int_0^\infty \hat{p}_{i,j}(t)dF(t) = \int_0^\infty \sum_{n=0}^\infty \bar{p}_{i,j}^n p(n; qt)dF(t).$$

We can now interchange integration and summation to find

$$\int_0^\infty \hat{p}_{i,j}(t)dF(t) = \sum_{n=0}^\infty \bar{p}_{i,j}^n \int_0^\infty p(n; qt)dF(t).$$

Hence, we first determine

$$v_n = \int_0^\infty p(n; qt)dF(t) \tag{10.17}$$

and then we calculate

$$\hat{p}_{i,j} = \sum_{n=0}^\infty \bar{p}_{i,j}^n v_n.$$

In this way, the integration need only be done once rather than for all $\hat{p}_{i,j}$. For many distributions $F(t)$, v_n can be obtained without integrations as shown in [26] and [34].

Once we have all $\hat{p}_{i,j}$, we can determine $p_{i,j}$ from equation (10.16). The $\pi_j = P\{X(t_n) = j\}$ are now the solution of the equilibrium equations

$$\pi_j = \sum_i \pi_i p_{i,j}.$$

This completes the calculation of all π_j.

The π_j are important for their own sake. They are used, for instance, for calculating waiting times. In particular, if the embedding events are arrivals, then the initial distribution of the waiting-time process is given by the π_j. For details, see Section 7.6.

In other cases, the $P_j = P\{X(t) = j\}$ are needed. We have

$$P_j = d \sum_i \pi_i \alpha_{i,j}. \tag{10.18}$$

Here, $\alpha_{i,j}$ is the expected time the inter-embedding process $\{Z(s),\ 0 \le s \le t_{n+1} - t_n^+\}$ spends in state j under the condition that $Z(0) = i$, and d is the reciprocal of $\sum_i P_i$, which ensures that the sum of all probabilities is 1.

We now provide an intuitive proof of equation (10.18). We look at the time interval from 0 to T, and we eventually let T go to infinity. We define $m(T)$ to be the number of events within this interval, and of these events, $m_i(T)$ happen when $X(t_n^+) = i$. Now, as $T \to \infty$, $\frac{m_i(T)}{m(T)}$ should approach π_i, and $\frac{m(T)}{T}$ approaches $E^{-1}(\tau)$, where $\tau = t_{n+1} - t_n$. Finally, let $S_{i,j,n}(T)$ be the actual time spend in state j when the inter-embedding process starts with $Z(0) = i$ for the nth time. Now, the average of the $S_{i,j,n}(T)$ should converge toward the expectation $\alpha_{i,j}$:

$$\lim_{T \to \infty} \frac{1}{m_i(T)} \sum_{n=1}^{m_i(T)} S_{i,j,n}(T) = \alpha_{i,j}.$$

Finally, P_j should be one over T multiplied by the limit of the total time the process spends in state j during the interval from 0 to T. The total time spent in state j is given by $\sum_i \sum_{n=1}^{m_i(t)} S_{i,j,n}(T)$, and we have

$$\frac{1}{T} \sum_i \sum_{n=1}^{m_i(T)} S_{i,j,n}(T) = \sum_i \frac{1}{m_i(T)} \left(\sum_{n=1}^{m_i(T)} S_{i,j,n}(T) \right) \frac{m_i(T)}{m(T)} \frac{m(T)}{T}.$$

Now, $\frac{1}{m_i(T)} \sum_{n=1}^{m_i(T)} S_{i,j,n}(T)$ should approach $\alpha_{i,j}$, the term $\frac{m_i(t)}{m(T)}$ should approach π_i, and $\frac{m(T)}{T}$ will approach $1/E(\tau)$. With this, equation (10.18) is proven, under the condition that all three fractions converge as stated. We will not go into details as to when this will happen. We merely mention that we need to assume that $\tau = t_{n+1} - t_n$ has a finite mean and variance. The method of letting averages converge to expectations, and proportions to probabilities is quite useful in many contexts. We used it already to prove Little's Law.

Given the inter-embedding time $\tau = t$, $\alpha_{i,j}$, the time $Z(t)$ spends in j, can be found using equation (7.8) of Chapter 7:

$$: \alpha_{i,j}(t) = \int_0^t \hat{F}_{i,j}(x)dx$$

$$= \sum_{n=0}^{\infty} \bar{p}_{i,j}^{(n)} \int_0^t p(n; qx)dx.$$

We uncondition this expression to obtain

$$\alpha_{i,j} = \int_0^\infty \alpha_{i,j}(t)dF(t)$$

$$= \sum_{n=0}^{\infty} \bar{p}_{i,j}^{(n)} \int_0^\infty \int_0^t p(n; qx)dx \, dF(t). \tag{10.19}$$

We use (7.9) of Chapter 7 to find the integral of the Poisson distribution and obtain

$$\int_0^\infty \int_0^t p(n; qx)dx \, dF(t) = \int_0^\infty \frac{1}{q} \sum_{m=n+1}^{\infty} p(m; qt)dF(t).$$

After interchanging the sum and the integral, and using the definition of v_n, this yields

$$\int_0^\infty \int_0^t p(n; qx)dx \, dF(t) = \frac{1}{q} \sum_{m=n+1}^\infty v_m.$$

We thus have

$$\alpha_{i,j} = \frac{1}{q} \sum_{n=0}^\infty \bar{p}_{i,j}^{(n)} \sum_{m=n+1}^\infty v_m, \tag{10.20}$$

and conclude

$$P_j = d \sum_i \pi_i \sum_{n=0}^\infty \bar{p}_{i,j}^{(n)} \sum_{m=n+1}^\infty v_m.$$

As before, d must be determined such that the sum of all $P_j = 1$. Note that q does not depend on i, j, or m and can therefore be ignored.

We have not yet discussed the case where the embedding event can be disabled, as is the case when using departures as embedding events: Departures are disabled when the system is empty. For the moment, we only address situations where the embedding event can only be disabled right after its occurrence, that is, the embedding event can only be disabled at time t_n^+. In this case, t_{n+1} is delayed until some other event, say an arrival, enables the event again. For instance, in the case of departures, the embedding event is disabled in state $i = 0$. In order to find $\hat{p}_{0,j}$, we have to find the state the process $\{Z(t)\}$ enters once the embedding event is enabled again. If this state is k, then we set $\hat{p}_{0,j} = \hat{p}_{k,j}$. Moreover, we have to find the expected time it takes to enable the embedding event again, which is $\alpha_{0,0}$. In this situation, $\alpha_{0,j} = \alpha_{k,j}$ for $j > 0$. Details of how to proceed are provided when we discuss our second example, which is a $M/G/1$ queue with varying arrival rates.

10.3.2 An Example where the Embedding Event is Never Disabled

We now present an example where the embedding event is never disabled. There is an inventory, and arrivals to the inventory are Poisson with rate λ. The size of the inventory is limited to N. Of interest is the number of items in stock at time t, which we denote by $X(t)$. At random times with distribution $F(t)$, a single part is taken from the inventory, and we refer to this event as "demand". If the inventory is empty during a demand, there is nothing to take, but the process generating demands is not interrupted, that is, the demand event is never disabled. Since the times between demands follow a general distribution, demands are used as embedding events.

In this example, the process $\{Z(t), \, t_n^+ \le t \le t_{n+1}\}$, is particularly simple: It is a Poisson process, except when the inventory level reaches N. Since the process starts in i,

$$\hat{p}_{i,j}(t) = p(j - i; \lambda t), \ 0 \le i \le j < N, \tag{10.21}$$

$$\hat{p}_{i,N}(t) = \sum_{n=N-i}^{\infty} p(n, \lambda t), \ 0 \le i \le N. \tag{10.22}$$

Therefore

$$\hat{p}_{i,j} = \int_0^\infty \hat{p}_{i,j}(t)dF(t) = \int_0^\infty p(j - i; \lambda t)dF(t) = v_{j-i}, \ 0 \le i \le j < N,$$

$$\hat{p}_{i,N} = \int_0^\infty \hat{p}_{i,N}(t)dF(t) = \int_0^\infty \sum_{n=N-i}^{\infty} p(n; \lambda t)dF(t) = \sum_{n=N-i}^{\infty} v_n, \ i > 0.$$

The transition probabilities of the embedded Markov chain $p_{i,j} = P\{X(t_{n+1}^+) = j \mid X(t_n^+) = i\}$ are found from $\hat{p}_{i,k} = P\{X(t_{n+1}) = k \mid X(t_n^+) = i\}$ by using $j = g(k) = \min(0, k - 1)$. Thus, for $j \ne 0$, $j = k - 1$. However, $j = 0$ is realized if either $k = 0$ or $k = 1$.

$$p_{0,0} = \hat{p}_{0,1} + \hat{p}_{0,0} = v_1 + v_0$$

$$p_{i,j} = \hat{p}_{i,j+1} = v_{j+1-i}, \ 0 \le i \le j + 1 \le N$$

$$p_{i,N-1} = \hat{p}_{i,N} = \sum_{v=N-i}^{\infty} v_v, \ 0 \le i. \tag{10.23}$$

The equilibrium equations can now be formulated and solved for the equilibrium probabilities π_j, $j = 1, 2, \ldots, N - 1$. Note that j ranges only to $N - 1$ rather than N because if there is a maximum of N elements in the system, then after a departure there are at most $N - 1$ left.

Equation (10.23) can also found directly as follows: If A is the number of arrivals between consecutive demands, we have as long as $0 \le X(t_{n+1}^+) \le N$:

$$X(t_{n+1}^+) = X(t_n^+) + A - 1,$$

and consequently, for $X(t_{n+1}) = j < N - 1$,

$$P\{X(t_{n+1}^+) = j \mid X(t_n^+) = i\} = p_{i,j} = v_{j-i+1}. \tag{10.24}$$

The cases of $j = 0$ and $j = N - 1$ can be found directly in a similar fashion.

To find the P_j we still need the $\alpha_{i,j} = \int_0^\infty \alpha_{i,j}(t)dF(t)$, where $\alpha_{i,j}(t) = \int_0^t \hat{p}_{i,j}(x)dx$. From equations (10.21) and (10.22), we find

$$\alpha_{i,j}(t) = \int_0^t p(j - i; \lambda x)dx = \frac{1}{\lambda} \sum_{n=j-i+1}^{\infty} p(n; \lambda t), \ 0 \le i \le j < N,$$

$$\alpha_{i,N}(t) = \int_0^t \sum_{n=N-i}^{\infty} p(n, \lambda x)dx = \frac{1}{\lambda} \sum_{n=N-i}^{\infty} \sum_{m=n+1}^{\infty} p(m, \lambda t), \ 0 \le i \le N.$$

When using $\alpha_{i,j} = \int_0^\infty \alpha_{i,j}(t)dF(t)$, this yields

$$\alpha_{i,j} = \frac{1}{\lambda} \sum_{n=j-i+1}^{\infty} \int_0^\infty p(n; \lambda t)dF(t) = \frac{1}{\lambda} \sum_{n=j-i+1}^{\infty} v_n, \ 0 \leq i \leq j < N,$$

$$\alpha_{i,N} = \frac{1}{\lambda} \sum_{n=N-i}^{\infty} \sum_{m=n+1}^{\infty} \int_0^\infty p(m, \lambda t)dF(t) = \frac{1}{\lambda} \sum_{n=N-i}^{\infty} \sum_{m=n+1}^{\infty} v_m, \ 0 \leq i \leq N.$$

The P_j can now be found from equation (10.18), and we are done.

10.3.3 An Example where the Embedding Event can be Disabled

To demonstrate how to proceed when the embedding event can be disabled, we consider the $M/D/1/N$ queue with varying arrival rates. For simplicity, we keep the example small, but larger examples can be solved in the same way. The embedding event is again the departure, and the service time is 1. Let X be the number in the system. When $X = 0$, the arrival rate is 1, when $X = 1$, the arrival rate is 0.5, and when $X = 2$, the arrival rate is 0.3. There are no arrivals for $X > 3$, which implies $X \leq 3$. Hence, the transition matrix Q of the inter-embedding process is as follows:

$$Q = \begin{bmatrix} -1 & 1 & 0 & 0 \\ 0 & -0.5 & 0.5 & 0 \\ 0 & 0 & -0.3 & 0.3 \\ 0 & 0 & 0 & 0 \end{bmatrix}.$$

We choose q to be 1, such that $\bar{P} = Q/q + I$ becomes

$$\bar{P} = \begin{bmatrix} 0 & 1 & 0 & 0 \\ 0 & 0.5 & 0.5 & 0 \\ 0 & 0 & 0.7 & 0.3 \end{bmatrix}.$$

In this matrix, we omitted the row of $X(t_n^+) = 3$ since $X(t_n^+) \leq 2$: This follows from $X(t_n) \leq 3$, and the fact that a departure is imminent. $\bar{P}^n$ can now be calculated readily. Also note that $v_n = \frac{1}{q}p(n; qt)$, with $q = 1$. The value of t is determined by the service time, which is 1. We also need $V_n = \sum_{m=n+1}^{\infty} v_m$, which we calculated, using $V_0 = 1 - p(0; 1)$, $V_{n+1} = V_n - p(n; 1)$, $n > 0$. We truncated the Poisson distribution at $n = 7$, where $\sum_{n=0}^{7} p(n; 1) = 0.99999$. Using these values, $\hat{P}$ evaluates to

$$\sum_{n=0}^{7} \bar{P}^n v_n = \hat{P} = \begin{bmatrix} 0.3679 & 0.4773 & 0.1387 & 0.0161 \\ 0.0000 & 0.6065 & 0.3357 & 0.0577 \\ 0.0000 & 0.0000 & 0.7408 & 0.2592 \end{bmatrix}.$$

We now have to derive $P = [p_{i,j}]$, where $p_{i,j} = P\{X(t_{n+1}^+) = j \mid X(t_n^+) = i\}$. The basis of this calculation is the matrix $\hat{P}$ above, with the entries $\hat{p}_{i,j} = P\{X(t_{n+1}) = j \mid X(t_n^+) = i\}$. Unless $X(t_n^+) = 0$, $X(t_{n+1}^+) = X(t_n) - 1$, that is, the columns of the matrix $\bar{P}$ move one step to the left. However, when $X(t_n^+) = 0$, then $X(t_{n+1}^+) = 0$ if either $X(t_{n+1}) = 0$ or 1, that is, $p_{0,0} = \hat{p}_{0,0} + \hat{p}_{0,1}$. We thus obtain the transition matrix:

$$P = \begin{bmatrix} 0.8452 & 0.1387 & 0.0161 \\ 0.6065 & 0.3357 & 0.0577 \\ 0.0000 & 0.7408 & 0.2592 \end{bmatrix}.$$

We can now formulate and solve the equilibrium equations for π_i, $i = 0, 1, 2$. The unnormalized results are

$$\pi_1 = 0.2552\pi_0, \quad \pi_2 = 0.04161\pi_0.$$

Next, consider equation (10.18) which becomes, if d is a constant to be determined such that $\sum_{i=0}^{3} P_i = 1$:

$$dP_0 = \pi_0\alpha_{0,0} \tag{10.25}$$

$$dP_1 = \pi_0\alpha_{0,1} + \pi_1\alpha_{1,1} \tag{10.26}$$

$$dP_2 = \pi_0\alpha_{0,2} + \pi_1\alpha_{1,2} + \pi_2\alpha_{2,2} \tag{10.27}$$

$$dP_3 = \pi_0\alpha_{0,3} + \pi_1\alpha_{1,3} + \pi_2\alpha_{2,3}. \tag{10.28}$$

Here, the $\alpha_{i,j}$ are the expected times during which $Z(t) = X(t_n^+ + t) = j$, given $Z(0) = i$. The process lasts 1 time unit if $i > 0$, but it lasts $1/\lambda_0 + 1$ time units if $i = 0$: Here, $1/\lambda_0$ is the expected time during which departures are disabled. Departures are enabled again at the point of the next arrival, and at this point, $Z(t) = 1$, and $\alpha_{1,j}$ must be used. The $\alpha_{i,j}$, $i = 1, 2$, can be found from equation (10.20), that is, $[\alpha_{i,j}] = \sum_{n=0}^{\infty} \bar{P}^n V_n$. We have

$$[\alpha_{i,j}, \ i = 1, 2, \ j = 1, 2, 3] = \begin{bmatrix} 0.7869 & 0.1925 & 0.0206 \\ 0.0000 & 0.8639 & 0.1361 \end{bmatrix}.$$

Substituting these values for $i > 0$, as well as $\alpha_{0,0} = 1/\lambda_0 = 1$, $\alpha_{0,j} = \alpha_{1,j}$ into equations (10.25) to (10.28) yields, after some calculation:

$$dP_0 = \pi_0 \times 1 = \pi_0$$

$$dP_1 = 0.9877\pi_0$$

$$dP_2 = = 0.2776\pi_0$$

$$dP_3 = = 0.0.0315\pi_0.$$

We now determine π_0/d such that $P_0 + P_1 + P_2 + P_3 = 1$, which yields $P_0 = 0.4354$, $P_1 = 0.4300$, $P_2 = 0.1209$ and $P_3 = 0.0137$.

10.3.4 The Embedded Markov Chains of $M/G/1/N$ and $GI/M/1/N$ Queues

Embedding is typically used to analyze the $GI/M/1$ and the $M/G/1$ queues. In both cases, X represents the number in the system. We first discuss the $M/G/1/N$ queue, then the $GI/M/1/N$. In the $M/G/1/N$ queue, the embedding events are departures, as in our two previous examples, and in the $GI/M/1/N$ queue, they are arrivals. In both models, we use λ for the arrival rate and μ for the service rate. In the $M/G/1/N$ queue, $F(t)$ is the cumulative service time distribution, and the probability of m arrivals during a service time given as

$$v_m = \int_0^\infty p(m; \lambda t) dF(t).$$

On the other hand, in the $GI/M/1/N$ queue, $F(t)$ is used for the cumulative inter-arrival time distribution, and v_m is defined as the number of service completion between two arrivals:

$$v_m = \int_0^\infty p(m; \mu t) dF(t).$$

In literature, the connection between the distributions of $X(t_n)$ and $X(t)$ is derived by a method that is tailor-made for $M/G/1$ and $GI/M/1$ queues. The method is based on the PASTA principle, as well as the connection between up and down moves. Details will follow in the next section.

If A is the number of arrivals between departures in an $M/G/1/N$ queue, we have for $X(t_{n+1}^+) + A < N$ and $X(t_n^+) > 0$:

$$X(t_{n+1}^+) = X(t_n^+) + A - 1. \tag{10.29}$$

We use this equation to find the transition probability $p_{i,j}$ of the embedded Markov chain, where

$$p_{i,j} = P\{X(t_{n+1}^+) = j \mid X(t_n^+) = i\}.$$

Note that $p_{i,j} = 0$ when $j < i-1$. Otherwise, one argues as follows. When $X(t_{n+1}^+) = j$ and $X(t_n^+) = i$, equation (10.29) becomes $j = i + A - 1$. Consequently, if $A = j - i + 1$, we make a transition from i to j, that is

$$p_{i,j} = v_{j-i+1}, \ i > 0, j < N.$$

When $X(t_n^+) = i = 0$, then j is equal to the number of arrivals after the first arrival following t_n^+, because only then, departures are possible. It follows that $p_{0,j} = p_{1,j}$, or

$$p_{0,j} = v_j.$$

Since the number in the system is restricted to N, $X(t_{n+1}) \leq N$, which implies that $X(t_{n+1}^+) \leq N - 1$. Any arrivals bringing $i + A$ above N are lost, that is, if $X(t_{n+1}) = i + A \geq N$, $X(t_{n+1}^+) = j = N - 1$. This argument implies

$$p_{i,N-1} = \sum_{m=N-i}^{\infty} v_m.$$

With this, all transition probabilities are determined, and the equilibrium probabilities π_i can now be calculated.

To find the π_i from the equilibrium equations, the state elimination algorithm is particularly efficient because the equation

$$p_{i,j}^{(n-1)} = p_{i,j}^{(n)} + \frac{p_{i,n}^{(n)} p_{n,j}^{(n)}}{\sum_{k=1}^{n-1} p_{n,k}^{(n)}}$$

simplifies considerably when $p_{i,j} = 0$ for $j < i - 1$. Indeed, since $i, j < n$, one has

$$p_{i,j}^{(n-1)} = \begin{cases} p_{i,j}^{(n)}, & j < n-1 \\ p_{i,n-1}^{(n)} + p_{i,n}^{(n)}, & j = n-1. \end{cases}$$

We call the systems given by the $p_{i,j}^{(n)}$ censored systems. As it happens, we obtain the same transition probabilities if we reduce the bound N to n. If we call the system with N replaced by n reduced bound system, then our claim is that the transition matrix of censored system is identical to the one of the reduced bound system. This is easily verified for $p_{i,j}$ with $i, j < n - 1$, that is, for all entries except the ones in the last column. Since the sum of the transition probabilities in each row must be 1, the last column must also be the same.

We now discuss the $GI/M/1/N$ queue. If D is the number of departures between arrivals, we have for $X(t_{n+1}^+) \geq 1$ and $X(t_n^+) < N$:

$$X(t_{n+1}^+) = X(t_n^+) - D + 1.$$

With $D \geq 0$, this equation implies $X(t_{n+1}^-) \leq X(t_n^+) + 1$. Moreover, $D = i - j + 1$ when $X(t_n^+) = i$ and $X(t_{n+1}^+) = j$:, and, as a consequence

$$p_{i,j} = P\{X(t_{n+1}^+) = j \mid X(t_n^+) = i\} = P\{D = i-j+1\} = v_{i-j+1}, \ 0 < j < i+1, \ i < N.$$

This equation also applies when $i = N$, except that

$$p_{N,N} = v_0 + v_1.$$

This is true, because when $X(t_n^+) = N$, we have $X(t_{n+1}) = N$ when there is either 0 or 1 departure. If there is no departure, then $X(t_{n+1})$ remains at N, and if there is one, then $X(t_{n+1}) = N - 1$, and $X(t_{n+1}^+) = X(t_{n+1}) + 1 = N$ as well.

If $D \geq i + 1$, then $j = 0$, that is

$$p_{i,0} = \sum_{m=i+1}^{\infty} v_m.$$

The equilibrium equations for the embedded Markov chain can now be formulated readily, and its equilibrium probabilities can be obtained. While executing the state elimination, it is better to eliminate π_N first, then π_{N-1}, and so on. In this way, the censored system looks just like the original system with N is reduced.

10.3.5 Finding Random Time Distributions from Embedding Point Distributions

To determine the random-time distributions from the embedding point distributions, one uses two principles. One is the PASTA principle, that is, Poisson Arrivals See Time Averages, a principle that was discussed in Section 7.6. The second principle is that in processes $\{X(t), t > 0\}$ where X can change by at most 1, the number of times that $X(t)$ goes from i to $i + 1$ can differ by at most 1 from the number of times $X(t)$ goes from $i + 1$ to i.

If $X(t)$ can only change by at most 1 at a time, then any move from $X = i$ to $X = i + 1$ must be matched by a move from $X = i + 1$ to $X = i$ when state i is visited the next time. Since all states are recurrent, they are visited infinitely often, and except for the last time that X moves from i to $i + 1$, all these moves can be matched by a move from $i + 1$ to i, that is, the moves from i to $i + 1$ can differ from the moves from $i + 1$ to i by at most 1 as claimed. More generally, the moves up from any state, minus the moves down, during the interval from 0 to T, is equal to $X(T) - X(0)$. Now let u_i be the probability that among the moves up, the move up starts at i, that is, u_i can be approximated by the proportion of moves from i to $i + 1$ among all the moves up. Similarly, let d_{i+1} be the proportion of moves from $i + 1$ to i among all moves down. We now have

$$u_i = d_{i+1}.$$

To see this, note that the total number of moves up from 0 to T minus the total number of moves down must be equal to the increase of $X(t)$ from 0 to T, and as $T \to \infty$, this difference can be neglected. Hence,

$$u_i = \frac{\#\text{going from } i \text{ to } (i + 1)}{\#\text{going up}} \approx \frac{\#\text{ going from } i + 1 \text{ to } i}{\#\text{going down}}.$$

Translated to queues, this means that u_i is the probability that just before an arrival that joins the queue, the number in the system is i, whereas d_{i+1} is the probability that just before a departure, the number in the system is $i + 1$. In the $M/G/1$ queue, π_i is the probability that the number in the system just after a departure is i, or $i + 1$ just before the departure which means that $\pi_i = d_{i+1}$, and then $d_{i+1} = u_i = \pi_i$. If there is no upper bound N, every arrival is a move from some i to $i + 1$, implying that u_i is the probability of i elements in line just before an arrival. Since arrivals see time averages, $u_i = P_i$, and consequently, $\pi_i = P_i$. If, however, there is an upper bound N, then the PASTA principle can only be used under the condition that $i < N$. In fact, $\pi_N = 0$ because this would imply that X just before a departure could be

$N + 1$. It follows that

$$\pi_i = P\{X = i \mid i < N\} = P_i/(1 - P_N). \tag{10.30}$$

Here, $1 - P_N$ can be found by using the fact that in the long run, the number of arrivals that join must be equal to the number of departures, that is

$$\lambda(1 - P_N) = \mu(1 - \pi_0).$$

Solving for $1 - P_N$ and substituting the resulting expression into (10.30) yields

$$P_i = \pi_i \frac{\mu(1 - \pi_0)}{\lambda}, \ i < N,$$

and

$$P_N = 1 - \frac{\mu}{\lambda}(1 - \pi_0).$$

Consider now the $GI/M/1/N$ queue. There, t_n is the time of the nth departure, and τ_ν is the time of the νth arrival, and we have

$$\pi_i = P\{X(t_n^+) = i\} = P\{X(t) = i \mid i > 0\} = P_i/(1 - P_0).$$

We now find $1 - P_0$ by exploiting the fact that departures must match the arrivals that join, which yields

$$(1 - P_0)\mu = \lambda(1 - \pi_N),$$

and

$$P_i = \pi_i(1 - P_0) = \pi_i \frac{\lambda(1 - \pi_N)}{\mu}, \ i > 0. \tag{10.31}$$

Moreover

$$P_0 = 1 - \frac{\lambda}{\mu}(1 - \pi_N).$$

With this, the connection between the random-time probabilities and the embedding point probabilities is established.

Problems

10.1 To reduce the number of states, discrete-time systems can be embedded at the time where the state changes. Write a matrix generator for the discrete-time $GI/G/1/N$ queue embedded at the times where the state changes. You can assume that the inter-arrival times are integers between 1 and h, and that the service times are integers between 1 and g. Assume that N, h, and g are input, and so are the inter-arrival and service time distributions.

10.2 Let X be defined as follows: When $X > 0$, then X is the number of units on stock in an inventory, and when $X < 0$, than X is the number of backorders. Demands, respectively backorders, are Poisson with a demand rate of λ. Once X reaches s, an order for Q units goes out, which arrives after a time T, where T has distribution $F_T(t)$. Once X reaches $-d$, where d is given, no backorders are accepted anymore. Formulate the Markov chain, with two embedding points: When an order goes out, and when an order arrives. Assume that when an order is outstanding, no other order goes out. Derive a formula for the average stock level for this system, and the average number of backorders.

10.3 Modify the formulas for v_n as given by equation (10.17) such that the integral is removed when $F(t)$ is an Erlang-k distribution. Do the same when $F(t)$ is a hyperexponential distribution.

10.4 Can you use the shortcut given by equation (10.31) in the $GI^X/M/1/N$ queue where arrivals are in groups of 1 or more. Justify your answer. Which principle is violated?

10.5 Given the arrival rate λ, the distribution of service time $F(t)$, and the buffer size N, write a program to find the expected number in a $M/G/1$ queue.

Chapter 11
Systems with Independent or Almost Independent Components

The complexity of a system increases exponentially with the number of state variables, and for this reason, it is advantageous to divide the system into several subsystems which can be dealt with separately. For instance, where a system includes d state variables $X_1, X_2, \ldots, X_d$, we can sometimes form two subsystems, with the first subsystem containing the first d_1 state variables, and the second subsystem containing the remaining state variables, such that the $X_i(t)$ placed into the first subsystem are independent of the state variables placed into the second subsystem. In a similar way, some systems can be divided into more than two independent subsystems.

We say that the subsystems are *fully independent* if for any initial condition, and for any t, state variables belonging to different subsystems are independent. If, however, state variables belonging to different subsystems are independent only as $t \to \infty$, then we talk about *steady-state independence*. Both types of independence will be discussed in this chapter. The steady-state independence is largely restricted to so-called *Jackson networks*, which will be analyzed in detail.

In a system divided into two subsystems, let $\pi_i^{(k)}$ be the probability that system k, $k = 1, 2$, is in state i. The states of the combined system are then given by the pairs i, j, where i is the state of system 1, and j the state of system 2. The probability of this state is $\pi_i^{(1)} \pi_j^{(2)}$. This type of solution is generally referred to as *product-form solutions*. Product-form solutions obviously also exist when dividing a system into more than two subsystems. The name is mainly used for systems that are steady-state independent, but we will use it also for fully independent subsystems.

Often, the dependence between subsystems is only weak, and in this case, one can, as a first approximation, treat them as independent. Possibly, this approximation can be improved in later steps. Thus, after dividing the system into several subsystems, and analyzing each subsystem separately, one has to combine them again in order to make statements about the system as a whole. To do this, we need so-called *Kronecker products*, which will be discussed in some detail. Kronecker products are also useful for formulating systems mathematically.

© The Author(s), under exclusive license to Springer Nature Switzerland AG 2022
W. Grassmann and J. Tavakoli, *Numerical Methods for Solving Discrete Event Systems*,
CMS/CAIMS Books in Mathematics 5, https://doi.org/10.1007/978-3-031-10082-6_11

The outline of this chapter is as follows: The first section discusses the advantages of dividing systems into independent subsystems and solving the subsystems separately. The next section discusses how to combine subsystems, using Kronecker products. The final section then deals with Jackson networks.

11.1 Complexity Issues when using Subsystems

We now provide a detailed analysis of the savings obtainable by dividing systems into subsystems. To set the stage, we first review the number of flops necessary for finding transient and equilibrium solutions. Since the iterative methods for finding equilibrium solutions are very similar to finding transient solutions, we combine their treatment. Thus, we first address the computational complexity of transient solutions and iterative methods, then the complexity of direct methods will be discussed.

The number of flops required for transient solutions is $2N^2$ per iteration when the sparsity is not taken into account. However, if there are only s non-zero elements in the transition (iteration) matrix, then the number of flops reduces to $2s$ flops. Of course, in iterative methods, the number of iterations is of importance. However, we observed in Section 9.3.5 that the dimensionality of the problem has only a minor effect on the number of iterations, and it may therefore be of minor importance when deciding to break a system into subsystems.

In the case of direct methods, handling a full matrix requires $\frac{2}{3}N^3$ flops. However, when the matrix is banded, with entries $p_{i,j}$ $(q_{i,j})$ greater than zero only if $-g \leq j - i \leq h$, then the number of flops decreases to $2ghN$. For simplicity, we assume $g = h$, which yields $2h^2N$ flops.

We now estimate the number of flops necessary in situations where the system can be divided into subsystems. For simplicity, we assume that there are only two state variables, X_1 and X_2, each of which may be obtained by combining several state variables. We also assume $1 \leq X_1 \leq N_1$ and $1 \leq X_2 \leq N_2$. Here, $X_1(t)$ and $X_2(t)$ are independent for all t. The number of states in the system is thus $N = N_1 N_2$.

First, we compare using the undivided system to the divided system when using the direct method for finding equilibrium solutions. While using the full matrix, the number of flops in the case where the system is not divided into subsystems is $\frac{2}{3}N_1^3 N_2^3$ flops, whereas when the system is divided into 2 parts yields $\frac{2}{3}N_1^3 + \frac{2}{3}N_2^3$ flops. The ratio of these two expressions becomes

$$\frac{N_1^3 N_2^3}{N_1^3 + N_2^3}.$$

This ratio is $N_1^3/2$ if $N_1 = N_2$, which shows that dividing a system into subsystems is very effective.

When using banded matrices, then the bandwidth of the whole system is larger than the bandwidth of the subsystems. To show this, we assume that for subsystem k $p_{i,j}^{(k)} = 0$ when $|j - i| > h_k$, $k = 1, 2$. It follows that one iteration of system k requires

$2N_k h_k^2$ flops for a total of $2N_1 h_1^2 + 2N_2 h_2^2$ flops. When combining both systems, the bandwidth increases. If the encoding of equation (3.1) is used, then when X_1 changes by 1, the index will change by $u_1 = N_2$, that is, the band is now $h^* = N_2 h_1$. Without going into details, we just note that the computational effort increases roughly by a factor $\Theta(N_1^3)$.

In the case of transient solutions and iterative methods with a full matrix, the complexity is $2N^2$ per iteration. When splitting the system into two subsystems, the flop count is reduced by the factor

$$\frac{N_1^2 N_2^2}{N_1^2 + N_2^2}.$$

If $N_1 = N_2$, this yields a factor of $N_1^2/2$. The increase in the flop count is still substantial, though not as drastic as in the case of equilibrium equations. Consequently, as the dimension increases, iterative methods become more competitive.

When using a sparse matrix representation with s non-zero entries, the number of flops is $2s$ per iteration. In a DTMC where the subsystems have s_i, $i = 1, 2$ non-zero entries, the system as a whole has $s_1 s_2$ non-zero entries because for each move from state i_1 to j_1 in subsystem 1, there is a move from i_2 to j_2 in subsystem 2. The gain from dividing the two systems into one is therefore given by the factor

$$\frac{s_1 s_2}{s_1 + s_2}.$$

If $s_1 = s_2$, this yields $s_1/2$.

The situation is better for CTMCs. Of course, in our algorithms, the CTMCs are converted to DTMCs, but this is done after combining the two subsystems. As long as one is working with a CTMC, only one event can happen at any time. However, all transitions of system 1 have to be repeated for each state of system 2, and all transitions of system 2 have to be repeated for each state of system 1, which yields $s = s_1 N_1 + s_2 N_2$ non-zero transitions for the combined system. The flop count for combining the two systems therefore increases by

$$\frac{s_2 N_1 + s_1 N_2}{s_1 + s_2}.$$

If $s_1 = s_2$ and $N_1 = N_2$, this yields an increase by a factor N_1.

We conclude: when possible, systems should be broken down into several subsystems. Even if the resulting subsystems are not completely independent, one can behave as if this is the case, and use the result as an approximation. This approximation can possibly be used as the starting point for iterative methods.

Dependencies between systems can have two forms: functional or synchronizing. In functional dependencies, the transition rates of one system are determined by the state of another system. An example of this is a central computer with two terminals. If only one of the terminals is active, the active terminal may complete tasks faster than when the other terminal is also active. This dependency can obviously be extended to more than two terminals. Synchronizing dependencies affect more than one system simultaneously. An example of a synchronizing event occurs when an entry moves from one subsystem to the other.

11.2 Mathematical Tools for Combining Independent Subsystems

In this section, we define and discuss Kronecker products and Kronecker sums. They are helpful for formulating systems that are formed by combining several subsystems. In particular, events in discrete event systems are generated by independent processes, often independent renewal processes, and their combined effect can be expressed by using Kronecker products. This is one of the reasons that Kronecker products are used extensively in the literature.

Here, we will use Kronecker products if two DTMCs are to be combined, whereas Kronecker sums are used for combining CTMCs.

11.2.1 Combining DTMCs via Kronecker Products

Consider two independent DTMCs $X_1(t)$ and $X_2(t)$, with transition matrices $P_1 = [p_{i,j,1}, \ i, j = 1, 2, \ldots, N_1]$ and $P_2 = [p_{i,j,2}, \ i, j = 1, 2, \ldots, N_2]$. Clearly,

$$P\{X_1 = j_1, X_2 = j_2 \mid X_1 = i_1, X_2 = i_2\} = p_{i_1,j_1,1} p_{i_2,j_2,2}.$$

If we bring this into lexicographic order, then we get the following matrix P of the combined process:

$$P = \begin{bmatrix} p_{1,1,1}P_2 & p_{1,2,1}P_2 & \cdots & p_{1,N_1,1}P_2 \\ p_{2,1,1}P_2 & p_{2,2,1}P_2 & \cdots & p_{2,N_1,1}P_2 \\ \vdots & \vdots & \vdots & \vdots \\ p_{n_1,1,1}P_2 & p_{n_1,2,1}P_2 & \cdots & p_{N_1,N_1,1}P_2 \end{bmatrix}.$$

We can abbreviate this by writing

$$P = [p_{i,j}P_2, \ i, j = 1, 2, \ldots, N_1].$$

This is the *Kronecker product* or *tensor product* of the matrices P_1 and P_2. Generally, the Kronecker product of two matrices $A = [a_{i,j}]$ and $B = [b_{i,j}]$ is defined as

$$A \otimes B = [a_{i,j}B].$$

Hence, the Kronecker product consists of the entries $a_{i_1,j_1} b_{i_2,j_2}$, arranged in lexicographic order. Consequently, the transition matrix that combines the two systems with transition matrices P_1 and P_2 is $P_1 \otimes P_2$. If P_1 has a dimension of N_1, and P_2 has a dimension of N_2, $P_1 \otimes P_2$ has dimension of $N_1 N_2$.

The main laws regarding Kronecker products are as follows: If $A = [a_{i,j}]$, $B = [b_{i,j}]$, $C = [c_{i,j}]$ and $D = [d_{i,j}]$, then we have

Associativity $A \otimes (B \otimes C) = (A \otimes B) \otimes C$.

Distributivity over addition $(A + B) \otimes (C + D) = A \otimes C + B \otimes C + A \otimes D + B \otimes D$, provided dimensions match.

Compatibility with multiplication $(A \otimes C)(B \otimes D) = AB \otimes CD$, provided the dimensions of the matrices allow the formation of all products involved.

Compatibility with matrix inversion $(A \otimes B)^{-1} = A^{-1} \otimes B^{-1}$.

To prove these identities, we have to show that the matrices on both sides of the equations have the same entries and that the entries are in lexicographic order. Here are the proofs.

1. In the law of associativity, we have the entries $a_{i_1,j_1} b_{i_2,j_2} c_{i_3,j_3}$ on both sides, where the entries on both sides are in lexicographic order.
2. To prove distributivity over addition, we have the entries $(a_{i_1,j_1} + b_{i_2,j_2})(c_{i_3,j_3} + d_{i_4,j_4})$ on the left, and $a_{i_1,j_1} c_{i_3,j_3} + b_{i_2,j_2} c_{i_3,j_3} + a_{i_1,j_1} d_{i_4,j_4} + b_{i_2,j_2} d_{i_4,j_4}$ on the right, and these expressions are obviously equal. Also, the entries are in the lexicographic order.
3. To prove the compatibility with ordinary matrix multiplication, the entries on the left are $\sum_{k_1} a_{i_1,k_1} b_{k_1,j_2} \sum_{k_2} c_{i_3,k_2} d_{k_2,j_4}$, and on the right, the entries are $\sum_{k_1} \sum_{k_2} a_{i_1,k_1} b_{k_1,j_2} c_{i_3,k_2} d_{k_2,j_4}$, and these sums are obviously equal. Also, both sides are in lexicographic order.
4. The compatibility with matrix inversion holds if $(A \otimes B)(A^{-1} \otimes B^{-1})$ is the identity matrix. Now, by the compatibility with matrix multiplication

$$(A \otimes B)(A^{-1} \otimes B^{-1}) = (AA^{-1}) \otimes (BB^{-1}),$$

and this is the identity matrix.

The associative law allows one to omit parentheses while writing products with more than two factors. Notice that the Kronecker product is not commutative, because although $A \otimes B$ and $B \otimes A$ have identical entries, the entries are in a different order. The distributivity over addition also holds for more than two factors, and so does the compatibility law over multiplication.

The compatibility with matrix inversion and matrix multiplication implies, for any integer n, that

$$(A \otimes B)^n = A^n \otimes B^n. \tag{11.1}$$

The proof of this is rather simple, and it is omitted.

If we have two DTMCs, with transition matrices P_1 and P_2, the transient solution $\pi^{(i)}(t)$, $i = 1, 2$ can be determined by the equation

$$\pi^{(i)}(t + 1) = \tau^{(i)}(t)P_i, \ i = 1, 2.$$

If $\pi(t)$ is the transient probability vector for the combined system, we have

$$\pi(t + 1) = \pi(t)(P_1 \otimes P_2).$$

One would now expect that $\pi(t)$ is $\pi^{(1)}(t) \otimes \pi^{(2)}(t)$, and really, if this equation holds for t, it also holds for $t + 1$. This follows from the compatibility of $\otimes$ with multiplication:

$$\pi(t + 1) = (\pi^{(1)}(t) \otimes \pi^{(2)}(t))(P_1 \otimes P_2) = (\pi^{(1)}(t)P_1) \otimes (\pi^{(2)}(t)P_2),$$

and the right side is $\pi^{(1)}(t + 1) \otimes \pi^{(2)}(t + 1)$, which proves

$$\pi(t + 1) = \pi^{(1)}(t + 1) \otimes \pi^{(2)}(t + 1)$$

as claimed.

Block-structured matrices in which the same block B appears in several places are of some importance, and to create such matrices, Kronecker products can be used. For instance, to create a block-structured matrix with the matrix B on the superdiagonal, we define A as

$$A = \begin{bmatrix} 0 & 1 & 0 & 0 & \ldots & 0 \\ 0 & 0 & 1 & 0 & \ldots & \vdots \\ \vdots & \ddots & \ddots & \ddots & \ddots & \vdots \\ 0 & 0 & \ldots & \ldots & 0 & 1 \\ 0 & 0 & \ldots & \ldots & 0 & 0 \end{bmatrix}.$$

Now, $A \otimes B$ is the block-structured matrix with B at the position where the entry of A is 1, that is

$$A \otimes B = \begin{bmatrix} 0 & B & 0 & 0 & \ldots & 0 \\ 0 & 0 & B & 0 & \ldots & \vdots \\ \vdots & \ddots & \ddots & \ddots & \ddots & \vdots \\ 0 & 0 & \ldots & \ldots & 0 & B \\ 0 & 0 & \ldots & \ldots & 0 & 0 \end{bmatrix}.$$

On the other hand, $B \otimes A$ yields a block-structured matrix where all blocks are formed by superdiagonal matrices, and the non-zero entries of the i, j block are all $b_{i,j}$.

Tridiagonal block-structured matrices are particularly important. If S_0 is the identity matrix, S_1 the superdiagonal matrix with all its non-zero entries equal to 1, and S_{-1} is the subdiagonal matrix with all its entries equal to 1, then the block-structured matrix with P_{-1} on its subdiagonal, P_0 on its diagonal, and P_1 on its superdiagonal can be written as

$$S_{-1} \otimes P_{-1} + S_0 \otimes P_0 + S_1 \otimes P_1.$$

In summary, Kronecker products are useful in many contexts.

11.2.2 CTMCs and Kronecker Sums

Let us now consider two independent CTMCs $X_1(t)$ and $X_2(t)$ having transition matrices Q_1 and Q_2 with sizes N_1 and N_2, respectively. Here, the essential fact is that as $h \to 0$, the probability that both systems change within an interval of length h time is $o(h)$. This means that only one system can change at any given time. The rates affecting only the first system are now given by the matrix $Q_1 \otimes I_{N_2}$, and the rates affecting only the second system are given by the matrix $I_{N_1} \otimes Q_2$. Consequently, the transition matrix Q of the combined process becomes

$$Q = Q_1 \otimes I_{N_2} + I_{N_1} \otimes Q_2.$$

To clarify this, let us consider the combination of the following systems

$$Q_1 = \begin{bmatrix} -\lambda_1 & \lambda_1 \\ \mu_1 & -\mu_1 \end{bmatrix}, \quad Q_2 = \begin{bmatrix} -\lambda_2 & \lambda_2 \\ \mu_2 & -\mu_2 \end{bmatrix}.$$

Here, λ_i is the rate of going from $X_i = 0$ to $X_i = 1$, and μ_i is the rate of going from $X_i = 1$ to $X_i = 0$ for $i = 1, 2$. The combined process now has four states, $[0, 0]$, $[0, 1]$, $[1, 0]$, $[1, 1]$. Since changes of X_2 do not affect X_1, Q_2 appears on the diagonal, that is, we obtain

$$\begin{bmatrix} -\lambda_2 & \lambda_2 & & \\ \mu_2 & -\mu_2 & & \\ & & -\lambda_2 & \lambda_2 \\ & & \mu_2 & -\mu_2 \end{bmatrix}.$$

This matrix can be expressed as $I_{N_1} \otimes Q_2$, where I_ν is the identity matrix of dimension ν. The following matrix catches the contribution of X_1.

$$\begin{bmatrix} -\lambda_1 & & \lambda_1 & \\ & -\lambda_1 & & \lambda_1 \\ \mu_1 & & -\mu_1 & \\ & \mu_1 & & -\mu_1 \end{bmatrix}.$$

This can be expressed as $Q_1 \otimes I_{N_2}$. Adding these two matrices together yields $Q = Q_1 \otimes I_{N_2} + I_{N_1} \otimes Q_2$ as claimed.

The expression

$$Q_1 \otimes I_{N_2} + I_{N_1} \otimes Q_2$$

is referred to as the *Kronecker sum* of the matrices Q_1 and Q_2, and is denoted by $Q_1 \oplus Q_2$. More generally, if A and B are two square matrices with dimensions N_1 respectively N_2, their Kronecker sum is

$$A \oplus B = A \otimes I_{N_2} + I_{N_1} \otimes B.$$

Kronecker sums are associative, but not commutative. We will not use them further.

11.2.3 Using Kronecker Products in Almost Independent Subsystems

The transition matrix of the combination of two almost independent subsystems may be written as $P = P_1 \otimes P_2 + C$, where C is a matrix with very few non-zero elements. In this case, one might suspect that it is advantageous to calculate transient solutions in the following manner:

$$\pi(t + 1) = \pi(t)(P_1 \otimes P_2 + C) = \pi(t)(P_1 \otimes P_2) + \pi(t)C.$$

If C has only very few non-zero elements, then the time for calculating $\pi(t)C$ can be neglected, and we can concentrate on efficient methods to find $\pi(t)(P_1 \otimes P_2)$. As shown earlier, if C vanishes, then we can treat the two subsystems defined by P_1 and P_2 separately. In this case, we would have two vectors $\pi^{(1)}(0)$ and $\pi^{(2)}(0)$, and could calculate $\pi^{(1)}(t + 1) = \pi^{(1)}(t)P_1$ and $\pi^{(2)}(t + 1) = \pi^{(2)}(t)P_2$. Then $\pi(t)$ would be $\pi^{(1)}(t) \otimes \pi^{(2)}(t)$. Also, $\pi(t + 1) = \pi(t)P$ would become

$$\pi^{(1)}(t + 1) \otimes \pi^{(2)}(t + 1) = (\pi^{(1)}(t) \otimes \pi^{(2)}(t))(P_1 \otimes P_2).$$

By applying the compatibility with multiplication to the right side of this equation, we get

$$\pi(t + 1) = \pi^{(1)}(t + 1) \otimes \pi^{(2)}(t + 1) = (\pi^{(1)}(t)P_1) \otimes (\pi^{(2)}(t)P_2).$$

Calculating $\pi(t + 1)$ in this fashion reduces the effort for 1 iteration from $2s_1s_2$ to $2s_1 + 2s_2$. Unfortunately, when we add a C to $P_1 \otimes P_2$, then the result $\pi(t + 1)$ can no longer be considered as a Kronecker product of two vectors, and the formula cannot be used except possibly for the first iteration. Are there any alternatives?

Let us consider the problem of finding $\pi(P_1 \otimes P_2)$. It is notationally simpler to replace P_1 by $A = [a_{i,j}]$, a square matrix of size N_1 and P_2 by $B = [b_{i,j}]$, a square matrix of size N_2. To be consistent, π must thus have a size of N_1N_2 to match the size of $A \otimes B$.

First, note that $A \otimes B$ is a block-structured matrix, which suggests structuring π consistently, that is, $\pi = [\pi_1, \pi_2, \ldots, \pi_{N_1}]$, where the π_i are vectors of size N_2. Now, $\pi(A \otimes B)$ can be written as

$$[\pi_1, \pi_2, \ldots, \pi_{N_1}] \begin{bmatrix} a_{1,1}B & a_{1,2}B & \ldots & a_{1,N_1}B \\ a_{2,1}B & a_{2,2}B & \ldots & a_{2,N_1}B \\ \ldots & \vdots & \vdots & \vdots \\ a_{N_1,1}B & a_{N_1,2}B & \ldots & a_{N_1,N_1}B \end{bmatrix}$$

$$= \left[\sum_{i=1}^{N_1} \pi_i a_{i,1}B \;\; \sum_{i=1}^{N_1} \pi_i a_{i,2}B \;\; \ldots \;\; \sum_{i=1}^{N_1} \pi_i a_{i,N_1}B \right] \tag{11.2}$$

$$= \left[\sum_{i=1}^{N_1} a_{i,1}y_i \;\; \sum_{i=1}^{N_1} a_{i,2}y_i \;\; \cdots \;\; \sum_{i=1}^{N_1} a_{i,N_1}y_i \right], \tag{11.3}$$

with

$$y_i = \pi_i B. \tag{11.4}$$

The difference between (11.2) and (11.3) is that in (11.2), the $\pi_i B$ are determined in each of the sums, whereas in (11.3), the $\pi_i B$ are calculated only once, and used in each of the N_1 sums. Notice that if either A or B are identity matrices, this advantage disappears. In the case $A = I_{N_1}$, (11.2) becomes

$$\left[\pi_1 B \; \pi_2 B \; \ldots \; \pi_{N_1} B \right]$$

and hence no saving results. If $B = I_{N_2}$, then $y_i = \pi_i$, and no saving occurs either. Since we use the identity matrix when representing CTMCs, this method only applies to DTMCs. Also, in some sparse matrices, the advantage gained may not be worth the extra effort of programming.

In conclusion, though initially, one has the impression that tremendous savings are possible, most savings disappear after a detailed investigation. However, the assumption of independence, with the concomitant product forms, can be used to select good initial conditions for iterative methods.

11.3 Jackson Networks

It is well known that tandem queues with Poisson arrivals and exponential service times often have a product-form equilibrium solution. This means that in equilibrium, the different line lengths are independent. Note, however, that transient solutions of tandem queues cannot be considered as independent. Indeed, if the system starts empty, it will take some time until the first arrival reaches the second line, and this causes dependencies. Independence is only observed when the system is in equilibrium. Consequently, we have what we called *steady-state independence*, but not *full independence*. We discuss steady-state independent systems, starting with the tandem queue, then we describe so-called *Jackson networks*, a class of networks for which the product-form solutions are valid. Indeed, tandem queues are special cases of Jackson networks.

11.3.1 Simple Tandem Queues

Consider the tandem queue given in Table 11.1. Note that we do not set any upper bounds, that is, the lines are not restricted to any finite length. We now write the

Table 11.1 A tandem queue

Event type	Event function	Event condition	Rate
Arrival	$X_1 + 1, X_2$		λ
From line 1 to 2	$X_1 - 1, X_2 + 1$	$X_1 > 0$	μ_1
Departure	$X_1, X_2 - 1$	$X_2 > 0$	μ_2

equilibrium equations, equating the flow into the state, with the flow out of the state. They are

$$\pi_{0,0}\lambda = \pi_{0,1}\mu_2 \tag{11.5}$$

$$\pi_{0,1}(\lambda + \mu_2) = \pi_{1,0}\mu_1 + \pi_{0,2}\mu_2 \tag{11.6}$$

$$\pi_{1,0}(\lambda + \mu_1) = \pi_{0,0}\lambda + \pi_{1,1}\mu_2 \tag{11.7}$$

$$\pi_{i,j}(\lambda + \mu_1 + \mu_2) = \pi_{i-1,j}\lambda + \pi_{i+1,j-1}\mu_1 + \pi_{i,j+1}\mu_2, \ i,j \geq 1. \tag{11.8}$$

The solution of these equations can be found as follows: We assume that the two lines are independent $M/M/1$ queues, in which case

$$P\{X_1 = i\} = \left(1 - \frac{\lambda}{\mu_1}\right)\left(\frac{\lambda}{\mu_1}\right)^i, \ P\{X_2 = j\} = \left(1 - \frac{\lambda}{\mu_2}\right)\left(\frac{\lambda}{\mu_2}\right)^j.$$

To find the $\pi_{i,j}$, we multiply these two probabilities to obtain

$$\pi_{i,j} = \left(1 - \frac{\lambda}{\mu_1}\right)\left(\frac{\lambda}{\mu_1}\right)^i\left(1 - \frac{\lambda}{\mu_2}\right)\left(\frac{\lambda}{\mu_2}\right)^j. \tag{11.9}$$

To show that this solution satisfies equations (11.5) to (11.8), we substitute (11.9) into the equilibrium equations. We start with equation (11.8). If we omit the factor

$$\left(1 - \frac{\lambda}{\mu_1}\right)\left(1 - \frac{\lambda}{\mu_2}\right),$$

which cancels out, this yields

$$\left(\frac{\lambda}{\mu_1}\right)^i\left(\frac{\lambda}{\mu_2}\right)^j(\lambda + \mu_1 + \mu_2) = \left(\frac{\lambda}{\mu_1}\right)^{i-1}\left(\frac{\lambda}{\mu_2}\right)^j\lambda + \left(\frac{\lambda}{\mu_1}\right)^{i+1}\left(\frac{\lambda}{\mu_2}\right)^{j-1}\mu_1 + \left(\frac{\lambda}{\mu_1}\right)^i\left(\frac{\lambda}{\mu_2}\right)^{j+1}\mu_2.$$

To show that this equation holds, divide both sides by $\left(\frac{\lambda}{\mu_1}\right)^i\left(\frac{\lambda}{\mu_2}\right)^j$ to obtain, after minor simplifications:

$$\lambda + \mu_1 + \mu_2 = \mu_1 + \mu_2 + \lambda. \tag{11.10}$$

This equation obviously holds. The equations (11.5), (11.6) and (11.7) can be checked in a similar way.

Equations (11.5) to (11.8) are called *global balance equations*, because they balance the flow into any state with the flow out of that state. Of importance are also the *local balance equations*. They indicate that in each state, the rate at which the state is left due to an increase of X_i must be equal to the rate the state is entered due to a decrease of X_i. Hence, there is one local balance equation for each state variable. In the tandem queue example, we have, for X_1 in state $i, j > 0$:

$$\pi_{i,j}\mu_1 = \pi_{i-1,j}\lambda. \tag{11.11}$$

Similarly, the local balance equation for X_2 in state $i, j > 0$ is

$$\pi_{i,j}\mu_2 = \pi_{i+1,j-1}\mu_1. \tag{11.12}$$

We add a third equation to these two, which balances, for each state, the rate of entering the system with the rate of leaving the system.

$$\pi_{i,j}\lambda = \pi_{i,j+1}\mu_2. \tag{11.13}$$

We treat this equation also like a local balance equation. It is easily verified that the $\pi_{i,j}$ as given in equation (11.9) satisfy all local balance equations.

Note that the sum of these three equations result in equation (11.8). Similarly, equation (11.5) is given by equation (11.13). Equation (11.6) is the sum of equations (11.13) and (11.11), and equation (11.7) is the sum of equations (11.12) and equation (11.13). Finally, equation (11.11) is the sum of equations (11.13) and (11.11). Since the local balance equations hold in our model, the global balance equations hold as well. In general, however, it is possible that the global balance equations hold, yet the local balance equations fail, but if the local balance equations hold, then it can be proven that a product-form solution exists.

The product form given by (11.9) does not hold for tandem queues with blocking. For the proof, we only need to show that a single equilibrium equation is not satisfied by (11.9). Logically, the equilibrium equation one would test first is the equation of $X_2 = N_2$, where N_2 is the size of X_2 which blocks server 1 from starting service. One has

$$\pi_{i,N_2}(\lambda + \mu_2) = \pi_{i-1,N_2}\lambda + \pi_{i+1,N_2-1}\mu_1. \tag{11.14}$$

If one enters the product given by equation (11.9) into this equation and simplifies the resulting equation, one arrives at $\lambda = \mu_1$. Only if $\lambda = \mu_1$ would the product form be valid, but then, the process would be divergent, and no equilibrium would exist. This shows that one cannot treat X_1 and X_2 as an independent.

One can also check for the local balance equations. Indeed, subtracting equation (11.12) with N_2 substituted for j from equation (11.14) yields

$$\pi_{i,N_2}\lambda = \pi_{i-1,N_2}\lambda,$$

and this can only be satisfied if $\pi_{i,N_2} = \pi_{i-1,N_2}$, which does not satisfy equation (11.9) unless $\lambda = \mu_1$.

Surprisingly, a slight modification of the tandem queue with blocking does allow the product to form solution given by (11.9). When $X_2 = N_2$, and server 1 finishes service, the entity in line 1 leaves the system rather than blocking server 1. The global equilibrium equation for i, N_2 then becomes

$$\pi_{i,N_2}(\lambda + \mu_1 + \mu_2) = \pi_{i-1,N_2}\lambda + \pi_{i+1,N_2-1}\mu_1 + \pi_{i+1,N_2}\mu_1, \quad i > 0. \tag{11.15}$$

This equation satisfies the product-form solution given by equation (11.9). This can be verified by substituting π_{i,N_2} with equation (11.9), which leads to an identity.

Alternatively, we can deduct equation (11.11) and (11.12) from equation (11.15) to obtain

$$\pi_{i,N_2}\lambda = \pi_{i+1,N_2}\mu_1$$

which is essentially equation (11.11).

We now check if one can limit the buffer size of line 1 and still has a product-form solution. For $i = N_1$ in (11.8), one has

$$\pi_{N_1,j}(\mu_1 + \mu_2) = \pi_{N_1-1,j}\lambda + \pi_{N_1,j+1}\mu_2.$$

If the product form is substituted for the equilibrium probabilities in question, the result is $\mu_2 = \lambda$, which is not an identity. Also, by deduction the local balance equation (11.11) from the equilibrium equation, one finds

$$\pi_{N_1,j}\mu_2 = \pi_{N_1,j+1}\mu_2.$$

This equation can only be satisfied if $\mu_2 = \lambda$. Consequently, the product form is not valid when arriving customers are lost and $\mu_2 \neq \lambda$.

We now generalize the tandem queue models by allowing the service rates to vary with the number of customers in the line they serve. However, we assume that λ does not depend on X_1 or on X_2 because this would destroy the product-form solution. The global balance equations are similar to the ones given by equation (11.8), except that μ_1 is replaced by $\mu_1(i)$, and μ_2 is replaced by $\mu_2(i)$. Therefore

$$\pi_{i,j}(\lambda + \mu_1(i) + \mu_2(j)) = \pi_{i-1,j}\lambda + \pi_{i+1,j-1}\mu_1(i+1) + \pi_{i,j+1}\mu_2(j+1). \quad (11.16)$$

Now X_1 and X_2 become birth–death processes, that is, if they were independent, then $\pi_i^{(k)}$, $k = 1, 2$ are, according to Section 2.4.3

$$\pi_{i+1}^{(1)} = \pi_i^{(1)}\frac{\lambda}{\mu_1(i+1)}, \quad \pi_{j+1}^{(2)} = \pi_j^{(2)}\frac{\lambda}{\mu_2(j+1)}. \quad (11.17)$$

The product form now becomes

$$\pi_{i,j} = \pi_i^{(1)}\pi_j^{(2)}.$$

We replace $\pi_{i,j}$ in the equilibrium equation (11.16) by this expression to check whether or not the above equation is satisfied. We have

$$\pi_i^{(1)}\pi_j^{(2)}(\lambda + \mu_1(i) + \mu_2(j)) = \pi_{i-1}^{(1)}\pi_j^{(2)}\lambda + \pi_{i+1}^{(1)}\pi_{j-1}^{(2)}\mu_1(i+1) + \pi_i^{(1)}\pi_{j+1}^{(2)}\mu_2(j+1),$$

or

$$\lambda + \mu_1(i) + \mu_2(j) = \frac{\pi_{i-1}^{(1)}}{\pi_i^{(1)}}\lambda + \frac{\pi_{i+1}^{(1)}}{\pi_i^{(1)}}\frac{\pi_{j-1}^{(2)}}{\pi_j^{(2)}}\mu_1(i+1) + \frac{\pi_{j+1}^{(2)}}{\pi_j^{(2)}}\mu_2(j+1),$$

From equation (11.17), we get

$$\frac{\pi_{i+1}^{(1)}}{\pi_i^{(1)}} = \frac{\lambda}{\mu_1(i+1)}, \quad \frac{\pi_{i-1}^{(1)}}{\pi_i^{(1)}} = \frac{\mu_1(i)}{\lambda},$$

and a similar relation can be found for $\pi_j^{(2)}$. Consequently

$$\lambda + \mu_1(i) + \mu_2(j) = \frac{\mu_1(i)}{\lambda}\lambda + \frac{\lambda}{\mu_1(i+1)}\frac{\mu_2(j)}{\lambda}\mu_1(i+1) + \frac{\lambda}{\mu_2(j+1)}\mu_2(j+1).$$

This is obviously an identity, and the product-form solution is still valid when the $\mu(i)$ depends on X_i.

11.3.2 General Jackson Networks

This section is about *Jackson networks*, a generalization of tandem queues. Instead of only two lines, there are many lines, and instead of going forward to the next line, elements randomly choose another line. To specify such a network, let there be M nodes, and let X_i be the number at node i, $i = 1, 2, \ldots, M$. Elements leave node i at a rate of $\mu_i(X_i)$, and the join node j with probability $p_{i,j}$. Also, arrivals occur at a rate of λ, and they join node j with probability $p_{0,j}$. Finally, with probability $p_{i,0}$ an element leaves the network after having received service at node i. Jackson networks have a product-form solutions, and as a result, it is relatively simple to calculate their equilibrium probabilities. However, as soon as blocking occurs, we no longer have a Jackson network, and the same is true if the choice of the next node an element will visit depends on the line length of this node.

To establish a product-form solution, we need, for each node, an arrival and a service rate. The service rate of node i is given as $\mu_i(X_i)$, The arrival rate λ_j for node j can be found as follows: Arrivals either come from the outside, and their rate is $\lambda p_{0,j}$, or they come from node i, $1 \le i \le M$, and the rate of elements moving from i to j is $\lambda_i p_{i,j}$. Consequently, we have

$$\lambda_j = \lambda p_{0,j} + \sum_{i=1}^{M} \lambda_i p_{i,j}, \ j = 1, 2, \ldots, M. \tag{11.18}$$

Note that this has the form of the equilibrium equations of a DTMC with states 0, 2, ..., M, except that π_i is replaced by λ_i, $i > 0$, and π_0 by λ. This stands to reason because as discussed in Section 9.2, by setting $\pi_0 = 1$, the solutions of (11.18) provide the number of visits to the different states between two visits of state 0. It follows that any algorithm available for solving equilibrium equations of DTMCs can be used here as well.

To prove the existence of a product-form solution, we need to formulate the equilibrium equations. To do this, let $\mathbf{e}_i$ be a vector with a 1 in position i, and a zero everywhere else. When an element coming from outside joins node i while the system is in state $X = [X_1, X_2, \ldots, X_M]$, X_i increases by 1, and the target state becomes $X + \mathbf{e}_i$. Similarly, when in state X, an element moves from node i to node j, the state changes from X to $X - \mathbf{e}_i + \mathbf{e}_j$. Finally, when an element leaves the system from node i the state changes from X to $X - \mathbf{e}_i$. If we have a product-form solution with the individual factors being birth-death processes, we must have

$$\pi_{X+\mathbf{e}_i} = \pi_X \frac{\lambda_i}{\mu_i(X_i + 1)}. \tag{11.19}$$

From this, we conclude

$$\pi_{X-\mathbf{e}_i} = \pi_X \frac{\mu_i(X_i)}{\lambda_i} \tag{11.20}$$

$$\pi_{X+\mathbf{e}_i-\mathbf{e}_j} = \pi_X \frac{\lambda_i}{\mu_i(X_i + 1)} \frac{\mu_j(X_j)}{\lambda_j}. \tag{11.21}$$

The system state X does not change if after finishing at node i, the element returns immediately to node i. This happens with probability $p_{i,i}$. Hence, at node i, a change only occurs at with a probability $1 - p_{i,i}$.

The equilibrium equation for the case where all X_i, $i = 1, 2, \ldots, M$ are non-zero is now

$$\pi_X \left(\lambda + \sum_{i=1}^{M} \mu_i(X_i)(1 - p_{i,i}) \right)$$

$$= \sum_{i=1}^{M} \pi_{X-\mathbf{e}_i} \lambda p_{0,i} + \sum_{i=1}^{M} \pi_{X+\mathbf{e}_i} \mu_i(X_i + 1) p_{i,0} \tag{11.22}$$

$$+ \sum_{i=1}^{M} \sum_{j \neq i} \pi_{X+\mathbf{e}_i-\mathbf{e}_j} \mu_i(X_i + 1) p_{i,j}.$$

In this equation, the first row provides the flow out of state X. The first sum of the second row gives the rate of entering the different nodes from the outside, and the second term is the rate of leaving the system from node i. The third row gives the moves from node i to node j, $i \neq j$. The eventuality that $X_i = 0$ can be handled by setting $\mu_i(0) = 0$ for $i < 0$.

We now substitute equations (11.19) to (11.21) for the $\pi_{X-\mathbf{e}_i}$, $\pi_{X+\mathbf{e}_i}$ and $\pi_{X+\mathbf{e}_i-\mathbf{e}_j}$ in equation (11.22) to obtain, after canceling π_X on both sides of the equilibrium equation:

$$\lambda + \sum_{i=1}^{M} \mu_i(X_i)(1 - p_{i,i})$$

$$= \sum_{i=1}^{M} \frac{\mu_i(X_i)}{\lambda_i} \lambda p_{0,i} + \sum_{i=1}^{M} \lambda_i p_{i,0} \qquad (11.23)$$

$$+ \sum_{i=1}^{M} \sum_{j=1, j \neq i}^{M} \frac{\lambda_i \mu_j(X_j)}{\lambda_j} p_{i,j}.$$

The term $\sum_{i=1}^{M} \lambda_i p_{i,0}$ is the rate of leaving the system, which must be equal to the arrival rate λ. The first term of the above equation, λ, therefore cancels the term $\sum_{i=1}^{M} \lambda_i p_{i,0}$ in row 2.

Consider now the last row of equation (11.23), which can be written as

$$\sum_{j=1}^{M} \sum_{i=1, i \neq j}^{M} \frac{\lambda_i \mu_j(X_j)}{\lambda_j} p_{i,j}.$$

Now we have from (11.18)

$$\sum_{i=1, i \neq j}^{M} \lambda_i p_{i,j} = \lambda_j(1 - p_{j,j}) - \lambda p_{0,j}.$$

Consequently, the last row of equation (11.23) becomes

$$\sum_{j=1}^{M} \frac{\mu_j(X_j)}{\lambda_j} \sum_{i=1, i \neq j}^{M} \lambda_i p_{i,j}$$

$$= \sum_{j=1}^{M} \mu_j(X_j)(1 - p_{j,j}) - \sum_{j=1}^{M} \frac{\mu_j(X_j)}{\lambda_j} \lambda p_{0,j}. \qquad (11.24)$$

Using this expression, equation (11.23) now becomes

$$\lambda + \sum_{i=1}^{M} \mu_i(X_i)(1 - p_{i,i}) = \sum_{i=1}^{M} \frac{\mu_i(X_i)}{\lambda_i} \lambda p_{0,i} + \sum_{i=1}^{M} \lambda_i p_{i,0}$$

$$+ \sum_{j=1}^{M} \mu_j(X_j)(1 - p_{j,j}) - \sum_{j=1}^{M} \frac{\mu_j(X_j)}{\lambda_j} \lambda p_{0,j},$$

and both sides of this equation are obviously identical. The product-form approach thus yields the correct result. Provided $\mu_i(0)$ is set to zero, this method even covers the case where some X_i are zero.

The main result is that in equilibrium, each node can be investigated on its own, without regard to the situation in the other nodes. It follows that instead of an M-dimensional problem, we have M independent problems which are much easier to solve. In other words, if X_i is the number of customers in node i, the distribution of the X_i can be treated like independent variables. Therefore, the joint distribution can easily be determined. If the expected number of elements in the system is required, one can first calculate the individual expectations, and the expected number in the system is then the sum of these expectations.

11.3.3 Closed Queueing Networks

Queueing networks with several nodes, but with no arrivals and departures are called *closed queueing networks*. In such networks, the number of elements in the system does not change, meaning that if there are N elements at the start, the number in the network remains N forever. An example of a closed queueing network is a company that owns a fixed number of trucks, which circulate between the processing plans of the company. The processing plants are then the nodes in this system.

The equilibrium solution for closed Jackson networks can be derived from the formulas of the open networks introduced in the previous section. One finds an appropriate arrival rate for each node, and given this arrival rate, product-form expressions for finding the joint probabilities can be obtained, just as in open networks. Naturally, one only needs expressions with N elements in the network. Once these expressions are obtained, they must be normed such that their sum is 1. As before, let $\mu_k(n)$ be the service rate when the number of node k is n. We are only covering the basics. For more dtails, see [12] and [87].

Since there are no outside arrivals, the traffic equation (11.19) becomes

$$\lambda_j = \sum_{j=0}^{M} \lambda_i p_{i,j}.$$

These equations are identical to the equilibrium equations of DTMCs. Like those, they only allow one to find λ_i up to a factor. To overcome this difficulty, let $\lambda_1 = \lambda$, and define v_i to be the number of visits to state i between two successive visits to state 1, in which case $\lambda_i = \lambda v_i$ and $v_1 = 1$. We calculate the probabilities $P\{X_k = n_k\} = \pi_{n_k}^{(k)}$ as in the open network, and form the products $\prod_{k=1}^{M} \pi_{n_k}^{(k)}$. In closed networks, these products are only needed for the case where $n_1 + n_2 + \cdots + n_M = N$. To simplify our notation, let S_N be the set of M non-negative integers with a sum of N. Note that we only need the products over array that are part of S_N. To find the equilibrium probabilities, these products must be normed such that their sum is equal to 1. This leads to the following equation:

$$P\{X_1 = n_1, X_2 = n_2, \ldots, X_M = n_M\} = \frac{\prod_{k=1}^{M} \pi_{n_k}^{(k)}}{\sum_{n_1, n_2, \ldots, n_M \in S_N} \prod_{k=1}^{M} \pi_{n_k}(k)}. \quad (11.25)$$

Here, the sum in the denominator has to be extended over all M-tuples $n_1 + n_2, \ldots n_M$ satisfying $n + 1 + n_2 + \ldots + n_M = N$.

As in the open network, we use the formulas for the birth-death process, that is

$$\pi^{(k)}_{n_k+1} = \pi^{(k)}_{n_k} \frac{\lambda_k}{\mu_k(n_k + 1)}.$$

We set $\pi^{(k)}_0 = 1$ for all k, and then

$$\pi^{(k)}_{n_k} = \frac{\lambda^{n_k} v_k^{n_k}}{\prod_{j=1}^{n_k} \mu_k(j)}.$$

Using this expression in equation (11.25), we find that both the nominator and the denominator have a factor λ^N, which cancel. Hence, if we define

$$f_k(n_i) = \frac{v_k^{n_k}}{\prod_{j=1}^{n_k} \mu_k(j)},$$

then we can substitute $f_k(n_k)$ for $\pi^{(k)}_{n_k}$ in equation (11.25) to obtain

$$P\{X_1 = n_1, X_2 = n_2, \ldots, X_M = n_M\} = \frac{\prod_{k=1}^{M} f_k(n_k)}{\sum_{n_1,n_2,\ldots,n_M \in S_N} \prod_{k=1}^{M} f_k(n_k)}.$$

With this, we have found the joint distribution of the X_k. If we denote the denominator by $G(N)$, then this yields

$$P\{X_1 = n_1, X_2 = n_2, \ldots, X_M = n_M\} = \frac{\prod_{k=1}^{M} f_k(n_k)}{G(N)}.$$

Here, $G(N)$ is the norming constant.

Of interest are also the marginal distributions, as well as the values for $\lambda_j = \lambda v_j$. Let us consider the marginal distribution of X_1. One has, if S_N^- is the set of $M - 1$ non-negative integers $n_2, n_3, \ldots n_M$ with a sum of N

$$P\{X_1 = n\} = \sum_{n_2,n_3,\ldots,n_M \in S_{N-n}^-} P\{X_1 = n, X_2 = n_2, X_3 = n_3, \ldots, X_M = n_M\}.$$

When using the expressions above, we obtain

$$P\{X_1 = n\} = \frac{f_1(n)}{G(N)} \sum_{n_2,n_3,\ldots,n_M \in S_{N-n}^-} \prod_{k=2}^{M} f_k(n_k).$$

This yields the marginal distribution of X_1. To simplify the notations, we define

$$G_1(N - n) = \sum_{n_2,n_3,\ldots,n_M \in S_{N-n}^-} \prod_{k=2}^{M} f_k(n_k).$$

$G_1(N - n)$ can be interpreted as the norming constant of a network consisting of nodes $2, 3, \ldots, M$, with $N - n$ elements in the system. Hence

$$P\{X_1 = n\} = \frac{f_1(n)G_1(N - n)}{G(N)}.$$

The marginal distributions of X_k, $k > 1$ can be obtained in a similar way.

We can now find λ as follows. Since $v_1 = 1$, $\lambda_1 = v_1\lambda = \lambda$, and λ_1 must be equal to the departure rate of node 1, that is

$$\lambda = \sum_{n=1}^{N} P\{X_1 = n\}\mu_1(n) = \frac{1}{G(N)} \sum_{n=1}^{N} f_1(n)G_1(N - n)f_1(n)\mu_1(n).$$

However, this expression can be simplified. First, note that with $v_1 = 1$,

$$f_1(n)\mu_1(n) = \frac{v_1^n}{\mu_1(1)\mu_1(2)\ldots\mu_1(n)}\mu_1(n) = f_1(n - 1).$$

Consequently

$$\lambda = \frac{1}{G(N)} \sum_{n=1}^{N} f_1(n - 1)G_1(N - n)$$

$$= \frac{1}{G(N)} \left(f_1(0)G_1(N - 1) + f_1(1)G_1(N - 2) + \cdots + f_1(N - 1)G_1(0)\right).$$

Now, $G_1(N - n - 1)$ contains all products corresponding to nodes $2, 3, \ldots M$, with a total number of $N - n - 1$ elements in these nodes, and $f_1(n)$ adds another n elements from node 1, meaning we have now all M nodes with $N - 1$ element in these nodes. We conclude

$$\lambda = \frac{G(N - 1)}{G(N)},$$

and $\lambda_j = \lambda v_j$. With this, we have found all joint probabilities, the marginal probabilities of X_1, and all $\lambda_i = \lambda v_i$.

For large networks, the number of terms of the sum forming $G(N)$ increase rather quickly as N and M increase, and efficient ways of calculating $G(N)$ become necessary. One can, for instance, modify the algorithm for finding the distributions of sums of random variables described in Chapter 1 to reduce the number of flops to find $G(N)$. There are, however, more sophisticated methods. In particular, we mention here the algorithm of Buzen [8], also described in Wikipedia under the heading "Buzen's algorithm", or in [87]. These sources also describe efficient algorithms to determine marginal distributions, using $G(N)$. Hence, even when M and N are large, one can find equilibrium probabilities readily.

11.4 Conclusions

In this chapter, we distinguished between two types of independence between subsystems: full independence and steady-state independence. The term steady-state independence is used for subsystems that are only independent when they are in a statistical equilibrium. In both types of independence, one has so-called *product-form solutions*. This means that the probability that $X_1 = i_1, X_2 = i_2, \ldots, X_d = i_d$ can be obtained as

$$P\{X_1 = i_1, X_2 = i_2, \ldots, X_d = i_d\} = \prod_{k=1}^{d} P\{X_k = i_k\},$$

Moreover, the distributions of the X_k, $k = 1, 2, \ldots, d$ can be found directly, which allows one to solve d problems with a small dimension rather than one problem with a large dimension. Since the number of states increases exponentially with the dimension of the problem, this leads to enormous savings.

Also, the distributions of the X_k are often of interest. If one has a model where the probability of the entire state vector $[X_1, X_2, \ldots, X_d]$ is obtained, the marginal probability $P\{X_k = i_k\}$ must be calculated, which requires additional flops. This step is obviously redundant in the case that $P\{X_k = i_k\}$ is calculated directly.

Can some of the advantages of substituting a number of small problems in the place of one large problem also be realized if a problem can be divided into sub-problems that are almost, but not completely independent? First, consider the equilibrium equations in Jackson networks. There, we noticed that changing a single equation destroys the product form. Generally, if one has two independent systems, one can combine them into a single system, call it system A. The transition matrix of system A is then given by the Kronecker product of the subsystems. Now, consider another system, we call it system B, which has almost the same transition matrix, except that a small number of entries are different, and which therefore has different equilibrium probabilities. In a system of linear equations for the variables $x_1, x_2, \ldots, x_N$, changing a single coefficient will normally change all x_i, and since the equilibrium equations form a system of linear equations, the same is true when changing a single transition rate, or, since the row sums must be 1, by changing two transition rates. The equilibrium probabilities of system B will thus differ from the ones of system A.

If a system can be divided into subsystems that are not quite independent, one can obtain an approximation by solving the subsystems obtained in this way using a product form. For details of this method, see [7]. If the resulting errors can be tolerated, then this is preferable. If not, one is tempted to try to improve the approximation obtained. For instance, one can use the results as a first approximation of an iterative algorithm. Another approach of exploiting the fact that systems are almost independent is as follows. One expresses the transition matrix as a Kronecker product, plus a matrix with very few non-zero entries that reflect any connection between the subsystems. If one can find efficient algorithms to perform vector/matrix multiplication for such transition matrices, the effort per iteration in an iterative method would be reduced. Unfortunately, what we found so far in literature in this respect did not persuade us to recommend such methods.

In this chapter, we also discussed closed Jackson networks.There, only certain state vectors are allowed: Any state $[X_1, X_2, \ldots, X_d]$ must satisfy $X_1 + X_2 + \cdots + X_d = N$, where N is the number in the system. Still, the solution has a product form, with each product involving only one state variable. The final probabilities can then be obtained by norming all these products such that their sum is 1. The question is now whether or not such product forms can be used as approximations in other systems. This may be an interesting topic for further research.

Problems

11.1 In a discrete-time $GI/G/1$ queue, formulate the Kronecker sum of the arrival and the service process. Use the remaining life process. The probability that the inter-arrival time is i is a_i, and the probability that the service time is i is s_i.

11.2 There are two independent DTMCs $X_1(t)$ and $X_2(t)$. Find the distribution of $Y(t) = X_1(t) + X_2(t)$, expressed in terms of the distributions of $X_1(t)$ and $X_2(t)$. Is $Y(t)$ a DTMC?

11.3 A repair facility has a station for diagnostics, which decides whether the repair is of type A or B. If the repair is of type A, it goes to station 1, and if it is of type B, it goes to stations 2. The arrival to the diagnostic station is Poisson. All stations have one server, and the service times are exponential. The arrival rates are $\lambda = 0.8$, and the service rates are μ_0 for the diagnostics station, $\mu_1 = 1$ for station 1, and $\mu_2 = 0.9$ for station 2. The probability of a type A repair is 0.6, and the probability of a type B repair is 0.4. After repair, the items leave the system. Find the expected line length for each station, as well as the total number in the system.

11.4 An assembly line has four stations, and all items must pass through all four stations. Each station can handle only one item at a time. To move items through the assembly line, four specialized containers are used. Whenever a container is free, it is loaded with a new item, which goes through the assembly line with the container. As soon as the item is processed, the container is released. The time spent by the item at each station is exponential with a rate of 1, and if the next station is occupied, the item, and the container, has to wait. The containers are thus part of a closed queueing network. Find the distribution of the waiting lines for all stations.

11.5 Provide formulas to determine the marginal distribution of $X_i, i > 1$ for a closed queueing network with a population of N and M nodes. Use an easily comprehensible notation.

Chapter 12
Infinite-state Markov Chains and Matrix Analytic Methods (MAM)

So far, we have restricted our attention to finite-state Markov chains. However, the majority of the standard queueing models do not put a bound on any of the queues, which means that they result in infinite-state Markov chains. This includes the $M/M/c$ queue, the $GI/M/c$ queue, and the $M/G/1$. The treatment of systems with no limit on some physical state variables is the topic of this chapter. We have to restrict our attention to cases where only one state variable can increase beyond any bound: We are not aware of any convenient, generally applicable numerical method to treat models in which several state variables have no upper bound.

When discussing infinite-state Markov chains, we will frequently use terms like "almost all states" or "almost all equilibrium equation". This refers to all but a finite number of states respectively equations. For instance, in the $M/M/c$ queue, the departure rate is $c\mu$ for almost all states, where μ is the service rate.

Infinite state systems can have properties not shared by any finite-state system. In particular, if $\{X(t), t \geq 0\}$ is a process where $X(t)$ can assume any value between 0 and infinity, the distribution of $X(t)$ can approach a normal distribution with increasing expectation and variance. In this case, the process approaches a so-called *Brownian motion* [78]. For instance, in a $M/M/1$ queue with arrival rate λ and service rate μ, where $\lambda > \mu$, the distribution of the line length $X(t)$ can be approximated by a normal distribution with expectation $t(\lambda - \mu)$ and a variance $t(\lambda + \mu)$. In this sense, the process does diverge. In the $M/M/1$ model, it is easy to determine when the process diverges, but this is no longer true in general. Hence, criteria must be developed for deciding whether or not the process is diverging. In diverging processes, we may be able to find limiting processes, such as the normal distribution with mean and variance increasing with t, but no equilibrium distribution exists, that is, for all finite value of i, the probability that $X(t) = i$ is zero.

If equilibrium probabilities exist, methods of obtaining them must be developed. One can, in principle, find them as limits of transient probabilities. Obviously, within a finite time, one can only reach a finite number of states. It follows that if equilibrium probabilities exist, and if the transient solution converges toward these equilibrium probabilities, then one can truncate the state space at a large enough level, and the equilibrium probabilities of the truncated system differ from the ones of the

© The Author(s), under exclusive license to Springer Nature Switzerland AG 2022
W. Grassmann and J. Tavakoli, *Numerical Methods for Solving Discrete Event Systems*, CMS/CAIMS Books in Mathematics 5, https://doi.org/10.1007/978-3-031-10082-6_12

infinite-state system by less than any prescribed $\epsilon > 0$. The resulting Markov chain could be very large, requiring large amounts of memory and long execution times to find the equilibrium probabilities. The question arises as to whether or not there are better methods to handle infinite-state systems. In certain cases, this is possible. In particular, there are special methods to handle transition matrices by repeating the rows. These methods can be extended to analyze systems where only one state variable, called the *level*, is unbounded, while all other state variables remain finite. These methods can only be used if, for almost all levels, the transition probabilities (rates) of the other state variables are not affected by the level.

The first method we describe to solve problems where there is only one unbounded state variable is an extension of the method of Crout as described in Section 9.1.6. While applying the Crout method to infinite-state Markov chains, we exploit certain regularities concerning the $a_{i,n} = p_{i,n}^{(n)}/(1 - p_{n,n}^{(n)})$ and $b_{n,j} = p_{n,j}^{(n)}$. The resulting extension will be called *extrapolating Crout method*.

Another group of well-known methods of solving systems where there is only one unbounded state variable are the so-called *matrix analytic methods*. It is standard to abbreviate Markov Analytic Method by MAM. As it turns out, the theory of MAMs can be derived from the extrapolating Crout method, as will be shown. We feel that this approach is easier to understand and it is frequently faster than the traditional methods developed for handling MAMs. For similar methods, see [35] and [99].

Before the introduction of MAMs, queueing problems were solved by methods based on characteristic roots for simpler models, and with eigenvalues for more complex models. We will explain these methods in some detail in this chapter. The idea is similar to the one used for transient solutions of Markov chains.

The outline of this chapter is as follows: we first discuss the problems arising when dealing with infinite-state Markov chains. Next, we deal with the extrapolating Crout method, solving first Markov chains where the rows repeat, then with Markov chains where the level is the only state variable that can assume arbitrarily high values. Next, MAMs will be discussed, and their relation to the extrapolating Crout method will be explored. Finally, we cover eigenvalue methods.

12.1 Properties Specific to Infinite-state Markov Chains

In contrast to finite-state Markov chains, infinite-state Markov chains can diverge, and we, therefore, need criteria to decide whether the process in question is diverging or converging. Also, whereas finite-state Markov chains with one communicating class have a unique solution, infinite-state Markov chains do not. This has implications for the numerical stability of algorithms. All these problems are addressed in this section.

12.1.1 Diverging and Converging Markov Chains

We consider an infinite-state non-periodic Markov chain with states numbered starting from 0 with one communicating class. We thus have a process $\{X(t), t \geq 0\}$, and we ask whether or not the process is diverging in the sense that its distribution converges to a distribution with parameters that change with t. Typically, convergence is toward the normal distribution with an increasing expectation. This suggests that we can rely on long-run behavior of $E(X(t))$ in deciding whether or not the process diverges. Specifically, if $E(X(t))$ of the limiting process keeps increasing once t is large enough, then the process is diverging. In the diverging process, $E(X(t))$ will normally eventually increase linearly with t. If, on the other hand, both $E(X(t))$ and $Var(X(t))$ approach a constant value, the process is said to be converging. A third case arises in processes where, in the limiting process, once the lower bound 0 of $X(t)$ is removed, $E(X(t))$ remains constant, but $Var(X(t))$ increases. In this case, the process spreads over a wider and wider area. If $X(t) \geq 0$, this implies that $E(X(t))$ increases because of the increasing spread. We will call these processes semi-diverging.

Typically, all states of diverging Markov chain are transient. By definition, state i is transient if the probability that a sample function starting in i will never return to i is greater 0, and it is recurrent if the probability of returning to i is 1. Also, since we assume that there is only one communicating class, all states of the Markov chain are either recurrent or transient.

Many results of infinite-state Markov chains apply to both DTMCs and CTMCs. However, the terminology differs: while we use "probability" when talking about DTMCs, we use "rate" in CTMCs. For simplicity, unless stated otherwise, when a result applies to both DTMCs and CTMCs, we merely use the term "probability", but the reader should understand in the case of CTMCs, the term "rate" should be used instead. We also apply the terms "stochastic" and "substochastic" to both DTMCs and CTMCs. Hence, a subgenerator will be called "substochastic".

To determine whether an infinite state Markov chain is transient or recurrent, we approach the infinite state Markov chains by a sequence of Markov chains of size N, and we let N go to infinity. This is done by cutting all rows and columns of the corresponding transition matrix above N. The remainder of the matrix is either left unchanged, making the transition matrix substochastic, or some of transition probabilities are changed in such a way that the transition matrix remains stochastic. Cutting without any change of the transition probabilities implies that no equilibrium probabilities exist. In this case, instead of π_i, we use f_i, which, in DTMCs, is the expected number of visits to state i on a path that starts in state 0, and ends as soon as state 0 is entered again. In CTMCs, f_i is the expected times spend in state i on such a path. As shown in Section 9.2.1.1, the f_i satisfy all equilibrium equations, except possibly for the one of state 0, which was not used in the proof of this result.

Indeed, the equilibrium equation of state 0 fails for non-recurrent Markov chains. The reason is that $\sum_{i=0}^{\infty} f_i p_{i,0}$ is the probability of eventually returning to state 0. We denote this probability by r_0:

$$r_0 = \sum_{i=0}^{\infty} f_i p_{i,0}. \tag{12.1}$$

If $r_0 < 1$, then state 0 is transient. By definition, $f_0 = 1$, that is, if $r_0 = 1$, then $f_0 = r_0$, and

$$1 = r_0 = f_0 = \sum_{i=0}^{\infty} f_i p_{i,0}. \tag{12.2}$$

This is exactly the equilibrium equation for state 0. If, however, $r_0 < 1$, then this equation fails. Note that when using state elimination, the right hand side of (12.2) reduces to $p_{0,0}^{(0)}$, which implies $r_0 = p_{0,0}^{(0)}$. Hence, state 0 is recurrent if $p_{0,0}^{(0)} = 1$, and transient if $p_{0,0}^{(0)} < 1$. Since the DTMC is assumed to have only 1 communicating class. all states are recurrent if $p_{0,0}^{(0)} = 1$, and transient otherwise.

In a CTMC with transition matrix $Q = [q_{i,j}, \ 0 \le i, j \ge 0]$, the f_i are the expected times in state i between two successive visits of state 0, with $f_0 = -1/q_{0,0}$. The rate to move from state i to state 0 is $q_{i,0}$, and with a time of f_i spent in state i, this yields an expected $f_i q_{i,0}$ transitions to state 0. Since there is at most 1 transition to state 0, $f_i q_{i,0}$ is also the probability of moving from state i to state 0. Consequently, the probability of ever moving to state 0 is

$$r_0 = \sum_{i=1}^{\infty} f_i q_{i,0}. \tag{12.3}$$

As before, if $r_0 = 1$, then state 0 is recurrent, and if $r_0 < 1$, then state 0 is transient. Since there is only one communicating class, the CTMC is recurrent if $r_0 = 1$, and transient if $r_0 < 1$. If $r_0 = 1$, then the equilibrium equation of state 0,

$$0 = f_0 q_{0,0} + \sum_{i=1}^{\infty} f_i q_{i,0}$$

reduces to $0 = -\frac{1}{q_{0,0}} q_{0,0} + r_0$, which holds if and only if $r_0 = 1$, where r_0 is given by equation (12.3). The relation between $q_{0,0}^{(0)}$ and r_0 is complicated and will not be discussed here. All we can say at this point is that whenever $q_{0,0}^{(0)} < 0$, then the Markov chain is not recurrent.

Let us use intuition to predict what happens to r_0 when the transition matrix is cut at N with no other changes and N increases. We assume that $X(t + 1) - X(t)$ remains finite, which is an assumption necessary to avoid jumps from states with a very high state number to states with a very low state number and vice versa. If $E(X(t))$ tends to decrease with t, the states with low state numbers will be visited more often than states with high state numbers. Consequently, as N increases, states with parts of their transitions to other states cut will be visited less frequently. The absorption probability r_0 will therefore decrease. As $N \to \infty$, r_0 is then likely to approach 1. It will be proven later for specific systems that indeed, $\lim_{N \to \infty} r_0 = 1$.

Once $r_0 = 1$, all equilibrium equations. including the one of state 0, are satisfied, and the f_i are proportional to the π_i.

If, however, $E(X(t))$ tends to increase with t, $X(t)$ will have the tendency to remain high and therefore close to N, where absorption is likely. In this situation, the probability for absorption will remain high for any N. Consequently, if the transition matrix is cut at N without changing any transition probabilities, r_0, the probability of ever returning to state 0, will remain below 1. It follows that all states are transient.

For the models that follow, we will first test if all states are transient or recurrent. We pick a single state, say state i, and test if the expected number of returns to state i is finite or infinite, and decide this way whether the process is converging or diverging. We will not consider the process limits of diverging Markov chains, such as the Brownian motion which are covered elsewhere [94]. We will only cover converging systems, and find their equilibrium distributions.

We have not yet discussed semi-diverging processes, that is, processes that spread over increasing area, where $E(X(t))$ increases only because of the increasing spread of $X(t)$. By the intuitive arguments presented above, one would expect that in such systems, r_0 converges to 1, but very slowly. What can be shown, however, is that the expected time to return to 0 is infinite. Recurrent processes with infinite return times are called *null-recurrent*. Recurrent processes that are not null-recurrent are *positive recurrent*. It can be proven that semi-diverging processes are null-recurrent.

12.1.2 Stochastic and Substochastic Solutions of Infinite-state Markov Chains

In infinite-state Markov chains, it was shown by Kemeny, Snell, and Knapp that the solution of the equilibrium equations is no longer unique [57]. Indeed, we showed in [39] that there are two solutions of interest: the stochastic and the substochastic solution. In diverging Markov chains, the substochastic solutions are obtained by cutting the transition matrix without changing any of the entries, while in converging systems, one changes the entries of the affected states in the transition matrix such that the matrix remains stochastic. The two types of solutions also exist for converging Markov chains.

To analyze transient Markov chains with states from 0 to N, we use again f_i instead of π_i. Given the f_i, the π_i become

$$\pi_i = \frac{f_i}{\sum_{j=0}^{\infty} f_j}. \tag{12.4}$$

This formula only works if $\sum_j f_j$ converges.

To demonstrate the concepts above, consider the $M/M/1$ queue. When the f_i replace the π_i, the equilibrium equations become

$$0 = -\lambda f_0 + \mu f_1 \tag{12.5}$$
$$0 = \lambda f_{i-1} - (\lambda + \mu) f_i + \mu f_{i+1} \ i \geq 1. \tag{12.6}$$

Clearly, if $\lambda > \mu$, then the server cannot handle the average flow, the expected line length increases, and the process is diverging. This is the natural view, and in this case, for reasons of symmetry, one would expect that all states are visited equally often, which implies that all f_i should have the same value. Let us now calculate the f_i recursively. First, we solve equation (12.5) for f_1 to obtain

$$f_1 = \frac{\lambda}{\mu} f_0. \tag{12.7}$$

Solving equation (12.6) for f_{i+1} yields

$$f_{i+1} = \frac{\lambda + \mu}{\mu} f_i - \frac{\lambda}{\mu} f_{i-1}.$$

We now use complete induction to show that

$$f_{i+1} = \frac{\lambda}{\mu} f_i. \tag{12.8}$$

This is true for $i = 1$ because of equation (12.7), and it is easy to show that if it is true for i, it is also true for $i + 1$. Note that if equation (12.5) holds, then this is the only solution. For $\lambda > \mu$, this violates our guess that all f_i should have the same value. Of course, setting f_i to 1 for all i will satisfy equation (12.6), but it does not satisfy equation (12.5). Indeed, by requiring that the equation for $X = 0$ holds, we force r_0 as given in equation (12.1) to be 1, ensuring in this way that the process is stochastic.

Note that equation (12.8) is the solution of the $M/M/1/N$ queue for $i \leq N$, and in this case, the π_i can be obtained by dividing the f_i by $\sum_i f_i$. This is the correct solution for any N, even for arbitrary large N. In all these cases, the transition matrices were stochastic, and what we got is the stochastic solution for the diverging $M/M/1/N$ queue. Note that the f_i keep increasing. This is what one would expect because the system spends most of its time close to its upper bound, even when this bound approaches infinity. This behavior is typical of the stochastic solution of diverging systems. In general, if there is an upper bound, the transition matrices are stochastic, and even as $N \rightarrow \infty$, the stochastic solution remains valid. Stochastic solutions are useful for exploring the dependence on upper bounds for state variables, such as queues.

We will encounter substochastic solutions even when the system is converging. For instance, in the case of the $M/M/1$ queue, the substochastic solution is $f_i = 1$ for all i, no matter what values λ and μ happen to have. Indeed, this solution satisfies the interior equilibrium equation (12.6) as is easily verified.

Let us see how this fits into the theory of Kemeny, Snell, and Knapp [57]. First of all, if we are looking at an infinite-state Markov chain with transition matrix $P = [p_{i,j}, i, j \geq 0]$, then the f_i can be obtained as follows, as shown in Section 9.2.1.

If $r = [p_{0,j}, j \geq 1]$ and $\bar{P} = [p_{i,j}, i, j \geq 1]$ then

$$f = [f_i, i \geq 1] = r(I - \bar{P})^{-1}.$$

Here, $(I - \bar{P})^{-1}$ is the fundamental matrix, yielding the number of visits to states j, given one starts in state 0 before entering state 0 again. Now, according to Kemeny, Snell, and Knapp, there are several inverses of $(I - \bar{P})$, and indeed, we identified two possible matrices that can be considered: the stochastic and the substochastic one. The authors say that one should take the minimal inverse of $I - \bar{P}$, and in diverging systems, the minimal solution is the one given by the substochastic matrix. We will show in Section 12.4 that in some cases, one can construct additional solutions, but these solutions have negative or complex entries in the transition matrix, which is not acceptable in our context.

12.1.3 Convergence to the Desired Solution

Since infinite-state Markov chains have two solutions, the question arises as to what happens when N is large, but not necessarily infinite. In particular, under which conditions is a convergence to the stochastic solution to be expected, and when will we have a convergence toward the substochastic solution? As we will see, what solution the elimination converges to is important for numerical analysis. Clearly, the GTH method ensures that the transition matrix always stays stochastic, which implies that the stochastic solution results. We thus assume that the GTH method is not used.

In converging systems, we can cut the transition matrix and change nothing else, and still obtain the stochastic solution. For instance, in the $M/M/1$ queue, cutting the transition matrix at N would result in the following equation for state N:

$$0 = \lambda \pi_{N-1} - (\lambda + \mu)\pi_N$$

instead of

$$0 = \lambda \pi_{N-1} - \mu \pi_N.$$

The first equation is the equation of the substochastic solution, whereas the second one is the equation for the stochastic solution. We will call the first equation "substochastic" and the second one "stochastic".

The formulas for state elimination given by equation (9.3) can now be applied, In fact, when state n is eliminated, only $q_{n,n}^{(n)}$ changes. One obtains

$$q_{n-1,n-1}^{(n-1)} = -(\lambda + \mu) - \frac{\lambda \mu}{q_{n,n}^{(n)}}.$$

Setting $q_{n,n}^{(n)} = s_n$, this yields

$$s_{n-1} = -(\lambda + \mu) - \frac{\lambda\mu}{s_n}. \tag{12.9}$$

Note that if $s_n = -\lambda$, then s_{n-1} is also $-\lambda$, and if $s_n = -\mu$, s_{n-1} is also $-\mu$. These are the only two values satisfying the condition $s_n = s_{n-1}$. To see this, set $s_n = s$ for all n, in which case one obtains a quadratic equation for s, and quadratic equations have only two solutions.

When starting with the stochastic equation for state N, $s_n = -\mu$, and then $s_{n-1} = -\mu$ for all n. Consequently,

$$0 = \lambda\pi_{n-1} - \mu\pi_n,$$

from which follows that $\pi_{n-1} = \frac{\lambda}{\mu}\pi_n$, which is the correct result. If we start with the substochstic equation, the result depends on whether $\lambda > \mu$ or $\lambda < \mu$. Consider first the case $\lambda < \mu$. For example, let $N = 40$, $\lambda = 0.9$ and $\mu = 1$. Calculations on a spreadsheet show that when starting with the substochastic solution $s_{40} = -(\lambda + \mu) = -1.9$, one has

$$s_{30} = -1.0457, \quad s_{20} = -1.0123, \quad s_{10} = -1.0039.$$

We thus observe a convergence to $-\mu = -1$ even when starting with the substochastic equation. In the limit, this leads again to the stochastic result, even though we used the substochastic equation to start with.

Next, consider the case where $\lambda > \mu$, say $\lambda = 1$ and $\mu = 0.9$. If we start again with $s_{40} = -(\lambda + \mu) = -1.9$, we get exactly the same values as above, that is, $s_{30} = -1.0457$, $s_{20} = -1.0123$, and so on. It follows that s_{40-i} converges to -1 as i increases, and -1 is now $-\lambda$. Hence, the reduced system converges to

$$0 = \lambda\pi_{n-1} - \lambda\pi_n,$$

implying $\pi_{n-1} = \pi_n$, which is the substochastic solution.

For approximating finite-state queues, we would like to obtain the stochastic solution. and this solution can be found by setting $s_{40} = -\mu = -0.9$. In this case, all s_i evaluate to -0.9, the correct result. However, if s_{40} is slightly below -0.9, then convergence is toward $-\lambda = -1$, which corresponds to the substochastic solution. For instance, for $s_{40} = -0.91$, $s_{10} = -0.972$. This indicates that convergence is toward the substochastic solution rather than to the stochastic one. Generally, if $\lambda > \mu$, the stochastic solution is unstable, and even slight deviations from the stochastic equation, which can be caused by rounding, can lead to rather large errors. To obtain the stochastic solution in this case, one must use the GTH method.

We now show that unless the GTH method is used, the results above imply that in converging systems, one should never start eliminating π_0 first, then π_1, then π_2, and so on, which is the way textbooks on linear algebra proceed. If we follow the textbook method, and eliminate the π_i in the order given by i, then the pivot elements become

$$s_{i+1} = -(\lambda + \mu) - \frac{\mu\lambda}{s_i}.$$

This is essentially the same equation as (12.9), with λ and μ interchanged. Consequently, using the same arguments as above, one can find experimentally that if $\lambda < \mu$, the recursion is numerically unstable.

More generally, eliminating the π_i in decreasing order of i, starting with $i = N$ is justified as long as the system is converging. It follows that eliminating the π_i with low values of i first is not optimal when dealing with converging systems. It may not be appropriate in other contexts either because the effect of rounding errors is reduced by dealing with the variables that have the smallest absolute values first.

12.2 Markov Chains with Repeating Rows, Scalar Case

There are many queueing systems with a single state variable X in which the columns of the transition matrix repeat from a certain point onward. Examples include the $M/D/c$ queue [44], [91] bulk queues [10], the $M/G/1$ and the $GI/M/c$ queues [44] as well as the waiting time of the discrete-time $GI/G/1$ queue [75]. Though different methods were used in literature to solve these problems, we found that the extrapolating Crout method in the form we will present next tends to be very efficient when compared to competing methods. Notice that since we are dealing with queueing systems, it makes sense to number the states starting with 0 rather than with 1.

We concentrate on DTMCs, but the derivation for treatment for CTMCs is very similar. We assume that there is no upper bound on X, and that for almost all i and j, $P\{X(t+1) = j \mid X(t) = i\} = p_{j-i}$. Stated mathematically,

$$P\{X(t+1) = j + k \mid X(t) = j\} = p_k.$$

This means, that for $X(t)$ outside the boundary, $X(t+1) = X(t) + D$, where $P\{D = k\} = p_k$. For simplicity, we also assume that the transition matrix is banded, with all non-zero transition probabilities of row i laying between $i - g$ and $i + h$. This implies that $p_k = 0$ unless $-g \le k \le h$. This restriction can be weakened, however. Thus, row i looks as follows for almost all i:

	column								
	$i-g$	$i-g+1$	$\ldots$	$i-1$	i	$i+1$	$\ldots$	$i+h-1$	$i+h$
row i	p_{-g}	p_{-g+1}		$\cdots\, p_{-1}$	p_0	p_1		$\cdots\, p_{h-1}$	p_h

Clearly, $\sum_{k=-g}^{h} p_k = 1$. Aside from the part where the rows repeat, there is often a *boundary*, where the $p_{i,j}$ may be different from p_{j-i}.

In this section, we first discuss whether $X(t)$ is a converging or diverging process. Next, the extrapolating Crout method is presented, which leads to an extremely efficient algorithm. The theory is then applied to solve the discrete-time $GI/G/1$ queue, and the $D/M/c$ queue.

12.2.1 Recurrent and Transient Markov Chains

We now develop a method to show whether the Markov chain is transient or recurrent. In order to do that, we pick a single state, and test if this state is recurrent. If the state is transient, then so is the Markov chain, provided, of course, that all states communicate. Similarly, if the state is recurrent, then so is the Markov chain.

The basic idea for our approach is that the $D(t) = X(t + 1) - X(t)$ are independent random variables with finite mean and variance. If we now consider a state $X_1(0) = x$, then $X_1(t) = x + \sum_{n=1}^{t} D(n)$. Since the $D(n)$ are independent random variables, the central limit theorem applies, and the distribution of $\sum_{n=1}^{t} D(n)$ converges toward the normal distribution. This fact allows us to show that if $E(D) > 0$, state x is transient. This is based on the following theorem:

Theorem 12.1 *Consider the process $\{X(n), n \geq 0\}$, defined by the relation $X(n + 1) = X(n) + D(n)$, where the $D(n)$ are independent, identically distributed random variables with finite expectation $E(D)$ and finite variance $Var(D)$. As long as all states of the process communicate, we have the following result: If $E(D) > 0$, then the Markov chain is transient, if $E(D) \leq 0$, then the Markov chain is recurrent.*

Proof We pick a state outside the boundary, say x, and we determine whether x is transient or recurrent. The process is transient whenever the expected number of returns to x is finite. Note that if x is high enough to never hit the boundary, we have

$$E(\text{returns to } x) = \sum_{t=1}^{\infty} P\{X(t) = x\} \leq \sum_{t=1}^{\infty} P\{X(t) \leq x\}.$$

It is thus sufficient to prove that $\sum_{t=1}^{\infty} P\{X(t) \leq x\} < \infty$. It follows from the definition of $X(n)$ that $X(n) = x + \sum_{i=1}^{n} D(i)$, and with the $D(i)$ being independent, $E(X(n)) = x + nE(D)$ and $Var(X(n)) = nVar(D)$. According to the central limit theorem, the probability that $X(n)$ is below x converges to:

$$\int_{-\infty}^{-z} \frac{1}{2\pi} e^{-t^2/2} dt = \frac{1}{2\pi} \int_{z}^{\infty} e^{-t^2/2} dt,$$

where

$$z = \frac{nE(D)}{\sqrt{nVar(D)}} = \sqrt{n}b, \ b = \frac{E(D)}{\sqrt{Var(D)}}.$$

The expected frequency of visiting states $X \leq x$ therefore converges to

$$\frac{1}{\sqrt{2\pi}} \sum_{n=1}^{\infty} \int_{\sqrt{n}b}^{\infty} e^{-t^2/2} dt.$$

Hence, we only need to prove that this sum converges. For this, we use the inequality $e^{-t^2/2} < e^{-t}$ for $t > 2$, which implies

$$\sum_{n=1}^{\infty}\int_{\sqrt{n}b}^{\infty}e^{-t^2/2}dt < \sum_{n=1}^{\infty}\int_{\sqrt{n}b}^{\infty}e^{-t}dt = \sum_{n=1}^{\infty}e^{-\sqrt{n}b}.$$

Now, if m is $\sqrt{n}$, rounded down to the closest integer, then

$$\sum_{n=1}^{\infty}e^{-\sqrt{n}b} < \sum_{n=1}^{\infty}(2m+1)e^{mb}.$$

This equation follows from the fact that as m increases by 1, $n = m^2$ increases by $(m+1)^2 - m^2 = 2m + 1$, which leads to the factor $2m + 1$ in the sum on the right. Thus, we obtain the arithmetic-geometric series which is known to converge. This settles the case $E(D) > 0$.

If $E(D) < 0$, then $E(X(t))$ is negative, and falling to a state at or below x is certain. But since we assumed that all states communicate, even if after the return to a state below x, we have a positive probability of hitting x later. Consequently, a return to x is certain. If $E(D) = 0$, then the probability that $X(t)$ is below x is 0.5, and the expected number of visits to state x is infinite. □

Note that the theorem does not make use of the fact that $-g \leq D \leq h$. Hence, it holds for any distribution of D. It only requires that the variance is finite. Of course, the variance is finite whenever D can only assume a finite number of values.

If $E(D) = 0$, then the probabilities spread over a wider and wider area, which implies that $E(X(t))$ increases. This increase is very slow, however. In fact, the variance increases at about the same rate, and the process is still recurrent. It can be shown, however, that the expected time between visits is infinite, making the process null-recurrent. We suspect that this is also true when $E(D) > 0$, but $Var(E(D))$ is infinite, but we found no proof in literature to this effect. Of course, if D can only assume a finite number of values, the variance is always finite.

12.2.2 The Extrapolating Crout Method

We first have to divide the transition matrix into a boundary and an interior part. In the interior part, the transition probabilities $p_{i,j}$ are equal to p_{j-i}, but in the boundary, this may not be the case. We assume that the boundary includes all entries of the transition matrix with $i < d_1$ or $j < d_2$.

In formulating the equilibrium equations, we have to concentrate on the columns, because the equilibrium equation for π_j is given by column j of the transition matrix. First, we need to determine in which rows of column j there are non-zero entries. Since $-g \geq j - i \geq h$, we have $j - h \leq i \leq j + g$, that is, the first row of column j inside the band is $j - h$, or 0 if $j - h < 0$, and the last entry of column j in the band is in row $j + g$. For $j < d_2$, the equilibrium equation is thus:

$$\pi_j = \sum_{i=0,j-h}^{j+g} \pi_i p_{i,j}, \ j < d_2.$$

In sums with two lower bounds, the higher one is to be used. In the sum above, $i = 0, j - h$ must thus be read $i = \max(0, j - h)$.

For $j \ge d_2$, we find, if $\pi_i = 0$ for $i < 0$:

$$\pi_j = \sum_{i=j-h}^{d_1-1} \pi_i p_{i,j} + \sum_{i=d_1,j-h}^{j+g} \pi_i p_{j-i}, \ j > 0. \tag{12.10}$$

Depending on j, one of the sums can be empty, that is, its upper bound is below its lower bound. The first sum is empty when $d_1 - 1 < j - h$, or $j > d_1 - 1 + h$, and then

$$\pi_j = \sum_{i=d_1,j-h}^{j+g} \pi_i p_{j-i}, \ j \ge d_1 + h. \tag{12.11}$$

This equation repeats forever.

The second sum of (12.10) is empty when $j + g < d_1$, and then

$$\pi_j = \sum_{i=j-h}^{d_1-1} \pi_i p_{i,j}, \ j < d_1 - g. \tag{12.12}$$

The equations given by (12.12) will be referred to as *boundary equations*. If $d_1 - g \le j < d_1 + h$, then equation (12.10) cannot be simplified.

The following theorem allows us to extend the Crout method to deal with infinite-state systems by using extrapolation, leading to a method we call *extrapolating Crout method*.

Theorem 12.2 Let $p_{i,j} = P\{X(t + 1) = j \mid X(t) = i\}$. If the $p_{i,j} > 0$ only if $-g \le j - i \le h$, and if there are values d_1 and d_2 such that all pairs $i \ge d_1$ and $j \ge d_2$, $p_{i,j} = p_{j-i}$, then for $i \ge d_1$ and $j \ge d_2$, we have $a_{i,n} = a_{n-i}$ and $b_{n,j} = b_{n-j}$.

Proof To prove the theorem, note that $p_{i,j}^{(n)}$, $i \ge d_1$, $j \ge d_2$, is determined by the probabilities of all paths going from i to j, passing only through states above n. Now any path such

$$X(t) = i, X(t + 1) > n, X(t + 2) > n, \dots, X(t + m - 1) > n, X(t + m) = j.$$

can be matched with the path

$$X(t) = i+k, X(t+1) > n+k, X(t+2) > n+k, \dots, X(t+m-1) > n+k, X(t+m) = j+k,$$

where k is a fixed positive integer. It follows that $p_{i,j}^{(n)} = p_{i+k,j+k}^{(n+k)}$. The theorem now follows from the fact that

$$b_{n,j} = p_{n,j}^{(n)}$$

$$a_{i,n} = \frac{1}{1 - b_0} p_{i,n}^{(n)}.$$

Note that the theorem does not settle the question of whether the a_{j-i} or the b_{i-j} are unique. Indeed, there are two solutions as proven in [36], the stochastic solution with $\sum_{j=0}^{g} b_j = 1$, and the substochastic solution with $\sum_{j=0}^{g} b_j < 1$.

Now let us look at a transition matrix of size N with repeating rows. Note if $p_{i,j}$ is 0 unless $-g \le j - i \le h$, then we only need the last h rows and the last g columns of the transition matrix. These rows and columns are independent of N. The other rows and columns can be added as the calculations progress. Consequently, nothing prevents us from letting N go to infinity. We now eliminate π_N by using the equilibrium equation of state N, add one row and one column, and eliminate π_{N-1}, and so we continue. What happens is that the elimination process either converges toward the stochastic or toward the substochastic solution, depending on whether the transition matrix is stochastic or substochastic. Of course, if we use the GTH method, we will always converge toward the stochastic solution.

To solve the equilibrium equations, we need the $a_{i,n}$ and $b_{n,j}$. To determine these, it is convenient to use the method of Crout, which is given in equations (9.16) and (9.17) of Chapter 9. We repeat it here for convenience:

$$a_{i,n} = \frac{p_{i,n} + \sum_{m=n+1}^{N} a_{i,m} b_{m,n}}{1 - b_{n,n}}, \ i < n, \tag{12.13}$$

$$b_{n,j} = p_{n,j} + \sum_{m=n+1}^{N} a_{n,m} b_{m,j}, \ j \le n. \tag{12.14}$$

It turns out to be convenient to replace the $a_{i,n}$ by a_{n-i}^{n} and the $b_{n,j}$ by b_{n-j}^{n}. Also, the upper bounds for our summations are often complicated, and to simplify the notation, we set $a_k^n = 0$ if $n > N$, or $k > h$, and $b_k^n = 0$ if $n > N$ or $k > g$. This allows us to use infinity as the upper bound for our summations. With these conventions, equations (12.13) and (12.14) become

$$a_{n-i}^{n} = \frac{1}{1 - b_0^n} \left(p_{n-i} + \sum_{m=n+1}^{\infty} a_{m-i}^{m} b_{m-n}^{m} \right), \ i < n, \tag{12.15}$$

$$b_{n-j}^{n} = p_{j-n} + \sum_{m=n+1}^{\infty} a_{m-n}^{m} b_{m-j}^{m}, \ j \le n. \tag{12.16}$$

When we use these formulas repeatedly, the a_i^n converge to a_i, and the b_j^n converge to b_j. Thus, we can stop iterating as soon as the difference between two iterations is less than some prescribed limit. At this point, we have to deal with the boundary equations.

The best way to explain how to handle the boundary equations is by means of an example. Suppose the first five rows and the first seven columns of the transition matrix are as follows:

$$\begin{bmatrix} p_{0,0} & p_1 & p_{0,2} & 0 & 0 & 0 & 0 \\ p_{1,0} & p_{1,1} & p_{1,2} & p_{1,3} & 0 & 0 & 0 \\ p_{2,0} & p_{2,1} & p_0 & p_1 & p_2 & 0 & 0 \\ 0 & p_{3,1} & p_{-1} & p_0 & p_1 & p_2 & 0 \\ 0 & 0 & p_{-2} & p_{-1} & p_0 & p_1 & p_2 \end{bmatrix}.$$

If we know the $a_{i,n}$ and $b_{n,j}$, then the $a_{i,n-1}$ and $b_{n-1,j}$ can be determined. Consequently, all entries of the fifth row, the one related to state 4, are fully determined, and it becomes $0, 0, b_2, b_1, b_0, a_1, a_2$. Similarly, the entries of column 5 and column 6 are given by a_2 and a_1. Also, $b_{3,1} = p_{3,1}$ according to equation (12.14). Moreover, by equation (12.13), $a_{1,3} = p_{1,3}/(1 - b_0)$. After this Crout iteration, the first four rows and columns of the transition matrix now look as follows:

$$\begin{bmatrix} p_{0,0} & p_1 & p_{0,2} & 0 \\ p_{1,0} & p_{1,1} & p_{1,2} & \frac{p_{1,3}}{1-b_0} \\ p_{2,0} & p_{2,1} & p_0 & a_1 \\ 0 & p_{3,1} & b_1 & b_0 \end{bmatrix}.$$

The next Crout iteration can now proceed. Note that p_1 in row 1 cannot be changed to a_1, but must be calculated.

To summarize: Once the boundary is hit, all p_i lead to a_i for $i > 0$, provided the entries of below p_i and to the right of p_i repeat. The other $a_{i,n}$ must be calculated. Similarly, the p_{-j} become b_j, provided the entries below and to the right of the entry in question repeat.

We now replace the upper limit "infinity" of the sums of equation (12.15). In the sum $\sum_{m=n+1}^{\infty} a_{m-i}^m b_{m-n}^m$, n is at most N, $m - i$ is at most h and $m - n$ is at most g. This implies that $m \le h + i$ and $m \le g + n$. A similar approach can be used to find the upper bounds of the sum $\sum_{m=n+1}^{\infty} a_{n-m}^m b_{m-j}^m$ of equation (12.16). One thus finds

$$a_{n-i}^n = \frac{1}{1 - b_0^n} \left(p_{n-i} + \sum_{m=n+1}^{N,h+i,g+n} a_{m-i}^m b_{m-n}^m \right), \quad i < n, \qquad (12.17)$$

$$b_{n-j}^n = p_{j-n} + \sum_{m=n+1}^{N,h+n,g+j} a_{m-n}^m b_{m-j}^m, \quad j \le n. \qquad (12.18)$$

The method of finding the a_i and b_j can be improved. As in the method of Gauss–Seidel, we can always use the most recent value for the calculation. In order to do so, we drop the m in equations (12.17) and (12.18), as well as any upper summation bounds. After further minor changes, we obtain

$$a_i = \frac{1}{1-b_0}\left(p_i + \sum_{j=1}^{h+i,g} a_{i+j}b_j\right) \tag{12.19}$$

$$b_j = p_{-j} + \sum_{i=}^{h,j+g} a_i b_{i+j}. \tag{12.20}$$

Note that once the a_i are known for $i = 1, 2, \ldots, h$, we can calculate the b_j as follows, where $m_b = \min(g, h)$:

$$b_g = p_{-g}$$
$$b_{g-1} = p_{-g+1} + a_1 b_g$$
$$\vdots$$
$$b_0 = p_0 + a_1 b_1 + a_2 b_2 + \cdots + a_{m_b} b_{m_b}.$$

Similarly, if the b_j are known for $j = 0, 1, \ldots, g$, the a_i, $i = 1, 2, \ldots, h$ can be found as follows, where $m_a = \max(h, g + 1)$:

$$a_h = \frac{1}{1-b_0} p_h$$
$$a_{h-1} = \frac{1}{1-b_0}(p_{h-1} + a_h b_1)$$
$$\vdots$$
$$a_1 = \frac{1}{1-b_0}(p_1 + a_2 b_1 + \cdots + a_{m_a} b_{m_a-1}).$$

The a_i and b_j can be obtained by successive approximation, that is, we obtain a_i^m and b_j^m as the mth approximation, starting with $a_i^0 = 0$, $i = 1, 2, \ldots, h$, then calculating the b_j^1, $j = g, g-1, \ldots, 0$, which can then be used to create a new approximation for a_i, $i = h, h-1, \ldots, 1$. This continues until the difference between two iterations is less than some prescribed value $\epsilon > 0$. Before we present the corresponding algorithm, note that there are two ways to determine $1 - b_0^m$: We can calculate b_0^m directly, or we can use the GTH method, resulting in $1 - b_0^m = \sum_{i=1}^{g} b_i^m$. We can use either value, but taking the average between the two values may be preferable, a hunch supported by numerical experiments. These ideas are implemented in the following algorithm.

Algorithm 32.

1. $a_i^0 = 0$, $i = 1, 2, \ldots, h$
2. For $m = 0, 1, 2, \ldots$ until $\max(|a_j^m - a_j^{m+1}|) < \epsilon$
   2.1 $b_j^m = p_{-j} + \sum_i a_i^{m-1} b_{i+j}^m$, $j = g, g-1, \ldots, 0$
   2.2 $S^m = 0.5\left(1 - b_0^m + \sum_{j=1}^{g} b_j^m\right)$
   2.3 $a_i^m = \left(p_i + \sum_j a_{i+j}^m b_j^m\right)/S^m$, $i = h, h-1, \ldots, 1$

In this algorithm, b_i^m is at least initially less than b_i, meaning that

$$\sum_{i=1}^{g} b_i^m < 1 - b_0 < 1 - b_0^m.$$

This again supports the statement that taking the average of the two bounds should improve the estimate for $1 - b_0$. Moreover, in all experiments reported in literature, such as [11], [36], [45], Algorithm 32 converged rapidly toward the stochastic solution.

Once the a_i and b_j are obtained, one can calculate all $a_{i,j}$ and $b_{i,j}$ of the boundary, as shown earlier. This then allows one to find all π_i recursively, except for a factor, using equation (9.18). This equation becomes

$$\pi_j = \begin{cases} \displaystyle\sum_{i=0}^{j-1} \pi_i a_{i,j}, & 0 < j < d_2 \\ \displaystyle\sum_{i=0}^{d_1-1,j-1} \pi_i a_{i,j} + \sum_{i=d_1}^{j-1} \pi_i a_{j-i}, & d_2 \leq j < d_1 + h \\ \displaystyle\sum_{i=j-h}^{j-1} \pi_i a_{j-i}, & j \geq d_1 + h. \end{cases} \tag{12.21}$$

There is no equation for π_0, and as usual, π_0 must be determined by the condition that the sum of all probabilities must be 1. To do this, one could calculate the $\tilde{\pi}_j$, $j \geq 0$ first, and add them all up. However, since there is an infinite number of π_j, one would have to truncate the sum, which leads to a truncation error. A better alternative is to use z-transforms, a method that will be discussed next.

12.2.3 Using Generating Functions for Norming the Probabilities

We now show how π_0 can be found by using z-transforms as they were discussed in Section 1.3.4. We have

$$P(z) = \sum_{i=0}^{\infty} \pi_i z^i. \tag{12.22}$$

Clearly,

$$P(1) = \sum_{i=0}^{\infty} \pi_i = 1.$$

We also will use $P(z)$ to find the expectation of X.

First, consider the case where $d_1 = 0$, when

$$\pi_j = \sum_{i=j-h}^{j-1} \pi_i a_{j-i}, \quad j > 0.$$

By substituting this expression into (12.22), we obtain

$$\sum_{j=0}^{\infty} \pi_j z^j = \pi_0 + \sum_{j=1}^{\infty} \sum_{i=j-h}^{j-1} \pi_i a_{j-i} z^j$$

$$= \pi_0 + \sum_{i=0}^{\infty} \pi_i z^i \sum_{j=i+1}^{i+h} a_{j-i} z^{j-i}$$

$$= \pi_0 + \sum_{i=0}^{\infty} \pi_i z^i \sum_{j=1}^{h} a_j z^j.$$

Substituting $\sum_{i=0}^{\infty} \pi_i z^i$ by $P(z)$ in the equation above, we get

$$P(z) = \pi_0 + P(z) \sum_{j=1}^{h} a_j z^j. \tag{12.23}$$

Since $P(1) = 1$, we conclude

$$1 = \pi_0 + \sum_{j=1}^{h} a_j,$$

hence

$$\pi_0 = 1 - \sum_{j=1}^{h} a_j. \tag{12.24}$$

Thus, instead of starting with $\tilde{\pi}_0 = 1$, we can use the real π_0 as given above to calculate all π_i directly, using equation (12.21).

To obtain $E(X)$, we take the derivative of (12.23) with respect to z:

$$P'(z) = P'(z) \sum_{j=1}^{h} a_j z^j + P(z) \sum_{j=1}^{h} j a_j z^{j-1}.$$

Consequently, by setting $z = 1$, and noting that $P'(1) = E(X)$, we find

$$E(X) = E(X) \sum_{j=1}^{h} a_j + \sum_{j=1}^{h} j a_j$$

or

$$E(X) = \frac{\sum_{j=1}^{h} j a_j}{1 - \sum_{j=1}^{h} a_j} = \frac{1}{\pi_0} \sum_{j=1}^{h} j a_j. \tag{12.25}$$

If $d_1 > 0$, then we proceed as follows. From equation (12.21), $j \geq d_1 + h$ we conclude

$$\pi_j = \pi_{j-1}a_1 + \cdots + \pi_{j-h}a_h, \ j \geq d_1 + h.$$

Now multiply these equations by z^j and add them to obtain

$$\sum_{j=d_1+h}^{\infty} \pi_j z^j = \sum_{j=d_1+h}^{\infty} \pi_{j-1}a_1 z^j + \cdots + \sum_{j=d_1+h}^{\infty} \pi_{j-h}a_h z^j$$

or

$$\sum_{j=d_1+h}^{\infty} \pi_j z^j = \sum_{j=d_1+h-1}^{\infty} \pi_j z^j a_1 z + \cdots + \sum_{j=d_1}^{\infty} \pi_j z^j a_h z^h.$$

One can replace any sum of the form $\sum_{j=k}^{\infty} \pi_j z^j$ by $P(z) - \sum_{j=0}^{k-1} \pi_j z^j$, where k ranges from d_1 to $d_1 + h$. This yields

$$P(z) - \sum_{j=0}^{d_1+h-1} \pi_j z^j = \left(P(z) - \sum_{j=0}^{d_1+h-2} \pi_j z^j\right)a_1 z + \cdots + \left(P(z) - \sum_{j=0}^{d_1-1} \pi_j z^j\right)a_h z^h.$$

We move all terms with $P(z)$ to the left, and all sums to the right, and after minor manipulations, we obtain

$$P(z)\left(1 - \sum_{i=1}^{h} a_i z^i\right) = \sum_{j=0}^{d_1+h-1} \pi_j z^j - a_1 \sum_{j=0}^{d_1+h-2} \pi_j z^{j+1} - \cdots - a_h \sum_{j=0}^{d_1-1} \pi_j z^{j+h}. \quad (12.26)$$

When $z = 1$, $P(1) = 1$, and we get

$$P(1)\left(1 - \sum_{i=1}^{h} a_i\right) = \sum_{j=0}^{d_1+h-1} \pi_j - a_1 \sum_{j=0}^{d_1+h-2} \pi_j - \cdots - a_h \sum_{j=0}^{d_1-1} \pi_j,$$

or

$$P(1) = \frac{\sum_{j=0}^{d_1+h-1} \pi_j - \sum_{i=1}^{h} a_i \sum_{j=0}^{d_1+h-1-i} \pi_j}{1 - \sum_{i=1}^{h} a_i}. \quad (12.27)$$

To find the equilibrium probabilities, we proceed as usual by setting $\tilde{\pi}_0 = 1$, in which case all $\tilde{\pi}_i$, $i > 0$ are determined. Clearly $\tilde{\pi}_i = c\pi_i$, and the $\tilde{\pi}_i$ therefore satisfy all equilibrium equations. The value of c can be found by replacing all π_i in equation (12.27) by $\tilde{\pi}_i$, resulting in $c = P(1) = \sum_i \tilde{\pi}_i$.

To find the expectation $E(X)$, take the derivative of equation (12.26):

$$P'(z)\left(1 - \sum_{i=1}^{h} a_i z^i\right) - P(z)\sum_{i=1}^{h} ia_i z^{i-1} =$$

$$\sum_{j=0}^{d_1+h-1} j\pi_j z^{j-1} - \sum_{i=1}^{h} a_i \sum_{j=0}^{d_1+h-1-i} (j+i)\pi_j z^{j+i-1}.$$

When $z = 1$, then $P(1) = 1$ and $P'(1) = E(X)$. The equation above then yields

$$P'(1) = E(X) = \frac{\sum_{i=1}^{h} i a_i + \sum_{j=0}^{d_1+h-1} j\pi_j - \sum_{i=1}^{h} a_i \sum_{j=0}^{d_1+h-1-i}(j+i)\pi_j}{1 - \sum_{i=1}^{h} a_i}. \tag{12.28}$$

The expectation can thus be determined once the π_j, $j \le d_1 + h - 1$ are known.

12.2.4 The Waiting-time Distribution of the $GI/G/1$ Queue

As an example of a process with repeating columns, we discuss how to find the waiting-time distribution of the $GI/G/1$ queue. The waiting time, W_n, of customer n is defined as the time from the moment that customer n arrives, up to the start of his/her service. To find the waiting -time distribution, let T_n be the time of the arrival of customer n, S_n her service time, and A_n the time between T_n and T_{n+1}.

The transition probabilities are

$$p_{i,j} = P\{W_{n+1} = j \mid W_n = i\}.$$

To find $p_{i,j}$, note that customer n leaves at time $T_n + W_n + S_n$. If $T_n + W_n + S_n \le T_{n+1}$, that is, if customer n leaves before customer $n + 1$ arrives, then $W_{n+1} = 0$, that is

$$W_{n+1} = 0 \text{ if } T_n + W_n + S_n - T_{n+1} \le 0.$$

Otherwise, customer $n + 1$ has to wait until $T_n + W_n + S_n$, the time when customer n leaves, and since he/she arrived at T_{n+1}, we have

$$W_{n+1} = T_n + W_n + S_n - T_{n+1}.$$

We combine these two formulas for W_{n+1}. With $T_{n+1} - T_n = A_n$, this yields

$$W_{n+1} = \max(0, W_n + S_n - A_n).$$

We now define U_n to be the random variable given by $S_n - A_n$. Consequently,

$$W_{n+1} = \max(0, W_n + U_n).$$

Since we are interested in the equilibrium solution, we can omit the subscripts n. If α_k is the probability that $A = k$, s_j is the probability that $S = j$, and p_k is the probability that $u = k$, one has

$$p_k = \sum_j s_{k+j}\alpha_j.$$

The transition matrix is thus:

$$
P = \begin{bmatrix}
\sum_{k=0}^{\infty} p_{-k} & p_1 & p_2 & p_3 & \cdots\cdots\cdots \\
\sum_{k=1}^{\infty} p_{-k} & p_0 & p_1 & p_2 & p_3 & \ddots & \ddots \\
\sum_{k=2}^{\infty} p_{-k} & p_{-1} & p_0 & p_1 & p_2 & p_3 & \ddots \\
\vdots & & \ddots & \ddots & \ddots & \ddots & \ddots & \ddots
\end{bmatrix}.
$$

We now use Algorithm 32 to calculate the a_i, $i \geq 1$. To apply our method, we require that p_k can be non-zero only if $-g \leq j - i \leq h$. Therefore, we must restrict A_n to be less than or equal to $g + 1$, and S_n to be less than or equal to $h + 1$. Once the a_k are found, $\pi_0 = \mathrm{P}\{W_n = 0\}$ is given by (12.24), π_n, $n > 0$ by (12.21), and $\mathrm{E}(W)$ by (12.25).

We now present two numerical examples to show how this works. In the first example, $\alpha_i = c/i$, $i = 1, 2, \ldots, 5$, and $s_i = d/i$, $i = 1, 2, 3, 4$, where c and d are determined such that the probabilities add up to 1. Applying Algorithm 32, we find $a_1 = 0.4182$, $a_2 = 0.2204$ and $a_3 = 0.0957$. From equation (12.24), we find $\pi_0 = 1 - 0.4182 - 0.2204 - 0.0957 = 0.2657$, $\pi_1 = \pi_0 a_1 = 0.1111$, $\pi_2 = 0.1050$, and so on. It took 8 iterations of Algorithm 32 to obtain the a_i.

Our second example involves a problem where inter-arrival times and service times are approaching long-range dependency. Specifically, we used truncated Pareto distributions: For the service time distribution, we used $\mathrm{P}\{S = i\} = \alpha i^{-(\alpha+1)}$ with $\alpha = 0.95$. The inter-arrival time distribution also was Pareto, except α was 0.9. Here, S and A are measured in seconds. We truncated both distributions at 1001, that is, both h and g were 1000. The traffic intensity in this case was $\rho = 0.8125$. The execution time was 0.85 seconds for 16 iterations, and the expected waiting time was 941 seconds. We should note that this problem was used mainly as a test problem for the quality of our algorithm. Indeed, as mentioned earlier, it is not clear that our theory can be extended to systems with long-range dependency.

12.2.5 The Line Length Distribution in a $GI/G/1$ Queue Obtained from its Waiting-Time Distribution

We compared several methods to find the distribution of the line length of the $GI/G/1$ queue [41], and we found that by far the fastest method of finding the distribution of the line length proceeds as follows: One first calculates the waiting-time distribution as discussed above, and then one uses the so-called *distributional Little's law* to obtain the distribution of the line length. We now present the details of this method.

Our aim is to find the distribution of $X_Q(t)$, the number of customers waiting for service. Let t_n be the time of the last arrival before t, that is, $t_n \leq t$ and $t_{n+1} > t$. Clearly, there are one or more customers waiting if customer number n, who arrived at time t_n, is still waiting. Similarly, there are tow or more customers waiting if the customer who arrived at time t_{n-1} is still waiting, because she is served before customer number n. Continuing in this fashion, one concludes that more than m customers are waiting if customer $n - m + 1$, the customer that arrived at time t_{n-m+1}

is still waiting. Now, customer $n - m + 1$ is still waiting at t if her arrival time plus her waiting time W exceeds t, that is, customer $n - m + 1$ is still waiting if $t_{n-m+1} + W > t$. Conversely, if there are m or more customers still waiting, then the waiting time of customer $n - m + 1$ must satisfy $t_{n-m+1} - W \geq t$. Consequently,

$$P\{t_{n-m+1} + W \geq t\} = P\{X_q(t) \geq m\}.$$

Also note that $X_Q(t) = 0$ if $t_n + W < t$, because in this case, customer n has finished waiting at time t. It suffices to determine $P\{t_{n-m+1} + W \geq t\}$ and $P\{t_n + W < t\}$ in order to find the complementary cumulative distribution of $X_Q(t)$.

To calculate the probabilities in question, note that $t - t_n$ is, by definition, the age at time t. We denote the age by A^* because A was used for the inter-arrival time. We have according to Section 4.1.2, equation (4.8)

$$P\{A^*(t) = i\} = \frac{1}{E(\tau)} \sum_{j=i}^{\infty} a_j.$$

Also note that since $A_n = t_n - t_{n-1}$ is the time between arrivals,

$$t - t_{n-m+1} = (t - t_n) + (t_n - t_{n-m+1}) = A^* + \sum_{k=n-m+2}^{n} A_k.$$

The A_k all have the same distribution, allowing us to replace $\sum_{k=n-m+2}^{n} A_k$ by $\sum_{k=1}^{m-1} A_k$. Now we have

$$P\{t_{n-m+1} + W \geq t\} = P\{t - t_{n-m+1} - W \leq 0\} = P\left\{W - A^* - \sum_{k=1}^{m-1} A_k \geq 0\right\},$$

that is

$$P\{X_Q \geq m\} = P\left\{W - A^* - \sum_{k=1}^{m-1} A_k \geq 0\right\}. \tag{12.29}$$

This formula is known as the *distributional Little's Law* [5] [59].

We thus have to determine the distribution of $Z(m) = W - A^* - \sum_{k=1}^{m-1} A_k$, which is given by $Z(0) = W - A^*$ and $Z(m + 1) = Z(m) - A$. If $\pi_i(m) = P\{Z(m) = i\}$ and $a_j^* = P\{A^* = j\}$, we have

$$\pi_j(0) = P\{Z(1) = j\} = \sum_{i=1}^{\infty} P\{W = i + j\}a_i^*, \; j \geq 0$$

$$\pi_j(t + 1) = P\{Z(t + 1) = j\} = \sum_{i=1}^{\infty} \pi_{i+j}(t)a_i, \; j \geq 0.$$

Now

$$P\{X_Q \geq m\} = P\{Z(m) \geq 0\} = \sum_{i=0}^{\infty} \pi_i(m), \ m > 0.$$

Also,

$$P\{X_Q = 0\} = P\{t_n + W < t\} = P\{W < A^*\} = \sum_{j=1}^{\infty} P\{W < j\}a_j^*.$$

To find $P\{X_Q = m\}$, one uses

$$P\{X_Q = m\} = P\{X_Q \geq m\} - P\{X_Q \geq m - 1\}, \ m > 0.$$

We now have the entire distribution of X_Q. If X represents the number in the system, including the one being served, one has

$$P\{X = m\} = P\{X_Q = m - 1\}, \ m > 1.$$

Also, for any $GI/G/1$ queue with arrival rate λ and service rate μ:

$$P\{X = 0\} = 1 - \frac{\lambda}{\mu}.$$

Since $X_Q = 0$ if $X \leq 1$, we have

$$P\{X = 1\} = P\{X_Q = 0\} - P\{X = 0\} = P\{X_Q = 0\} + \frac{\lambda}{\mu} - 1.$$

With this, we have the entire distribution of the number of elements in the system.

12.2.6 The $M/D/c$ Queue

In the $M/D/c$ queue, there are c servers, arrivals are Poisson with a rate of λ, and the service time is constant. Without loss of generality, we assume the service time has a length of 1. In order to ensure that the server can handle the average flow, we assume that $\lambda < c$. If at time t, the number in the system is $X(t)$, and if $X(t) \geq c$, then all elements in the system at time t have received service by time $t + 1$, and if H elements arrive from time t to time $t + 1$, the number in the system is $X(t) + H - c$. Hence, if $X(t) = i, X(t+1) = j = i + H - c$, and since arrivals are Poisson, $P\{H = n\} = p(n; \lambda)$. With $H = j - i + c$, we obtain

$$p_{i,j} = p(j - i + c; \lambda) = p_{j-i}, \ i \geq c.$$

If $i \leq c$, then all elements in the system are served, but the arrivals between t and $t + 1$ are still in the system, that is,

$$p_{i,j} = p(j; \lambda), \ i \le c.$$

Clearly, X can decrease as much as c, that is $g = c$. To find h, the Poisson distribution has to be truncated. As in the randomization method, we set the upper bound for the Poisson distribution as $h = \lambda + z_\alpha \sqrt{\lambda} + r$, with $z_\alpha = 4.26$ and $r = 4$. Algorithm 32 can now be used to calculate the $a_i, i = 1, 2, \ldots, h$ and the $b_j, 0 \le j \le c$.

We now have to consider the boundary where $X(t) < c$. Once the $a_i, i \ge 0$, and $b_j, j \ge 0$ are known, all $a_{i,j}, i < c$ can be found by the standard Crout method, with $b_{i,j} = b_{j-i}$ and $a_{i,j} = a_{j-i}$ for $i \ge c$. At this point, the $\tilde{\pi}_j, j \le h$ must be evaluated, which allows us to obtain π_0 from equation (12.27). Once this is done, all π_j can be obtained from equation (12.21), and so can the expected line length from (12.28).

Actually, there is another way to deal with the initial conditions. When $i \le c$, the probability to go from i to j is $p(j; \lambda)$, which is independent of i. Consequently, all states less than or equal to c can be combined into a single state, say state $\bar{c}$, and the probability of this state becomes

$$\pi_{\bar{c}} = \sum_{i=0}^{c} \pi_i.$$

Now, state $\bar{c}$ is the first state, followed by state $c+1, c+2, \ldots$. The column of $\bar{c}$, being the first column, is not used for state elimination, and it can therefore be ignored. The transition matrix of the new matrix is thus, when $\sum$ denoted the probability of entering state $\bar{c}$:

	$\bar{c}$	$c+1$	$c+2$	$c+3$	$\ldots$
$\bar{c}$	$\sum$	$p(c+1; \lambda)$	$p(c+2; \lambda)$	$p(c+3; \lambda)$	$\ldots$
$c+1$	$\sum$	$p(c; \lambda)$	$p(c+1; \lambda)$	$p(c+2; \lambda)$	$\ldots$
$c+2$	$\sum$	$p(c-1; \lambda)$	$p(c; \lambda)$	$p(c+1; \lambda)$	$\ldots$
$c+3$	$\sum$	$p(c-2; \lambda)$	$p(c-1; \lambda)$	$p(c; \lambda)$	$\ldots$
$\vdots$	$\vdots$	$\ddots$	$\ddots$	$\ddots$	$\ddots$

As long as the first column is ignored, all the rows are the same except for a shift, and consequently, equations (12.24) to (12.25) can be applied. In particular, if $\pi_{\bar{c}}$ is the probability to be in state $\bar{c}$,

$$\pi_{\bar{c}} = 1 - \sum_{i=1}^{\infty} a_i.$$

Moreover

$$\pi_{c+1} = \pi_{\bar{c}} a_1$$

$$\pi_{c+1} = \pi_{\bar{c}} a_2 + \pi_{c+1} a_1$$

$$\vdots \quad = \quad \vdots .$$

Note that $\pi_{\bar{c}}$ is the probability that there is no queue, and that π_{c+i} is the probability that i elements are waiting. Also, if X_Q is the queue length,

$$E(X_Q) = \frac{\sum_{i=1}^{\infty} i a_i}{1 - \sum_{i=1}^{\infty} a_i}.$$

The probabilities $\pi_0, \pi_1, \ldots, \pi_c$ can be found from the following simple equation:

$$\pi_i = b_{c-i} * \pi_{\bar{c}}.$$

To prove this equation, note that the $p_{i,j}^{(c)} = b_{c-j}$, $i = 0, 1, \ldots, c$, $j = 1, 2, 3, \ldots$, that is, the $p_{i,j}^{(c)}$ are independent of i as long as $i \le c$. The equilibrium equations for the system reduced to the states from 0 to c therefore become

$$\pi_i = b_{c-i}(\pi_0 + \pi_1 + \cdots \pi_c), \ i = 0, 1, \ldots, c.$$

By definition, $\pi_{\bar{c}} = \pi_0 + \pi_1 + \cdots \pi_c$, and the result follows.

Note that the solution method used for the $M/D/c$ queue cannot be used for the $M/D/c/N$ queue. The reason is that we do not know how many arrivals are lost. For instance, when $X(t) = N$, and an arrival occurs before the first service completion, it is lost, whereas if it occurs after the first service completion, it is not.

12.2.7 Increase of X limited to 1, and Decrease of X limited to 1

In some cases, X can only decrease by 1, that is, $g = 1$. For example, if X is the number in the system in the $M/G/1$ queue immediately after a service completion, then $X(t + 1) = \max(0, X(t) + H - 1)$, where H is the number of arrivals during a service time. The change of X immediately after service is $k = H - 1 \ge -1$. It follows that $g = 1$ and $p_k = P\{H - 1 = k\}$, $k \ge -1$. As shown in Section 10.3, we have, if $F_H(t)$ is the distribution of H:

$$P\{H = i\} = \int_0^{\infty} p(i; \lambda t) dF_H(t).$$

In other cases, X can increase by at most 1. If $X(t)$ is the number in the system immediately before an arrival in a $GI/M/1$ queue, then $X(t_{n+1}) = \max(0, X(t_n) - D+1)$, where D is the number of service completion between two successive arrivals. Using arguments similar to the ones used for the $M/G/1$ queue, we have

$$p_{-k} = P\{D - 1 = k\} = \int_0^{\infty} p(k + 1; \mu t) dF_D(t),$$

where $F_D(t)$ is the distribution of D.

In the case where $g = 1$, equations (12.19) and (12.20) become

$$a_i = \frac{1}{1 - b_0}(p_i + a_{i+1}b_1) \tag{12.30}$$

$$b_1 = p_{-1}. \tag{12.31}$$

We thus know the value of b_1, and since $b_0 + b_1 = 1$, we also know b_0. The a_i can therefore be calculated recursively as

$$a_h = \frac{1}{1 - b_0}p_h, \quad a_i = \frac{1}{1 - b_0}(p_i + a_{i+1}b_1).$$

This recursion allows us to find all a_i, $i = h, h - 1, \ldots, 1$, and since $b_1 = p_{-1}$ and $b_0 = 1 - b_1 = 1 - p_{-1}$, the π_j outside the boundary can be expressed as

$$\pi_j = \sum_{i=1}^{F} a_i \pi_{j-i}.$$

When $h = 1$, equations (12.19) and (12.20) simplify to

$$a_1 = \frac{1}{1 - b_0}p_1 \tag{12.32}$$

$$b_j = p_{-j} + a_1 b_{j+1}. \tag{12.33}$$

Moreover,

$$\pi_{i+1} = \pi_i a_1.$$

We still need to find a_1. In contrast to the case where $g = 1$, all algorithms to calculate a_1 are iterative. Of course, one way to find a_1 is to use Algorithm 32. Alternatively, one can find a_1 from the following equation, which we will prove now:

$$a_1 = \sum_{j=-1}^{g} a_1^{j+1} p_{-j}. \tag{12.34}$$

There is software to solve this polynomial equation for a_1. To prove it, we apply equation (12.33) repeatedly:

$$b_g = p_{-g}$$
$$b_{g-1} = p_{-g+1} + a_1 b_g = p_{-g+1} + a_1 p_{-g}$$
$$b_{g-2} = p_{-g+2} + a_1 b_{g-1} = p_{-g+2} + a_1 p_{-g+1} + a_1^2 p_{-g}$$
$$\vdots \ .$$

This suggests

$$b_{g-i} = \sum_{j=0}^{i} a_1^j p_{-g+i-j},$$

a guess that can readily be verified by complete induction. Setting $i = g$ leads to

$$b_0 = \sum_{j=0}^{g} a_1^j p_{-j}.$$

To finish the proof, we solve $a_1 = \frac{p_1}{1-b_0}$ for b_0, which yields

$$b_0 = 1 - \frac{p_1}{a_1}.$$

Consequently,

$$1 - \frac{p_1}{a_1} = \sum_{j=0}^{g} a_1^j p_{-j}.$$

This equation is equivalent to equation (12.34) as required.

12.3 Matrices with Repeating Rows of Matrix Blocks

Many models lead to block-structured transition matrices where outside the boundary, the rows repeat, with the rows consisting of matrices. Instead of the p_k, which are scalars, we have thus matrices, which we denote by P_k, where k ranges from $-g$ to h. It is customary to combine all boundary states into a set of states, and row 0 then contains all transition out of states of this set, and column 0 contains all the transitions ending in a state of this set. To denote the resulting matrices, we use $P_{0,j}$ for the matrices of row 0, and $P_{i,0}$ for matrices of column 0. We assume that $P_{0,j} = 0$ if $j > h$, and $P_{i,0} = 0$ for $i > g$. The size of the matrix $P_{0,0}$ is denoted by m_0, and the matrices P_k all have size m. The transition matrix thus looks as follows:

$$P = \begin{bmatrix} P_{0,0} & P_{0,1} & P_{0,2} & \cdot & P_{0,h} & 0 & 0 & \cdots \\ P_{1,0} & P_0 & P_1 & P_2 & \cdots & P_h & 0 & \ddots \\ P_{2,0} & P_{-1} & P_0 & P_1 & P_2 & \cdots & P_h & \ddots \\ & & & & & & & \\ P_{g,0} & P_{-g+1} & P_{-g+2} & \cdots & & \cdots & \cdots & P_h & 0 \\ 0 & P_{-g} & P_{-g+1} & P_{-g+2} & \cdots & & \cdots & \cdots & P_h \\ \vdots & \ddots & \ddots & & \ddots & & \ddots & \ddots & \ddots & \ddots \end{bmatrix}. \quad (12.35)$$

The objective is to find the vectors π_i, which contain the equilibrium probabilities for states with $X_1 = i$. Thus, if $\pi_{i,j}$ is the equilibrium probability that $X_1 = i$, $X_2 = j$, $\pi_i = [\pi_{i,j}, j = 1, 2, \ldots, m]$.

In the transition matrix of CTMCs, the probabilities would have to be replaced by rates, and to indicate this, we would write Q_k, $-g \le k \le h$ instead of P_k, and a similar change of notation is used when dealing with the boundaries.

Models with transition matrices of the form given by (12.35) originally arose in the analysis of queueing problems, and the level was a queue length, or sometimes a line length. In these models, two events are prominent: arrivals, which increase the line length, and departures, which decrease the line length. To illustrate, consider the tandem queue, with X_1 being the length of the first line. We now assume that there is no bound on X_1, while the lengths of the other lines remain bounded. In this case, the sublevel, X_2, is the combination of all other line lengths. Now, Q_1 includes the rates of all events that increase X_1, such as arrivals, while Q_0 includes the rates of all events that leave X_1 unchanged, and Q_{-1} includes all events that decrease X_1. We conclude that $h = 1$ and $g = 1$. Such models are called *quasi birth-death* processes, abbreviated by QBD. processes. The tandem queue model with X_1 not bounded is thus a QBD process. Other QBD processes include the $PH/PH/1$ queue and even the $PH/PH/c$ queue. The discrete-time $GI/G/1$ queue can also be formulated as a QBD process, but the resulting solution procedures are slower than the one presented earlier. More generally, any process where X_1 can change at most by 1 is a QBD process.

Another important type of process is the $GI/M/1$-type process, where X_1 can increase at most by 1, that is, $h = 1$, but it can decrease by any amount. We will assume, however, that the decrease is limited by g, an assumption that is needed when using numerical methods. $GI/M/1$ type processes were introduced by Neuts [72], who coined the term $GI/M/1$ paradigm. $GI/M/1$ type processes arise when embedding a queueing system at the times of arrivals, with X_1 used to represent the length of the line. In this way, the line length, X_1, can increase by at most 1 at a time. Examples of $GI/M/1$ type processes include the $GI/PH/c$ queue embedded at arrival points, as well as models with server breakdowns. Processes where X_1 cannot decrease by more than 1 at any time are referred to as $M/G/1$ type processes. $M/G/1$ processes arise when embedding a queueing process at the times of departure. They include the $PH/G/1$ queue embedded at departure points, and processes with MAP arrivals and general service time distributions.

The most general process is the $GI/G/1$ process, where both h and g can have arbitrary values. The name $GI/G/1$ paradigm was coined by Grassmann and Heyman [35]. $GI/G/1$ type processes arise when arrivals to a queue are in groups, or bulk, or when departures are in groups. Examples would include gondolas, where arrivals occur in groups, and several people are transported in the same gondola. The $GI/G/1$ paradigm, being the most general one, is ideally suited to explain the underlying concepts.

Theoretically, any $GI/G/1$ type process with g and h limited to finite values can be converted into a QBD process. We will not elaborate on this method because it is usually computationally inefficient.

The outline of this section is as follows. We start with the $GI/G/1$ paradigm to lay the groundwork of our future discussion. In particular, we cover norming in connection with the $GI/G/1$ paradigm. Next, we show how an extension of the state elimination method can be used to find equilibrium solutions of QBD processes. We then cover the $GI/M/1$ and the $M/G/1$ paradigms.

12.3.1 Recurrent and Transient $GI/G/1$ Type processes

As in the scalar case, we use the central limit theorem to decide whether or not the Markov chain is recurrent. We therefore need $E(X_1(t))$ and $Var(X_1(t))$. As before, we define $D(t) = X_1(t+1) - X_1(t)$, such that $X_1(t) = X_1(0) + \sum_{n=1}^{t} D(n)$. As long as the block rows repeat, and this is true outside the boundary, the process $X_2(t)$, the sublevel, does not depend on X_1. Consequently, $X_2(t)$ forms a Markov chain on its own as long as the boundary is avoided, and its transition matrix is $B = \sum_{i=-g}^{h} P_i$, where $B = [b_{i,j}, 1 \le i, j \le m]$.

If the process $\{X_1(t), t \ge 0\}$ stays outside the boundary, transient probabilities $\beta_i(t) = P\{X_2(t) = i\}$ can be obtained, and, as long as the process $\{X_2(t), t \ge 0\}$ is ergodic and non-periodic, the $\beta_i(t)$ converge to their equilibrium probabilities β_i. The expectations of $D(t)$ now depend on $X_2(t)$, and we therefore need $E(D \mid X_2 = i)$. The vector $[E(D(t) \mid X_2(t) = i), i = 1, 2, \ldots, m]$ can be found as $\sum_{k=-g}^{h} k P_k$.

We now select a state x, that is, we select a value for X_1 and X_2, and we develop criteria for deciding whether or not state x is transient, that is, whether or not the process $\{X(t) = [X_1(t), X_2(t)], t \ge 0\}$, when starting in state x, will eventually return to state x. This return can be far in the future. We can therefore assume that the process $\{X_2(t), t \ge 0\}$ reaches equilibrium, when $E(D)$ becomes

$$E(D) = \sum_{i=-1}^{m} \beta_i E(D \mid X_2 = i).$$

To find $E(D \mid X_2 = i)$, note that

$$P\{D = k \mid X_2 = i\} = \sum_{i=1}^{m} p_{i,j}^k,$$

where $p_{i,j}^k$ are the entries of the matrix P_k, and m is the dimension of P_k. It follows that

$$E(D \mid X_2 = i) = \sum_{k=-g}^{h} k \sum_{i=1}^{m} p_{i,j}^k.$$

It is well known that the central limit theorem also applies to Markov chains, that is, if the process $\{D(t), t \ge 0\}$ forms a Markov chain, then $\sum_{n=1}^{t} D(n)$ approaches a normal distribution. We still need the parameters of the distribution of $x + \sum_{n=1}^{t} D(n)$, which are expectation and variance. For the expectation, we find

$$E\left(x + \sum_{n=1}^{t} D(n)\right) = x + \sum_{n=1}^{t} E(D(n)) = x + tE(D).$$

Hence, $E(X_1(t)) - x$ is proportional to t as long as $X_1(t)$ remains outside the boundary.

It is known that the variance of $X_1(t)$ increases linearly with t. We will not prove this here, and refer to the literature on the variance of time averages in Markov chains, in particular [29], [69], [77]. In these sources, it is shown that the variance of an average like $\frac{1}{t} \sum_{n=1}^{t} D(n)$ is V/t, where V is a constant. The details of how to find V will not be discussed here. What is important is that if the variance of the average of t terms is of the form V/t, then the variance of the sum of t terms must be tV. Hence, like in the scalar case, both the expectation and the variance show an asymptotic increase or decrease that is proportional to t. This reduces the matrix case to the scalar case, except that instead of $\mathrm{Var}(D(t))$, one must use V. Everything else stays the same.

In conclusion: We first calculate $\beta = [\beta_i, i = 1, 2, \ldots, m]$ from the equilibrium equation $\beta = \beta \sum_{k=-g}^{h} P_k$, then we obtain

$$E(D) = \sum_{i=1}^{m} \beta_i E(D \mid X_2 = i) = \sum_{i=1}^{m} \beta_i \sum_{k=-g}^{h} k \sum_{j=1}^{m} p_{i,j}^k.$$

In matrix form, this can be written as

$$E(D) = \beta \sum_{k=-g}^{h} k P_k \mathbf{e}, \qquad (12.36)$$

where $\mathbf{e}$ is the column vector with all its entries equal to 1.

We now have the well-known result [63] that all states are transient if $E(D) > 0$ and they are recurrent if $E(D) \leq 0$. Without proof, we also state that the system is null-recurrent if $E(D) = 0$. We claim that these results only hold when V is finite, which is true as long as D can only assume a finite number of values, and the number of sublevels is finite. If V is infinite, all states are null-recurrent. We should mention that in literature, the role of the variance is ignored.

Equation (12.36) also holds for CTMCs, provided P_k is replaced by the rate matrices Q_k, $k = -g, -g + 1, \ldots, h$. This is allowed because P_0 is not needed when calculating $\sum_{k=-g}^{h} k P_k$, and aside from P_0, respectively Q_0, the transition matrices of DTMCs and CTMCs have the same structure.

12.3.2 The $GI/G/1$ Paradigm

If the transition matrix is given as in (12.35), the equilibrium equations become

$$\pi_0 = \sum_{i=-h}^{g} \pi_i P_{i,0} \tag{12.37}$$

$$\pi_j = \pi_0 P_{0,j} + \sum_{i=1}^{j-1} \pi_i P_{j-i}, \ j \leq h \tag{12.38}$$

$$\pi_j = \sum_{i=j-h}^{j+g} \pi_i P_{j-i}, \ j > h. \tag{12.39}$$

Processes with this structure are called *GI/G/1 type processes*.

Using the method discussed in Section 12.1.1, we now calculate E(D), and if E(D) < 0, then the Markov chain is recurrent.

We now show that if $P_{i,j} = P_{j-i}$ for all $i, j > 0$, then $A_{i,j} = A_{j-i}$ and $B_{i,j} = B_{j-i}$, where $A_{i,j}$ and $B_{i,j}$ are defined in Section 9.1.6 for the block version of the Crout method. Consider a path starting in $X_1 = i_1 \leq n$ and $X_2 = i_2$, and ending in $X_1 = j_1$ and $X_2 = j_2$, and never passing through any state with $X_i \leq n$. When $n > 0$, this path can be matched with a path starting in $X_1 = i_1 + k$, $X_2 = i_2$ and ending in state $X_1 = j_1 + k$ and $X_2 = j_2$ and never passing through any state with $X_1 \leq n + k$. Since censoring maintains the paths, it follows that $P_{i,j}^{(n)} = P_{i+k,j+k}^{(n+k)}$. Consequently, $A_{i,n} = P_{i,n}^{(n)}(I - P_{n,n}^{(n)})^{-1}$ is equal to $A_{i+k,n+k}$. In both cases, the difference between the second and the first subscript remain the same, that is $A_{i,n} = A_{i+k,n+k} = A_{n-i}$. The proof that $B_{n,j} = B_{n-j}$ follows the same lines. The argument is thus very similar to the ones used in the scalar case, except that scalars are replaced by matrices.

We now explore the use of the extrapolating Crout method. For this purpose, we define $A_{n-i}^n = A_{i,n}$ and $B_{n-j}^n = B_{n,j}$. With this notation, equation (9.24) and (9.25) become after manipulations similar to the ones used in the scalar case:

$$A_{n-i}^n = \left(P_{n-i} + \sum_{v=n+1}^{\infty} A_{v-i}^v B_{v-n}^v \right)(I - B_0^v)^{-1}, \ i < n \tag{12.40}$$

$$B_{n-j}^n = P_{j-n} + \sum_{v=n+1}^{\infty} A_{v-n}^v B_{v-j}^v, \ j \leq n. \tag{12.41}$$

The A_i^n converge toward A_i, and the B_j^n converge toward B_j When convergence is achieved, the superscripts can be dropped, and after some slight changes involving the subscripts, one obtains

$$A_i = \left(P_i + \sum_{j=1}^{h+i,g} A_{i+j} B_j \right)(I - B_0)^{-1}, \ i \geq 1 \tag{12.42}$$

$$B_j = P_{-j} + \sum_{i=1}^{h,j+g} A_i B_{i+j}, \ j \geq 0. \tag{12.43}$$

Note that for each iteration of equations (12.40) and (12.41), we need to store only the P_k with k between $-g$ and h, the A_i^v with i between 1 and h and v between i and $n + 1$, as well as the B_j^v with j between 0 and g, and v between $n + 1$ and $n + 1 + g$. All these values can be stored in a matrix of size $4gh$, and the matrix is updated after each iteration. After some iterations, the A_k^n and B_k^n converges to A_k and B_k, respectively. We therefore can stop as soon as the results of two successive iterations do not differ by more than a prescribed value $\epsilon > 0$. To do the iterations, one can use an algorithm similar to Algorithm 22. When using such an algorithm, one can either calculate the diagonal elements directly or use the GTH method, Taking the average of these two methods is also possible.

Once the A_k and B_k are known, the problem can be solved by using the standard Crout method based on matrices. We demonstrate this with a model where h=2 and $g = 3$. Once the A_k and B_k are known, the initial 6 rows and 5 columns can be written as follows:

$$\begin{bmatrix} P_{0,0} & P_{0,1} & P_{0,2} & 0 & 0 \\ P_{1,0} & B_0 & A_1 & A_2 & 0 \\ P_{2,0} & B_1 & B_0 & A_1 & A_2 \\ P_{3,0} & B_2 & B_1 & B_0 & A_1 \\ 0 & B_3 & B_2 & B_1 & B_0 \end{bmatrix}.$$

We now calculate the $A_{0,i}$, which replace the $P_{0,j}$, $j = 1, 2$. One finds, using the Crout equation (9.24):

$$A_{0,2} = P_{0,2}(I - B_0)^{-1}$$
$$A_{0,1} = \left(P_{0,1} + A_{0,2}B_1\right)(I - B_0)^{-1}.$$

Similarly, the $B_{0,i}$ become

$$B_{3,0} = P_{3,0}$$
$$B_{2,0} = P_{2,0} + A_1 B_{0,3}$$
$$B_{1,0} = P_{1,0} + A_1 B_{0,2} + A_2 B_{0,3}.$$

Once all $A_{0,j}$ and $B_{i,0}$ are found, we obtain

$$B_{0,0} = P_{0,0} + \sum_{i=1}^{2} A_{0,i} B_{i,0}.$$

Now, $B_{0,0}$ is $P_{0,0}^{(0)}$, the matrix with all states except the ones of level 0 censored. Therefore,

$$\pi_0 = \pi_0 B_{0,0}.$$

This allows us to find π_0 up to a factor. How to find this factor will be discussed in the next section. For $\pi_1, \pi_2, \pi_3, \dots$, the following equations can be used:

$$\pi_1 = \pi_0 A_{0,1}$$
$$\pi_2 = \pi_0 A_{0,2} + \pi_1 A_1$$
$$\pi_3 = \pi_1 A_2 + \pi_2 A_1.$$

For all other π_j, one can use

$$\pi_j = \pi_{j-1} A_1 + \pi_{j-2} A_2.$$

In general, the formulas for π_j are as follows: First, π_0 is found as

$$\pi_0 = \pi_0 \left(P_{0,0} + \sum_i A_{0,i} B_{i,0} \right).$$

The π_j, $j > 0$ can then be obtained as

$$\pi_j = \pi_0 A_{0,j} + \sum_{i=1}^{j-1} \pi_i A_{j-i}, \ j \le h \tag{12.44}$$

$$\pi_j = \sum_{i=1}^{h} \pi_{j-i} A_i, \ j > h. \tag{12.45}$$

Using these equations, one can find all π_j recursively, up to a factor. The determination of this factor is the subject of the next section.

As stated in textbooks on linear algebra, one should avoid using inverses, and use *LU* factorizations instead. This has to be kept in mind.

12.3.3 Generating Functions

As in the scalar case, generating functions can be used when dealing with matrices for the purpose of norming the probabilities such that their sum is 1. They also allow one to find expectations. The generating function approach is easiest to use when $P_{0,j} = P_j$. This case is done first.

Clearly,

$$P(z) = \pi_0 + \sum_{j=1}^{\infty} \pi_j z^j.$$

Note that $\pi_i = [\pi_{i,k}, \ k = 1, 2, \ldots, m]$, and therefore,

$$P(z) = \left[\sum_{i=0}^{\infty} \pi_{i,k} z^k, \ k = 1, 2, \ldots, m \right]. \tag{12.46}$$

Now, equation (12.45) can be used for $j \ge 1$, which yields

$$P(z) = \pi_0 + \sum_{j=1}^{\infty} \sum_{i=1}^{h} \pi_{j-i} A_i z^j$$

$$= \pi_0 + \sum_{i=1}^{h} \sum_{j=1}^{\infty} \pi_{j-i} z^{j-i} A_i z^i$$

$$= \pi_0 + \sum_{i=1}^{h} \sum_{j=0}^{\infty} \pi_j z^j A_i z^i.$$

Hence,

$$P(z) = \pi_0 + \sum_{j=0}^{\infty} \pi_j z^j \sum_{i=1}^{h} A_i z^i,$$

or

$$P(z) = \pi_0 + P(z) \sum_{i=1}^{h} A_i z^i. \tag{12.47}$$

Solving for $P(z)$ yields

$$P(z) = \pi_0 \left(I - \sum_{i=1}^{h} A_i z^i \right)^{-1}.$$

When $z = 1$, we obtain $P(1)$, which also represents the marginal distribution of X_2:

$$P(1) = \left[\sum_{i=0}^{\infty} \pi_{i,j}, \ j = 1, 2, \ldots, m \right] = \pi_0 \left(I - \sum_{i=1}^{h} A_i \right)^{-1}. \tag{12.48}$$

The sum of the components of $P(1)$ must be 1, and this sum can be used for norming. As in the scalar case, we first find all $\tilde{\pi}_i$ $i \geq 0$, with one $\tilde{\pi}_{i,j}$ set to 1. Now, if $\mathbf{e}$ is the column vector with all its entries equal to 1, then $P(1)\mathbf{e}$ is the sum of all $\pi_{i,j}$, or, if all π_i in equation (12.48) are replaced by $\tilde{\pi}_i$, the sum of all $\tilde{\pi}_{i,j}$. Once this sum is known, say it is c, then $\pi_{i,j} = \tilde{\pi}_{i,j}/c$.

To find $E(X_1)$, we take the derivative of equation (12.47):

$$P'(z) = P(z) \sum_{i=1}^{h} A_i z^i + P(z) \sum_{i=1}^{h} i A_i z^i. \tag{12.49}$$

Set $z = 1$ and solve for $P'(1)$:

$$P'(1) = P(1) \sum_{i=1}^{h} i A_i \left(I - \sum_{i=1}^{h} A_i \right)^{-1}.$$

From equation (12.46), one concludes that $P'(1)$ yields the expectations of X_1 under the condition that $X_2 = k$:

$$P'(1) = \left[\sum_{i=0}^{\infty} i\pi_{i,k}, \ k = 1, 2, \ldots, m \right].$$

Consequently, $E(X_1) = P'(1)\mathbf{e}$, that is,

$$E(X_1) = P'(1)\mathbf{e} = P(1) \left(I - \sum_{i=1}^{h} A_i \right)^{-1} \mathbf{e}.$$

We now consider the more general case where $P_{0,j}$ is different from P_j, and instead of the dimension m, it has dimension m_0. To simplify our notation, we define

$$\bar{P}(z) = \sum_{j=1}^{\infty} \pi_j z^j.$$

Clearly, $\bar{P}(1)\mathbf{e}$ is no longer 1, but it is $1 - \sum_{k=0}^{m_0} \pi_{0,k}$.

Now, according to equation (12.45), we have for $j > h$:

$$\pi_j = \pi_{j-1}A_1 + \pi_{j-2}A_2 + \cdots + \pi_{j-h}A_h.$$

Multiplying these equations by z^j and taking the sum from $h + 1$ to infinity yields

$$\sum_{j=h+1}^{\infty} \pi_j z^j = \sum_{j=h+1}^{\infty} \pi_{j-1}A_1 z^j + \sum_{j=h+1}^{\infty} \pi_{j-2}A_2 z^j + \cdots + \sum_{j=h+1}^{\infty} \pi_{j-h}A_h z^j.$$

Like in the scalar case, we can write this equation as

$$\bar{P}(z) - \sum_{j=1}^{h} \pi_j z^j = \left(\bar{P}(z) - \sum_{j=1}^{h-1} \pi_j z^j \right) A_1 z$$

$$+ \left(\bar{P}(z) - \sum_{j=1}^{h-2} \pi_j z^j \right) A_2 z^2$$

$$+ \ldots$$

$$+ \left(\bar{P}(z) - \sum_{j=1}^{0} \pi_j z^h \right) A_h z^h.$$

We bring all the terms involving $\bar{P}(z)$ to the left, and all other terms to the right to obtain

$$\bar{P}(z) \left(1 - \sum_{i=1}^{h} A_i z^i \right) = \sum_{j=1}^{h} \pi_j z^j - \sum_{i=1}^{h} \sum_{j=1}^{h-i} \pi_j A_i z^{i+j}. \tag{12.50}$$

Consequently,

$$\bar{P}(1)\left(1 - \sum_{i=1}^{h} A_i\right) = \sum_{j=1}^{h} \pi_j - \sum_{i=1}^{h}\sum_{j=1}^{h-i} \pi_j A_i. \tag{12.51}$$

To use this equation, we set a particular probability equal to 1, which allows us to find all the other probabilities with $X_1 \leq h$, except for a factor, let us say c. If $\tilde{\pi}_{i,j}$ are the values found this way, then $\tilde{\pi}_{i,j} = c\pi_{i,j}$, Next, we determine $c = \sum_i \sum_j \tilde{\pi}_{i,j}$. We use equation (12.51), with all $\pi_{i,j}$ replaced by $\tilde{\pi}_{i,j}$. If $\bar{c}$ is the value of $\bar{P}(1)\mathbf{e}$ obtained after the substitution, then we have

$$c = \bar{c} + \sum_{k=1}^{m_0} \tilde{\pi}_{0,k}.$$

The $\pi_{i,j}$ are then $\tilde{\pi}_{i,j}/c$, and with this, all $\pi_{i,j}$ are determined.

We now determine $E(X_1)$, the average level. We take the derivative of equation (12.50) to obtain

$$\bar{P}'(z)\left(1 - \sum_{i=1}^{h} A_i z^i\right) - \bar{P}(z)\sum_{i=1}^{h} i A_i z^{i-1} = \sum_{j=1}^{h} j\pi_j z^{j-1} - \sum_{i=1}^{h}\sum_{j=1}^{h-i}(i+j)\pi_j A_i z^{i+j-1}.$$

When $z = 1$, this yields

$$\bar{P}'(1)\left(1 - \sum_{i=1}^{h} A_i\right) - \bar{P}(1)\sum_{i=1}^{h} i A_i = \sum_{j=1}^{h} j\pi_j - \sum_{i=1}^{h}\sum_{j=1}^{h-i}(i+j)\pi_j A_i.$$

We can solve this equation for $\bar{P}(1)$, and $E(X_1) = \bar{P}'(1)\mathbf{e}$. With this, we have solved the problem.

12.3.4 The QBD Process

$GI/G/1$ type processes with g and h equal to 1 are called *QBD process*. In the QBD process, the equation (12.40) and (12.41) specialize to

$$B_1 = P_{-1}, \quad B_0 = P_0 + A_1 B_1, \quad A_1 = P_1(I - B_0)^{-1}.$$

These equations were obtained from the block version of Crout given by equation (9.23), which can be written as

$$P_{i,j}^{(n)} = P_{j-i} + \sum_{m=n+1}^{N} P_{i,m}^{(m)}\left(I - P_{m,m}^{(m)}\right)^{-1} P_{m,j}^{(m)}, \quad i,j \leq n. \tag{12.52}$$

Since the block matrix is banded with a bandwidth of 1, to be non-zero, $m - i \leq 1$ and $m - j \leq 1$. This implies as one can easily verify

$$P_{n-1,n}^{(n)} = P_1, \quad P_{n,n-1}^{(n)} = P_{-1}.$$

Moreover, for $i = j = n$, we obtain

$$P_{n-1,n-1}^{(n-1)} = P_0 + P_1 \left(I - P_{n,n}^{(n)}\right)^{-1} P_{-1}. \tag{12.53}$$

This equation essentially reduces the Crout version to the standard block elimination given by equation (9.14), that is, the advantages of using Crout disappear. Now, according to our arguments, $P_{n,n}^{(n)}$ converges to B_0 as n decreases. Also, $P_{n,n-1}^{(n)}$ is always P_{-1}, and it thus converges trivially to B_1. Finally, $P_1(I - P_{n,n}^{(n)})^{-1}$ converges to A_1. Once convergence is achieved, no further iterations are needed, and we can move to level 1, where we find the following matrix:

$$P^{(1)} = \begin{bmatrix} P_{0,0} & P_{0,1} \\ P_{1,0} & P_{1,1} \end{bmatrix}.$$

Hence, we have

$$B_{0,0} = P^{(0)} = P_{0,0} + P_{0,1}(1 - B_0)^{-1} P_{1,0}.$$

Also,

$$A_{0,1} = P_{0,1}(I - B_0)^{-1}.$$

When using MATLAB$^{©}$ or Octave, this leads to the following algorithm:

Algorithm 33.

```
A = 0
Aold = P(1)
while not yet converged
      Aold = A
      B = P0 + A * Pm
      A = P1/(eye(m) - B)
end while
A01 = P01/(eye(m)- B)
B00 = P00 + A01* P10
```

Here, A stands of A_1, B for B_0, P1 for P_1, P0 for P_0, and Pm for P_{-1}. The other symbols are to be understood in a similar fashion. Note that eye(m) is the identity matrix of size m, and A = P1/(eye(m) -B) can be used instead of P1*inv(eye(m) - B). In fact, P1/(eye(m) - B) uses LU factorization, which is more efficient than matrix inversion.

There are several ways to check for convergence. Here, we use the Octave statement sum(sum(abs((Aold-A))))< epsilon to check for convergence, where epsilon is a small enough number.

The next step is to find the π_i, $i \geq 0$. As usual, we use $\tilde{\pi}_{i,j}$, which is $\pi_{i,j}c$, where c is the sum of all $\tilde{\pi}_{i,j}$. We select a sublevel j of level 0, and set $\tilde{\pi}_{0,j} = 1$. The vector $\tilde{\pi}_0 = [\tilde{\pi}_{0,j}, \ j = 1, 2, \ldots, m_0]$ can then be obtained from the following equation, where $\tilde{\pi}_{0,1} = 1$:

$$\tilde{\pi}_0 = \tilde{\pi}_0 B_{0,0}.$$

Once $\tilde{\pi}_0$ is known, we have

$$\tilde{\pi}_1 = \tilde{\pi}_0 A_{1,0}$$
$$\tilde{\pi}_{n+1} = \tilde{\pi}_n A_1, \ n > 0.$$

The factor c, which is the sum of all $\tilde{\pi}_{i,j}$, can now be found from equation (12.51), which in the case of QBD simplifies to

$$\bar{P}(1)(I - A_1) = \pi_1.$$

We calculate $\bar{P}(1)$, except that π_i is replaced by $\tilde{\pi}_i$. If $\bar{c}$ is $\bar{P}(1)\mathbf{e}$ with $\bar{P}(1)$ modified as indicated, then

$$c = \bar{c} + \sum_{k=1}^{m_0} \tilde{\pi}_{0,k}.$$

With this, the problem is solved.

Before doing all these calculations, one should check if the process is recurrent. For this, one can use equation (12.36), which specializes to

$$E(D) = \beta(P_1 - P_{-1})\mathbf{e},$$

where β is the equilibrium solution of the DTMC with transition matrix $P_{-1}+P_0+P_1$. When using Octave, we write "beta" for β, where beta can be found as

```
PT = Pm + P0 + P1 - eye(m,m)
beta = [1, -PT(1,2:m)/PT(2:m, 2:m)]
beta = beta/sum(beta)
```

$E(D)$ (ED) then becomes

```
ED = beta*sum((P1 - Pm)')'
```

Here, the prime (') indicates transposition, and sum in MATLAB calculates the column sums when applied to a matrix. The expected down-drift $E(D)$ is negative for positive recurrent DTMCs.

The formulas for $P(1)$ and $P'(1)$ simplify considerably. In particular, for the case where $P_{1,0} = P_1$, equations (12.48) and (12.49) simplify to

$$P(1) = \pi_0(I - A)^{-1} \tag{12.54}$$
$$P'(1) = P(1)A(I - A)^{-1}. \tag{12.55}$$

Since we will apply the QBD process mostly to CTMCs, we discuss our approach for this situation as well. In this case, the $P_{i,j}^{(n)}$ must be replaced by $Q_{i,j}^{(n)}$, except for

$i = j$, where $-Q_{i,i}^{(i)}$ corresponds to $I - P_{i,i}^{(i)}$. Also, P_i must be replaced by $Q_i, i = 1, -1$. Thus, instead of equation (12.53), we have

$$Q_{n-1,n-1}^{(n-1)} = Q_0 + Q_1 \left(-Q_{n,n}^{(n)}\right)^{-1} Q_{-1}.$$

Hence, the essential statements of the while loop become

B = Q0 + A * Qm

A = –Q1 / B

Also, the last two statements become

A01 = –Q01/B

B00 = Q00 + Q10*A01

When determining the drift, one finds β from the equilibrium equations with the transition matrix $Q_{-1} + Q_0 + Q_1$. The formula for the drift does not change. The equations for $P(1)$ and $P'(1)$ also remain unchanged.

As an alternative to block elimination, one can use the state elimination given by Algorithm 30 to obtain B_0 as well as the $\pi_i, i \geq 0$. This method works for both DTMCs and CTMCs, but we explain it for DTMCs. We start with column N, where N is assumed to be infinite, but we only keep the last two matrix rows and the last two matrix columns. The lower right corner of the transition matrix is now:

$$P^{(1)} = \begin{bmatrix} P_0 & P_1 \\ P_{-1} & B^{(N)} \end{bmatrix}.$$

Here, $B^{(N)}$ is an initial value for B_0, such as P_0. Now, one applies state elimination using Algorithm 30, but only until all m states of level N have been eliminated. After that, the lower right corner looks as follows, where B^{N-1} is the new approximation of B_0:

$$\begin{bmatrix} B^{(N-1)} & A^{(N-1)} \\ C^{(N-1)} & D^{(N-1)} \end{bmatrix}.$$

Here, $B^{(N-1)}$ is really $P_0 + P_1 \left(I - B^{(N)}\right)^{-1} P_{-1}$ because we get the same result when we eliminate all sublevels of level N at once, or when we eliminate the sublevels one by one.

Obviously, the entire procedure can be repeated. Generally, once all levels above n are eliminated, one has

$$\begin{bmatrix} P_0 & P_1 \\ P_{-1} & B^{(n)} \end{bmatrix}.$$

After the eliminations of the sublevels of level n, one finds

$$\begin{bmatrix} B^{(n-1)} & A^{(n-1)} \\ C^{(n-1)} & D^{(n-1)} \end{bmatrix}.$$

As n decreases, this matrix converges to

$$\begin{bmatrix} B & A' \\ C & D \end{bmatrix}.$$

Note that though the $B^{(n)}$ converge to B_c, the $A^{(n)}$, $C^{(n)}$ and $D^{(n)}$ converge to values that have no equivalent in the matrix-based algorithm. Once $B^{(n)}$ does not change by more than $\epsilon > 0$, we obtain the following matrix for levels 0 and 1:

$$P^{(1)} = \begin{bmatrix} P_0 & P_{0,1} \\ P_{1,0} & B \end{bmatrix}.$$

We eliminate all states of level 1 from this matrix to find $B_{0,0}$. Now, π_0 can be determined, up to a factor, from the equation $\pi_0 = \pi_0 B_{0,0}$. This is followed by the back-substitution step in the normal way. In particular, for $j > 1$, one finds, if $A' = [a'_{i,j}, i, j = 1, 2, \ldots, m]$ and $D = [d_{i,j}, i, j = 1, 2, \ldots, m]$:

$$\pi_{j,n} = \sum_{v=1}^{n-1} \pi_{j,v} d_{v,n} + \sum_{v=1}^{m} \pi_{j-1,v} a'_{v,n}.$$

This equation allows one to find all $\pi_{t,j}$ recursively. If $D' = [d'_{i,j}]$ is the upper triangular part of D, that is, $d'_{i,j} = d_{i,i}$ if $i < j$, and zero otherwise, the above equation can be written as

$$\pi_j = \pi_j D' + \pi_{j-1} A'.$$

For finding the sum of all probabilities, we need A_1. To find A_1, we solve the equation above for π_j, which yields

$$\pi_j = \pi_{j-1} A' (I - D')^{-1}.$$

It follows that $A_1 = A'(I - D')^{-1}$.

There are minor changes to this approach when dealing with a CTMC. These changes should be obvious, however. This concludes our discussion of finding the equilibrium probabilities of a QBD process.

The advantage of the direct application of state elimination is that one can exploit the structure of the matrices. For instance, if P_1 or P_{-1} are diagonal matrices, this can be exploited, and the same is true when these matrices contain rows or columns of zero. For further information on this, see [24].

12.3.4.1 The Example of the Tandem Queue

We now formulate the two-station tandem queue as a QBD process, with the lengths of line i denoted by X_i, $i = 1, 2$. X_1 is unbounded, but X_2 is always at or below N_2, because when X_2 reaches N_2, server 1 is blocked. In the QBD process, there are three matrices, Q_1, Q_0 and Q_{-1}. The Q_i, $i = -1, 0, 1$ contain the rates of events that change X_1 by i. For simplicity, we set $N_2 = 3$, but extensions to any N_2 are easy. Consequently, our matrices Q_i are square matrices of size 4, with one row for each value of X_2. As discussed earlier, there are three events: "Arrival", "Move" and "Departure". An arrival increases X_1 by 1, leaving X_2 unchanged. Since no other event increases X_1, we have

$$Q_1 = \begin{bmatrix} \lambda & 0 & 0 & 0 \\ 0 & \lambda & 0 & 0 \\ 0 & 0 & \lambda & 0 \\ 0 & 0 & 0 & \lambda \end{bmatrix}.$$

The event "Move" causes X_1 to decrease by 1, and X_2 to increase by 1. No other event decreases X_1. Hence:

$$Q_{-1} = \begin{bmatrix} 0 & \mu_1 & 0 & 0 \\ 0 & 0 & \mu_1 & 0 \\ 0 & 0 & 0 & \mu_1 \\ 0 & 0 & 0 & 0 \end{bmatrix}.$$

Finally, "Depart" leaves X_1 unchanged, while decreasing X_2. Moreover, we have to add the diagonal, which is equal to the sum of all rates affecting the state in question. Thus,

$$Q_0 = \begin{bmatrix} -(\mu_1 + \lambda) & 0 & 0 & 0 \\ \mu_2 & -(\mu_1 + \mu_2 + \lambda) & 0 & 0 \\ 0 & \mu_2 & -(\mu_1 + \mu_2 + \lambda) & 0 \\ 0 & 0 & \mu_2 & -(\lambda + \mu_2) \end{bmatrix}.$$

Algorithm 33, with the changes as indicated, can now be applied, starting with the lower right corner

$$\begin{matrix} Q_0 & Q_1 \\ Q_{-1} & B. \end{matrix}$$

Once convergence of this algorithm is achieved, one has to consider the boundary equations. Note that the rows differ only when $X_1 = 0$. In this case the event "Move" cannot occur, but arrivals and departures can. It follows that

$$Q_{0,0} = \begin{bmatrix} -\lambda & 0 & 0 & 0 \\ \mu_2 & -(\mu_2 + \lambda) & 0 & 0 \\ 0 & \mu_2 & -(\mu_2 + \lambda) & 0 \\ 0 & 0 & \mu_2 & -(\lambda + \mu_2) \end{bmatrix}.$$

However, $Q_{0,1} = Q_1$ and $Q_{1,0} = Q_{-1}$. It follows that once convergence is achieved, the matrix with all states above level 2 eliminated becomes

$$\begin{bmatrix} Q_{0,0} & Q_1 \\ Q_{-1} & B \end{bmatrix}.$$

The transition matrix $P^{(0)} = B_{0,0}$ becomes

$$B_{0,0} = Q_{0,0} + Q_1 (B)^{-1} Q_{-1}.$$

We now find π_0 up to a factor, using

$$0 = \pi_0 B_{0,0}.$$

This is a system of equilibrium equations that can be solved readily for π_0, except for a factor. To find this factor, we take the sum of all π_j, which is, according to equation (12.54):

$$\sum_{j=0}^{\infty} \pi_j = P(1) = \pi_0(I - A_1)^{-1}.$$

This is a vector with 4 entries, say $[s_0, s_1, s_2, s_3]$, with the s_j being the probability that $X_2 = j$. Now, $s_0 + s_1 + s_2 + s_3 = 1$, and this relation can be used for norming.

We did all the calculations with MATLAB, using $\lambda = 1$, $\mu_1 = \mu_2 = 1$. As a solution of $0 = \pi_0 B_{0,0}$, we found

$$\pi_0 = [1.00000, 0.69997, 0.45931, 0.23876].$$

Next, the unnormed value of $P(1)$ was

$$P(1) = \pi_0(I - A)^{-1} = [7.2867, 6.2866, 5.5866, 5.1273].$$

The norming constant is thus the sum of the entries of $P(1)$, which is 24.287. The normed version of π_0 and $P(1)$ are thus

$$\pi_0 = [0.0411738, 0.0288203, 0.0189117, 0.0098308]$$
$$P(1) = [0.30002, 0.25884, 0.23002, 0.21111].$$

Note that $P(1)$ provides the distribution of X_2. In particular, the probability that $X_2 = 3$ is 0.21111, and in this situation, server 2 is blocked. This means that server 1 cannot work for 21% of the time, which reduces its service rate from 1 to $1 - 0.21111 = 0.78889$. Hence, whereas without blocking, the traffic intensity for server 1 is $\rho = \lambda/\mu_1 = 0.7$, it would increase to 0.78889, which is substantial. For an $M/M/1$ queue, this change would increase the expected line length from 2.333 to 7.875. We now discuss how this will work for the tandem queue.

First, we find $P'(1)$ according to equation (12.55). We have

$$P'(1) = P(1)A(I - A)^{-1} = [2.9076, 2.6975, 2.5652, 2.5019].$$

We add the entries of this vector to find $E(X_1) = 10.672$. This is even worse than in an $M/M/1$ queue with $\rho = 0.78889$. The reason is that as X_1 increases, so does the blocking probability. The values obtained from our calculation indicate that when $X_1 = 0$, the blocking probability is 0.09957, but when $X_1 = 10$, it increases to 0.2363. Blocking probabilities that increase with X_1 affect the $P\{X_1 = i\}$: It slows its decrease with i. This, in turn, increases $E(X_1)$.

12.3.5 The Classical MAM Paradigms

There is an extensive body of literature describing the so-called *matrix analytic methods*, or MAMs, where methods are discussed for finding equilibrium solutions of $M/G/1$, $GI/M/1$ and QBD type processes. At first sight, the approach used by the MAM community seems to be different from ours, but on closer inspection, MAM methods are closely related to ours, as will be shown in this section. For further information about MAM, the reader should consult the following books, which are listed in order of their appearance: [72], [73], [62], [6], [46]. We should stress, however, that with the possible exception of the so-called logarithmic reduction algorithm [62], [6], our methods are at least as efficient as the corresponding MAMs. In fact, the final formulas are sometimes identical, though they were obtained by using different arguments.

12.3.5.1 The $GI/M/1$ Paradigm

The foundations of the $GI/M/1$ paradigm were developed by Wallace [93], and put on a mathematical sound basis by Neuts [72]. The main formulas developed for the $GI/M/1$ paradigm are

$$\pi_{j+1} = \pi_j R.$$

where R is the solution of the matrix equation

$$R = \sum_{v=0}^{\infty} R^v P_{1-v}. \tag{12.56}$$

Actually, Neuts [72] uses the symbols A_v instead of P_{1-v}, but we use A_v for different purposes. He also assumed that g can be infinite. We now show that these equations immediately follow from equations (12.45), (12.42) and (12.43), with $g \to \infty$.

We now show that the $GI/M/1$ paradigm can be obtained from our equations (12.42) and (12.43), which when $h = 1$, simplify to

$$A_1 = P_1(I - B_0)^{-1}$$
$$B_j = P_{-j} + A_1 B_{j+1}, \ j \geq 0.$$

Also, equation (12.45) becomes for $h = 1$:

$$\pi_{j+1} = \pi_j A_1.$$

It follows that $R = A_1$. Next, we assume that we cannot decrease the level by more than g, which implies the $P_k = 0$ for $k > g$. Since sums have to be truncated anyhow, using finite values for g does not pose a serious restriction. Clearly, for $h = 1$ and $A_1 = R$, (12.43) becomes

$$B_g = P_{-g}$$
$$B_j = P_{-j} + RB_{j+1}, \ 0 \le j < g.$$

Consequently,

$$B_g = P_{-g}$$
$$B_{g-1} = P_{-g+1} + RB_g = P_{-g+1} + RP_{-g}$$
$$B_{g-2} = P_{-g+2} + RB_{g-1} = P_{-g+2} + RP_{-g+1} + R^2 P_{-g}.$$

Continuing in this fashion, we get

$$B_0 = P_0 + RP_{-1} + R^2 P_{-2} + \cdots + R^g P_{-g} = \sum_{v=0}^{g} R^v P_{-v}. \tag{12.57}$$

Also, (12.42) becomes for $h = 1$, $R = A_1 = P_1(I - B_0)^{-1}$, which yields $R(I - B_0) = P_1$, and with equation (12.57), we obtain

$$R\left(I - \sum_{v=0}^{g} R^v P_{-v}\right) = P_1.$$

This is equivalent to equation (12.56), which is therefore proven. It follows that the method of Neuts can be derived from the extrapolating Crout method.

Neuts [72] also shows that the i, j entry of R is equal to the expected number of visits to state j of level $n + 1$ on a path starting in state i of level n before returning to a level of n or below. This result can also be derived from the extrapolating Crout method. Indeed, $(I - B_0)^{-1}$ is the fundamental matrix of B_0, and its i, j entry $t_{i,j}$ is the expected number of visits to state j before leaving the present level n, given the level was entered in sublevel i. Also, $p_{i,j}^1$, the i, j entry of P_1, is the probability to move from level $n - 1$ to level n. Consequently the i, j entry of $R = P_1(I - B_0)^{-1}$, given by $\sum_{k=1}^{m} p_{i,k}^1 t_{k,j}$, represents the expected number of visits to state j of level n, given the process starts in sublevel i of level $n - 1$. This is equivalent to the interpretation of R given by Neuts.

Neuts suggests to solve the matrix equation for R by successive approximation, with R_n being the nth approximation. He uses $R_0 = 0$ as his initial approximation and calculates

$$R_{n+1} = \sum_{v=0}^{\infty} R_n^v P_{1-v}.$$

In each iteration, R_n increases, converging eventually to R. He also suggested the following improvement to his algorithm: By writing equation (12.56) as

$$R(I - P_0) = P_1 + \sum_{v=2}^{\infty} R_n^v P_{1-v},$$

he obtains

$$R_{n+1} = \left(P_1 + \sum_{v=2}^{\infty} R_n^v P_{1-v} \right) (I - P_0)^{-1}.$$

This matrix equation can also be solved by successive approximation.

Another approach has been suggested by Latouche [61, Theorem 3.1]. Using our notation, he defines

$$U = \sum_{v=0}^{\infty} R^v P_{-v}.$$

and finds $R = P_1(I - U)^{-1}$. This approach is identical to ours, with $U = B_0$. What is important is that Latouche found experimentally that his algorithm outperforms the ones of Neuts discussed above. Now, if g is finite, and the polynomial is evaluated by Horner's method [49], which is the standard method for evaluating polynomials [49], then the method of Latouche and our method are identical, as the reader may verify.

12.3.5.2 The $M/G/1$ Paradigm

In the $M/G/1$ paradigm, $g = 1$, and h can assume any value. To solve this problem, Neuts [73] used a matrix G, which he finds by solving the following matrix equation which, in our notation becomes

$$G = \sum_{v=0}^{\infty} P_{v-1} G^v. \tag{12.58}$$

To find G, Neuts suggests using a successive approximation, similar to the one he used for R. We now claim that

$$G = (I - B_0)^{-1} P_{-1}.$$

We prove this first by using an interpretation of G suggested by Neuts, who found that $g_{i,j}$, the (i, j) entry of G, is the probability that starting in sublevel i of level $n + 1$, the first time the process falls to level n, it goes to sublevel j. We already stated that the entries of $(I - B_0)^{-1}$, we called them $t_{i,j}$, are the expectations of the number of visits to sublevel j of level n when starting in sublevel i of level n. From this, it is easy to show that the entries of $(I - B_0)^{-1} P_{-1}$ are really the probabilities that starting in sublevel i of level n, the first time the process falls below level n, it goes to sublevel j.

We now prove that the matrix equation for G can be derived form equations (12.42) and (12.43), which simplify to

$$A_h = P_h$$
$$A_i = (P_i + A_{i+1} B_1)(I - B_0)^{-1}, \ i = 1, 2, \ldots, h - 1$$
$$B_{-1} = P_{-1}$$
$$B_0 = P_0 + A_1 B_1.$$

From this, we find

$$A_i(I - B_0) = P_i + A_{i+1}P_{-1} = P_i + A_{i+1}(I - B_0)(I - B_0)^{-1}P_{-1}.$$

If we set $\hat{A}_i = A_i(I - B_0)$, and use $G = (I - B_0)^{-1}P_{-1}$, then this yields

$$\hat{A}_i = P_i + \hat{A}_{i+1}G.$$

As a consequence,

$$
\begin{aligned}
\hat{A}_h &= P_h \\
\hat{A}_{h-1} &= P_{h-1} + \hat{A}_h G = P_{h-1} + P_h G \\
\hat{A}_{h-2} &= P_{h-2} + \hat{A}_{h-1}G = P_{h-2} + P_{h-1}G + P_h G^2 \\
&\vdots \\
\hat{A}_1 &= P_1 + \hat{A}_2 G = P_1 + P_2 G + \cdots + P_h G^{h-1}.
\end{aligned}
$$

Also, since $B_1 = P_{-1}$:

$$B_0 = P_0 + A_1(I - B_0)(I - B_0)^{-1}B_1 = P_0 + \hat{A}_1 G = \sum_{\nu=0}^{h} P_\nu G^\nu.$$

From $G = (I - B_0)^{-1}P_{-1}$, we obtain $(I - B_0)G = P_{-1}$, and

$$G = P_{-1} + B_0 G = \sum_{\nu=-1}^{h} P_\nu G^{\nu+1}.$$

This equation is equivalent to equation (12.58), which shows that the paradigms of Neuts can be derived from the extrapolating Crout equations.

12.4 Solutions using Characteristic Roots and Eigenvalues

Equilibrium equations with repeating rows are frequently solved by using characteristic roots. Indeed, before MAM methods gained acceptance, this was the standard method.. A related solution method, which is used quite often, is based on generating functions. Both methods are covered in this section, and their relation to the extrapolating Crout method is shown. In the case of the block-structured matrices with repeating rows discussed in Section 12.3, the characteristic roots are replaced by so-called *generalized eigenvalues*.

12.4.1 Solutions based on Characteristic Roots

When rows repeat, the repeating steady-state equation can be written as in (12.11):

$$\pi_j = \pi_{j-h}p_h + \pi_{j-h+1}p_{h-1} + \cdots + \pi_{j+g}p_{-g}. \tag{12.59}$$

We are looking for solutions of this equation in the form $\pi_j = x^j$. Substituting x^j for π_j into the equation above yields

$$x^j = x^{j-h}p_h + x^{j-h+1}p_{h-1} + \cdots + x^{j+g}p_{-g}.$$

To convert this into a polynomial equation, we multiply this equation by x^{h-j} to obtain

$$x^h = p_h + p_{h-1}x + \cdots + p_{-g}x^{h+g}. \tag{12.60}$$

This equation is called *characteristic equation*, and its roots are the *characteristic roots*. The characteristic equation is a polynomial equation of degree $g + h$, and it yields $g + h$ roots for x, say $x_1, x_2, \ldots, x_{g+h}$. One can show by using Rouché's theorem, that of the $g + h$ values for x, exactly h are within the unit circle (that is, their absolute value is less than 1). Since for x on and outside the unit circle, $\pi_j = x^j$ does not converge as $j \to \infty$, we can discard these roots. It can also be shown that the largest root inside the unit circle is real and positive.

We now have to form a linear combination of the powers of these roots such that the initial conditions are satisfied. Hence, we set

$$\pi_j = \sum_{k=1}^{h} c_k x_k^j. \tag{12.61}$$

The c_k must be determined in such a way that the boundary equations are satisfied.

Note that the largest root x_1 will eventually dominate. This root is positive and has a multiplicity of 1, which implies that for large enough n $\pi_{n+1} \approx \pi_n x_1$, that is, the π_n eventually decrease geometrically with n.

If rows repeat for i greater or equal to $d_1 > 0$, then the equilibrium equations for $j \geq d_1 + h$ are all of the form

$$\pi_j = \sum_{i=j-h}^{j+g} \pi_i p_{j-i}, \quad j \geq d_1 + h. \tag{12.62}$$

To show that this equation is valid for any $j \geq d_1 + h$, note that $p_{i,j} = p_{j-i}$ for $i \geq d_1$, that is

$$\pi_j = \sum_{i=j-h}^{j+g} \pi_i p_{i,j}$$

becomes equation (12.62) once $j - h \geq d_1$. However, if $j < d_1 + h$, then equation (12.62) is no longer valid. All π_i occurring in equation (12.62) must satisfy equation

(12.61), that is, π_j is given by equation (12.61) $j \geq d_1$, but for $j < d_1$, π_j must be found by other means. When we want expressions for all π_j, have to determine the c_k, $k = 1, 2, \ldots, h$, as well as the π_j, $j = 0, 1, \ldots, d_1 - 1$. We have thus $d_1 + h$ variables, which must be determined, and for doing this, we need $d_1 + h$ independent equations.

First, there are the equilibrium equations of the states $0, 1, \ldots, d_1 + h - 1$. In these equilibrium equations, all π_i, $i \geq d$, are substituted by $\sum_k c_k x_k^i$. One equation depends on the other equilibrium equations. To account for this, we drop the equilibrium equation for state 0. Hence, we have only $d_1 + h - 1$ equilibrium equations for $d_1 + h$ variables. The missing equation is supplied by the fact that the sum of all probabilities must be 1, that is

$$1 = \sum_{i=0}^{\infty} \pi_i = \sum_{i=0}^{d_1-1} \pi_i + \sum_{i=d_1}^{\infty} \sum_{k=1}^{h} c_k x_k^i = \sum_{i=0}^{d_1-1} \pi_i + \sum_{k=1}^{h} c_k \frac{x_k^{d_1}}{1 - x_k}.$$

With this, we have all the equilibrium equations we need.

Since the x_i may be negative, or even complex, the result is subject to rounding errors. As an alternative, we suggest finding the a_i, $i = 1, 2, \ldots, h$ as given by equation (12.19), through applying the following theorem.

Theorem 12.3 *Let* $x_1, x_2, \ldots, x_h$ *be the roots of equation (12.60) inside the unit circle, and let the* a_i, $i = 1, 2, \ldots, h$ *be the values defined by equation (12.19). We then have*

$$(x - x_1)(x - x_2) \ldots (x - x_h) = x^h - a_1 x^{h-1} - a_2 x^{h-2} - \cdots - a_h.$$

Proof We have

$$\pi_j = \sum_{i=1}^{h} \pi_{j-i} a_i.$$

We substitute x^i for π_i in this equation to obtain

$$x^j = \sum_{i=1}^{h} x^{j-i} a_i$$

which leads to the equation:

$$x^h = \sum_{i=1}^{h} a_i x^{h-i}, \tag{12.63}$$

or

$$x^h - a_1 x^{h-1} - a_2 x^{h-2} - \cdots - a_h = 0.$$

Since the solution of the equilibrium equation is unique, except for a factor, the x_i obtained from this equation must also satisfy (12.60), which completes the proof. $\square$

The a_i can thus be found by expanding $(x - x_1)(x - x_2) \cdots (x - x_h)$. For instance, if $h = 2$, $x_1 = 0.9$ and $x_2 = -0.5$, then

$$(x - 0.9)(x + 0.5) = 1 - 0.4x - 0.45.$$

Consequently, $a_1 = 0.4$, $a_2 = 0.45$. Once the a_i are known, one can find the π_i as in the extrapolating Crout method, and the z-transform is given by equation (12.26).

12.4.1.1 The Wiener-Hopf Factorization

We now discuss the *Wiener-Hopf factorization*, which for our models can be formulated as follows, where the a_i and b_j as defined in equations (12.19) and (12.20).

$$\left(1 - \sum_{i=1}^{h} a_i x^{-i}\right)\left(1 - \sum_{j=0}^{g} b_j x^i\right) = \left(1 - \sum_{j=-h}^{g} p_{-j} x^j\right). \tag{12.64}$$

To prove this equation, note that the left side of (12.64) can be written as

$$\left(1 - \sum_{i=1}^{h} a_i x^{-i}\right)\left(1 - \sum_{j=0}^{g} b_j x^j\right) = 1 - \sum_{i=1}^{h} a_i x^{-i} - \sum_{j=0}^{g} b_j x^j + \sum_{i=1}^{h}\sum_{j=0}^{g} a_i b_j x^{j-i}.$$

Using the expression at the right of this equation in (12.64) yields

$$\sum_{i=1}^{h} a_i x^{-i} + \sum_{j=0}^{g} b_j x^j - \sum_{i=1}^{h}\sum_{j=0}^{g} a_i b_j x^{j-i} = \sum_{j=-h}^{g} p_{-j} x^j.$$

We now compare the coefficients of x^k on both sides of this expression. First, suppose $k \geq 0$. On the left, we find the coefficient of x^k in the second sum when $j = k$, and in the third when $j - i = k$. This must be equal to the coefficient of x^k on the right side, which is p_{-k}. It follows that

$$b_k - \sum_{i=1}^{h} a_i b_{i+k} = p_{-k}.$$

Using also $b_j = 0$ for $j > g$, we find

$$b_k = p_{-k} + \sum_{i=1}^{h,g-k} a_i b_{i+k},$$

which is exactly the value for b_k given by equation (12.20).

For $k < 0$, we find x^k in the first sum on the left when $i = k$, and when $j - i = k$ in the third sum. Consequently,

$$a_k - \sum_{j=0}^{g,h-k} a_{j+k} b_j = p_k.$$

or, since a_k appears on both sides of this equation,

$$a_k(1 - b_0) = p_k + \sum_{j=1}^{g,h-k} a_{j+k} b_j.$$

This matches equation (12.19), and equation (12.64) is proven.

It is clear that the right side of equation (12.64) has $g + h$ zeros. From equation (12.63), and the fact that the π_j converge as $j \to \infty$, we conclude that $1 - \sum_{i=1}^{h} a_i x^{-i}$ contains the zeros inside the unit circle. This implies that the remaining zeros, the zeros of $1 - \sum_{j=0}^{g} b_j x^j$ must be outside or on the unit circle. It follows that the Wiener Hops factorization separates the zeros inside the unit circle from the ones not inside the unit circle.

Incidentally, $A(z) = \sum_{i=1}^{h} a_i z^i$, $B(z) = \sum_{j=0}^{g} b_j z^j$, and $T(z) = \sum_{j=-g}^{h} p_j z^j$ are the z-transform of a_i, b_j and p_k, respectively. By using these transforms, we can formulate the Wiener–Hopf factorization as

$$(1 - A(z))(1 - B(1/z)) = (1 - T(z)).$$

The Wiener Hops factorization provides a factorization of the expression at the left of equation (12.64). There are many other such factorizations. Indeed, one can use any h roots of the expression on the right to construct the factor on the left involving the a_i. In particular, by exchanging the root $x = 1$ of the polynomial involving the b_j with the largest root for x of the expression involving the a_i, one converts the stochastic solution into the substochastic solution. From a numerical point of view, the question arises as to whether or not Algorithm 32 converges to the correct solution. In literature, no case is reported where Algorithm 32 converge to the substochastic solution while $E(D) < 0$, and we think such cases are extremely rare. Of course, one can always force it to converge to the stochastic solution by setting S^m in Algorithm 32 to $\sum_{j=1}^{g} b_j^m$.

12.4.1.2 Characteristic Roots and z-Transforms

The classical methods to find the π_i are through the z-transform. This method is used extensively, especially in the literature on bulk queues [10]. However, it is very susceptible to rounding errors, and we feel there are better methods.

To find the z-transform $P(z)$, we replace the π_j in $P(z) = \sum_{i=0}^{\infty} \pi_j z^j$ by the values given by equations (12.10), (12.11) and (12.12) to obtain by setting $\pi_i = 0$ for $i < 0$:

$$P(z) = \pi_0 + \sum_{j=1}^{\infty} z^j \left(\sum_{i=j-h}^{d_1-1} \pi_i p_{i,j} + \sum_{i=\max(d_1, j-h)}^{j+g} \pi_i p_{j-i} \right).$$

We now present the transform method in outline, without going into details. One finds, where $T(z) = \sum_{i=-g}^{h} p_i z^i$:

$$P(z) = P(z)T(z) + N(z, \pi_j, 0 \le j \le d_1 + g). \tag{12.65}$$

Here, $N(z, \pi_j, 0 \le j \le d_1 + g)$ becomes after lengthy calculations we omit

$$N(z, \pi_i, i = 0, 1, \ldots d_1 + g - 1) = \pi_0 + \sum_{j=1}^{d_1+h-1} z^j \sum_{i=j-h}^{d_1-1} \pi_i p_{i,j}$$

$$+ \sum_{i=d_1}^{d_1+g-1} \pi_i z^i \sum_{j=d_1-g}^{i+h} p_{j-i} z^{j-i} - \sum_{i=0}^{d_1+g-1} \pi_i z^i \sum_{j=-g}^{h} p_j z^j.$$

Solving (12.65) for $P(z)$ yields

$$P(z) = \frac{N(z, \pi_j, 0 \le j \le d_1 + g)}{1 - T(z)}. \tag{12.66}$$

Clearly, if $1-T(z) = 0$, then $P(z)$ would be infinite unless $N(z, \pi_j, 0 \le j \le d_1+g) = 0$. We know that $P(z)$ is finite for any z with $|z| \le 1$. To prove this, note that for any sum of constants v_i, $|\sum_i v_i| \le \sum |v_i|$, which implies

$$|P(z)| = \left| \sum_{j=0}^{\infty} \pi_i z^i \right| \le \sum_{j=0}^{\infty} \pi_i |z^i| < 1, \ |z| < 1.$$

Of course, $P(z)$ also exists for $z = 1$. Hence, for any z with $|z| > 1$, if $1 - T(z) = 0$, then $N(z, \pi_j, 0 \le j \le d_1 + g) = 0$. Now, equation (12.60) is equivalent to $T(1/x) = 1$. This implies that the characteristic roots outside the unit circle match with the roots of $T(z) = 1$ inside the unit circle. We therefore have exactly g values of z with $|z| < 1$. This provides g equations, which, in addition to boundary equations, and $P(1) = 1$, provide enough equations to find all π_j, $0 \le j \le d_1 + g$. The remaining equations can then be found from equation (12.61). Unfortunately, this method is numerically unstable except for small values of g, and not recommended in general. Instead, we suggest to calculate the a_i from the z_i outside the unit circle (or the x_i inside the unit circle) by methods discussed in Section 12.4, and to solve the problem this way.

12.4.2 Using Eigenvalues for Block-Structured Matrices with Repeating Rows

Given a block-structured matrix with repeating rows, one can find the equilibrium probability vectors π_j as follows. One sets

$$\pi_j = v x^j,$$

where the vector v and the scalar x have to be determined. The dimension of v is given by the number of sublevels. From (12.39), one obtains

$$vx^j = \sum_{i=j-h}^{j+g} x^i P_{j-i}.$$

After dividing this equation by x^{j-h} and bringing vx^h to the right, one finds

$$0 = \sum_{i=0}^{g+h} x^i P_{i-g} - vx^h. \tag{12.67}$$

In case of CTMCs, this equation must be modified as follows:

$$0 = \sum_{i=0}^{g+h} x^i Q_{i-g}.$$

We now briefly interrupt our discussion to present so-called lambda matrices, which are polynomials with coefficients that are matrices, that is, a lambda matrix has the form

$$A(\lambda) = \sum_{i=0}^{n} A_i \lambda^i,$$

where the A_i are matrices, and λ is a scalar. Consequently, the polynomial in equation (12.67) is a lambda matrix. If there is a value λ and a vector v satisfying

$$vA(\lambda) = v \sum_{i=0}^{n} A_i \lambda^i = 0, \tag{12.68}$$

then λ is a *generalized eigenvalue* and v is a *generalized eigenvector*. The name *lambda matrices* [60], [95] is due to the fact that eigenvalues are typically denoted by λ, that is, instead of x as in equation (12.67), the symbol λ is used. Other names for lambda matrices are matrix pencils [56] and matrix polynomials [23]. The term matrix polynomial is also used for expressions like $\sum_{i=1}^{n} c_i A^i$, with c_i being scalars, and A being a matrix. The term "pencil" is sometimes only used when $n = 1$, that is, only expressions like $A_0 + A_1 \lambda$ are called pencils. Note that if $A(\lambda) = A_0 - I\lambda$, then

$$vA(\lambda) = v(A_0 - I\lambda) = 0$$

is a standard eigenvalue problem.

If A_n in the lambda matrix $\sum_{i=0}^{n} A_i \lambda^i$ is the identity matrix, then the generalized eigenvalue problem given by equation (12.68) can be converted into a standard eigenvalue problem and solved as such. To do this, let $u = [u_0, u_2, \ldots, u_{n-1}]$ and

$$L = \begin{bmatrix} 0 & I & 0 & \ldots\ldots & 0 \\ 0 & 0 & I & \ddots\ \ddots & 0 \\ \vdots & \ddots & \ddots\ \ddots\ \ddots & & \vdots \\ 0 & 0 & \ldots\ldots & 0 & I \\ -A_0 & -A_1 & \ldots\ldots\ldots & & -A_{n-1} \end{bmatrix}. \tag{12.69}$$

One can now solve the standard eigenvalue problem $uL = u\lambda$, and set $v = u_{n-1}$. The eigensolution is then (λ, v). In other words, this eigensolution satisfies $u_{n-1}A(\lambda) = 0$. To prove this, expand $uL = u\lambda$ to obtain

$$-u_0\lambda - u_{n-1}A_0 = 0$$
$$u_0 - u_1\lambda - u_{n-1}A_1 = 0$$
$$u_1 - u_2\lambda - u_{n-1}A_2 = 0$$
$$\vdots \qquad \vdots \quad \vdots$$
$$u_{n-2} - u_{n-1}\lambda - u_{n-1}A_{n-1} = 0.$$

By multiplying the ith equation by λ^i, $i = 0, 1, 2, \ldots, n - 1$ and then adding all equations together, one finds

$$u_{n-1}\sum_{i=0}^{n-1} A_i\lambda^i + u_{n-1}\lambda^n = 0.$$

This is equation (12.68) with $u_{n-1} = v$.

With this, standard eigenvalue routines can be used to solve the generalized eigenvalue problem as long as A_n is the identity matrix. If A_n is not the identity matrix, then the procedure above can still be used as long as A_n is invertible. In this case, one forms $A_iA_n^{-1}$ and uses these values instead of the A_i. Also, if A_n is not invertible, but A_0 is, one sets $z = \lambda^{-1}$, and if (12.68) holds, so does

$$v\sum_{i=0}^{n} A_{n-i}z^i = 0.$$

One can now post-multiply all A_{n-i} by A_0^{-1}, and proceed as before. If neither A_0 nor A_h is invertible, other methods must be used. We will mention some of these methods later. At the moment, we refer to Stewart [84], who shows how to find eigenvalues and eigenvectors of the type $vA = vB\lambda$. Methods similar to the one given above can be used to convert every generalized eigenvalue problem into such a pencil as the reader may verify.

There are nm generalized eigenvalues, where m is the dimension of the matrices A_i. This can be seen from the fact that the generalized eigenvalue problem can be converted into a standard eigenvalue problem with a matrix of dimension nm. This result holds generally, even when neither A_n nor A_0 is invertable. Obviously, if A_0 is singular, then $\lambda = 0$ is a generalized eigenvalue because $v\sum_{i=1}^{n} A_i = 0$ reduces to

$vA_0 = 0$. Note also that $\lambda = \infty$ can be a generalized eigenvalue [84]. For simplicity, we occasionally omit from now on the word "generalized" because this is what we are mainly using from now on.

We now return to the solution of equation (12.67), which is a generalized eigenvalue problem. First note that if $\pi_i = vx^i$, and π_i must converge, then only values of x with absolute value less than 1 are acceptable. There are hm such eigenvalues. To prove this, note that according to equation (12.45), $\pi_j = \sum_{i=j-h}^{j-1} \pi_{j-i} A_i$, and if $\pi_j = vx^j$, then $v \sum_{i=j-h}^{j-1} A_i x^{j-i} = vx^j$. This is a generalized eigenvalue problem, and it therefore has mh eigensolutions, where m is the size of the matrices A_i. Moreover, of the eigenvalues $|x| < 1$, the eigenvalue with the largest absolute value must be positive because otherwise, π_j would turn negative for some j, which is impossible. The consequence of this is that the distribution of X_1, the level, has a geometric tail, with the parameter given by the largest eigenvalue.

12.4.3 Application of the Generalized Eigenvalue Problem for Finding Equilibrium Solutions

We now describe how to proceed in order to find the equilibrium probabilities when the transition matrix has repeating rows of blocks, as indicated in equation (12.35). The equilibrium equations are therefore:

$$\pi_0 = \sum_{i=0}^{g} \pi_i P_{i,0} \tag{12.70}$$

$$\pi_j = \pi_0 P_{0,j} + \sum_{i=1}^{j+g} \pi_i P_{j-i}, \; j \le h \tag{12.71}$$

$$\pi_j = \sum_{i=j-h}^{j+g} \pi_i P_{j-i}, \; j > h. \tag{12.72}$$

First, we replace the π_i in equation (12.72) by vx^i, where v is a vector and x is a scalar:

$$vx^j = \sum_{i=j-h}^{j+g} vx^i P_{j-i}.$$

This can be converted into a generalized eigenvalue problem, and solved as such. If the matrices P_k are of size m, then there are mh eigensolutions (v_k, x_k) with $|x| < 1$. Any of these eigensolutions satisfies equation (12.72), and so does any linear combination of them. We therefore set

$$\pi_j = \sum_{k=1}^{mh} c_k v_k x^j.$$

Here, the c_k are determined such that the boundary conditions given by equations (12.70) and (12.71) are satisfied. Replacing π_i, $i > 1$ by $\sum_{k=1}^{mh} c_k v_k x^i$ yields

$$\pi_0 = \pi_0 P_{0,0} + \sum_{i=0}^{g} \sum_{k=1}^{h} c_k v_k x_k^i P_{i,0} \tag{12.73}$$

$$\pi_j = \pi_0 P_{0,j} + \sum_{i=1}^{j+g} \sum_{k=1}^{h} c_k v_k x_k^i P_{j-i}, \; j \le h. \tag{12.74}$$

If level 0 has m_0 sublevels, then the expression (12.73) contains m_0 equation, and expression (12.74) contains mh equations. One of these equations can be derived from the other ones, that is, there are $m_0 + mh - 1$ independent equations. These equations must be used to determine the mh values of c_k and the m_0 values of $\pi_{0,j}$. An additional equation can be found from the condition that the sum of all probabilities must be 1, that is:

$$1 = \sum_{j=1}^{m_0} \pi_{0,j} + \sum_{i=1}^{\infty} \pi_i \mathbf{e} = \sum_{j=1}^{m_0} \pi_{0,j} + \left(\sum_{k=1}^{h} \frac{c_k x_k v_k}{1 - x_k} \right) \mathbf{e}.$$

Here, $\mathbf{e}$ is the column vector with all its entries equal to 1. We conclude that as soon as the eigenvalues and eigenvectors are determined, the problems are solved. Note, however, that for large m, solving the initial equations may be problematic because of rounding errors.

When $P_{i,j} = P_{j-i}$ for all $j > 0$ and for all i, the method above can be simplified. It is then convenient to set all $\pi_j = 0$ for $j < 0$. In this way, equation (12.72) holds for all j, even for $j < h$. We now have the following equations to find the c_k and π_0:

$$\pi_0 = \pi_0 P_{0,0} + \sum_{i=1}^{g} \sum_{k=1}^{h} c_k v_k x_k^i P_{0,i}$$

$$0 = \sum_{i=-h}^{g} \sum_{k=1}^{h} c_k v_k x_k^i P_{j-i}, \; j < 0.$$

The norming condition is now

$$1 = P_0 \left(\sum_{k=0}^{h} \frac{c_k v_k}{1 - x_k} \right) \mathbf{e}.$$

The question remains of how to find the mh eigenvalues x_k and the corresponding eigenvectors v_k. If P_h is invertable, the problem can be converted into an eigenvalue problem and solved as such. Also, if the matrix $\sum_{i=-g}^{h} P_i x^i$ is tridiagonal, then all eigenvalues are real as shown in [31]. The proof for this is similar to the one used for the standard eigenvalue problem discussed in Section 8.2.11. When all eigenvalues are real, then one can just pick two values of x between 1 and -1, and use the secant method to minimize $\left| \sum_{i=-g}^{h} P_i x^i \right|$.

12.4.4 Eigenvalue Solutions Requiring only one Eigenvalue

There are two important situations where only one eigenvalue is required, which must be positive and less than 1. This can happen for two different reasons: In some models, only one eigensolution exists with $x \neq 0$, whereas in others, more than one eigensolution with $x \neq 0$ exists, but the initial conditions can be satisfied by using just one eigensolution. In the latter case, any change in the initial conditions will require additional eigenvalues.

Consider now the case where there is only one non-zero eigenvalue. This happens when $h = 1$ and P_1 contains only one non-zero row. To show this, note that according to equation (12.42), when $h = 1$ then $A_1 = P_1(I - B_0)^{-1}$, and all rows of P_1 that are 0 lead to rows of A_1 that are 0. Now

$$\pi_{j+1} = \pi_j A_1.$$

By setting $\pi_n = v x^n$, this problem turns into a standard eigenvalue problem. If A_1 has only one non-zero row, then all eigenvalues of A_1 are 0, except for one non-zero eigenvalue. The eigenvalue $x = 0$ has $m - 1$ different independent eigenvectors because A_1, having only one non-zero row, has a rank of 1. In conclusion, there is only one non-zero eigenvalue.

A model where there is only one non-zero eigenvalue is the shorter queue problem, and this model will now be analyzed. The case where the dominant eigensolution satisfies the initial conditions is the tandem queue, which is used as our second example.

12.4.4.1 The Shorter Queue Problem

In the shorter queue problem, there are two servers, each with its own line, and arrivals join the shorter line. When both lines have the same length, they join line 1. If the difference between the lines is greater m, the last customer switches from the longer to the shorter line. The difference between lines is thus at most m. Arrivals are Poisson with a rate of λ, and the service times are exponential, with both servers having a service rate of μ. To formulate the system, we could use the line lengths as state variables, but this would mean that we have two unbounded state variables. Instead, we denote X_1 as the longer line, and L_2 as the shorter line. However, as the sublevel, we do not use L_2, but $X_2 = X_1 - L_2$. Hence, X_2 indicates the difference between the lines. Since jockeying occurs as soon as X_2 exceeds m, we have $0 \leq X_2 \leq m$.

There are now three events, arrivals, departures from the longer line, and departures from the shorter line, and we denote the corresponding event functions by $f_A(X_1, X_2)$, $f_{D_1}(X_1, X_2)$ and $f_{D_2}(X_1, X_2)$. An arrival will increase X_1 only when $L_1 = L_2$, and then $X_2 = 0$. Otherwise, arrivals will leave X_1 unchanged, and decrease X_2. Consequently

$$f_A(X_1, X_2) = \begin{cases} [X_1 + 1, X_2 + 1] & \text{if } X_2 = 0 \\ [X_1, X_2 - 1] & \text{otherwise.} \end{cases}$$

A departure from the longer line will decrease both X_1 and X_2 by 1, that is

$$f_{D_1}(X_1, X_2) = [X_1 - 1, X_2 - 1].$$

A departure from the shorter line leaves X_1 unchanged, but increases X_2, except when $X_2 = m$. When $X_2 = m$, then there is a switch from the longer line to the shorter line, decreasing X_1 by 1. Since L_2 remains unchanged, $X_1 - L_2$ also decreases by 1. Consequently

$$f_{D_2}(X_1, X_2) = \begin{cases} [X_1, X_2 + 1] & \text{if } X_2 < m \\ [X_1 - 1, X_2 - 1] & \text{if } X_2 = m. \end{cases}$$

We are now ready to formulate the interior equilibrium equations. One has, if $m > 3$:

$$\pi_{i,0}(\lambda + 2\mu) = \lambda\pi_{i,1} + \mu\pi_{i+1,1} \tag{12.75}$$

$$\pi_{i,1}(\lambda + 2\mu) = \lambda\pi_{i-1,0} + 2\mu\pi_{i,0} + \lambda\pi_{i,2} + \mu\pi_{i+1,2} \tag{12.76}$$

$$\pi_{i,j}(\lambda + 2\mu) = \lambda\pi_{i,j+1} + \mu\pi_{i,j-1} + \mu\pi_{i+1,j+1}, \ 2 \le j < m-1 \tag{12.77}$$

$$\pi_{i,m-1}(\lambda + 2\mu) = \lambda\pi_{i,m} + 2\mu\pi_{i+1,m} + \mu\pi_{i,m-2} \tag{12.78}$$

$$\pi_{i,m}(\lambda + 2\mu) = \mu\pi_{i,m-1}. \tag{12.79}$$

If $m = 3$, then equation (12.77) disappears. When $m = 2$, we still have equations (12.75) and (12.79). In addition, we have for $j = 1$

$$\pi_{i,1}(\lambda + 2\mu) = \lambda\pi_{i-1,0} + \lambda\pi_{i,2} + 2\mu\pi_{i+1,0} + 2\mu\pi_{i+1,2}.$$

From now on, we assume that $m > 3$. In order to solve these equations, we set $\pi_{i,j} = g_i x^j$, and obtain after dividing by x^i and simplifying:

$$g_0(\lambda + 2\mu) = g_1(\lambda + \mu x) \tag{12.80}$$

$$g_1(\lambda + 2\mu) = g_0(\lambda x^{-1} + 2\mu) + g_2(\lambda + \mu x) \tag{12.81}$$

$$g_j(\lambda + 2\mu) = g_{j+1}(\lambda + \mu x), +g_{j-1}\mu, \ 2 \le j < m-1 \tag{12.82}$$

$$g_{m-1}(\lambda + 2\mu) = g_m(\lambda + 2\mu x) + g_{m-2}\mu \tag{12.83}$$

$$g_m(\lambda + 2\mu) = \mu g_{m-1}. \tag{12.84}$$

Once we know x, we have $m + 1$ homogeneous equations, that is, if there is a non-zero solution, then this solution can only be determined up to a factor. This allows us to set $g_0 = 1$. After that, we can find g_1 from equation (12.80), g_2 from (12.81),

g_{j+1} from (12.82), $j = 3, 4, \ldots, m - 1$, and g_m from (12.83). The value of x can be found by using a slight modification of the method described in Section 8.2.11.

As it happens, in the case of the model under investigation, the value of x that satisfies equation (12.84) can be shown to be

$$x = \left(\frac{\lambda}{2\mu}\right)^2 = \rho^2. \tag{12.85}$$

Here, $\rho = \frac{\lambda}{2\mu}$. We will prove later that this is the required eigenvalue. Given x, we can calculate all g_i, and hence the $\pi_{i,j}$ that satisfy the interior equations. For instance, for $m = 4$, $\lambda = 1.8$ and $\mu = 1$, $x = 0.81$, we found the following values

$$\begin{array}{ccccc}
g_0 & g_1 & g_2 & g_3 & g_4 \\
1.000 & 1.456 & 0.502 & 0.173 & 0.046
\end{array}$$

This solution satisfies equation (12.84), as claimed.

The proof of equation (12.85) is as follows. If one discounts the boundary, and if L_1 is the length of the longer line, and L_2 the length of the shorter line, then $L_1 + L_2$ form the same process as an $M/M/2$ queue as long as both L_1 and L_2 are greater 0, which is satisfied when $X_1 > m$. Consequently,

$$r_{k+1} = P\{L_1 + L_2 = k + 1\} = r_k \rho. \tag{12.86}$$

Since $L_1 = X_1$ and $L_2 = X_1 - X_2$, $L_1 + L_2 = 2X_1 - X_2$. If X_1 increases by 1, $L_1 + L_2$ must therefore increase by 2, and $\pi_{i,j} = g_j x^i$ must increase by ρ^2 as i increases by 1. From this, it follows that $x = \rho^2$ as claimed.

If $m > 4$, the equations (12.80) to (12.84) form a set of difference equations for g_i, and this can be exploited to save a few flops, but since these savings are minor, we will not discuss this method further. For a similar approach, see [51].

We now come to the boundary equations. Note that equations (12.84) fails for $i = m$ because the rate of leaving state (m, m) is $\lambda + \mu$ rather than $\lambda + 2\mu$. Hence, we consider all states with $i \leq m$ as boundary equations. Once x and g_j, $j = 0, 1, \ldots, m$ are known, we can satisfy the boundary equations as follows. Note that all equilibrium probabilities appearing in any interior equation have the form $\pi_{i,j} = g_j x^i$, even when they appear in a boundary equation. We can therefore replace any $\pi_{i,j}$ appearing in a boundary equation as well as in an equilibrium equation by $g_j x^i$. This yields a system of equations that can be solved for the remaining $\pi_{i,j}$, except for a factor. This factor can be obtained as follows. Let

$$P_m = \sum_{(i,j:2i-j=m)} \pi_{i,j}.$$

The sum of all probabilities is now:

$$\sum_{i,j} \pi_{i,j} = \sum_{i,j:2i-j<m} \pi_{i,j} + P_m \frac{1}{1 - \rho}.$$

The reason is that $L_1 + L_2 = 2X_1 - X_2$, and

$$P\{L_1 + L_2 = n + 1\} = P\{L_1 + L_2 = n\}\rho,$$

or

$$P\left\{\sum_{(i,j):2i-j=n+1} \pi_{i,j}\right\} = P\left\{\sum_{(i,j):2i-j=n} \pi_{i,j}\right\}\rho.$$

It follows that

$$\sum_{i,j:2i-j\geq m} \pi_{i,j} = \sum_{n=m}^{\infty} P_m\rho^{n-m} = P_m\frac{1}{1-\rho}.$$

For our example, we found

$$P_0 = \pi_{0,0} = 0.0434$$
$$P_1 = \pi_{1,1} = 0.0560$$
$$P_2 = \pi_{2,2} + \pi1,0 = 0.0847$$
$$P_3 = \pi_{3,3} + \pi_{2,1} = 0.0805$$
$$P_4 = \pi_{4,4} + \pi_{3,2} + \pi_{2,0} = 0.07354.$$

Also, with $m = 4$,

$$P\{L_1 + L_2 \geq 4\} = P_4\frac{1}{1-\rho} = 0.7354.$$

The shorter queue model can be expanded to more than two lines. In addition to that, arrival need not be Poisson. For details, see [51], [97], [98].

12.4.4.2 A Tandem Queue with No Blocking

In order to demonstrate a system in which the initial conditions are satisfied by the eigenvector of x_1, we consider a tandem queue similar to the one used in Table 11.1, except that now there is no blocking. In detail, we have a tandem queue with Poisson arrivals and exponential service times. Customers arrive at server 1 at a rate of λ, where they are served at a rate μ_1. They then proceed to server 2, who has a service rate of μ_2. After completing service with the second server, customers leave the system. The waiting space between servers, the buffer, provides space for only N customers, and once this space is full, any customers completing service with server 1 leaves the system. The state variables are X_1 and X_2, the numbers of customers in the two lines, including the ones in service. It is required to find the equilibrium probabilities $\pi_{i,j}$ for this system, where

$$\pi_{i,j} = P\{X_1 = i, X_2 = j\}.$$

The equilibrium equations are as follows. First, consider the boundary equations, which are the equations of the states with $X_1 = i = 0$:

$$0 = -\pi_{0,0}\lambda + \pi_{0,1}\mu_2 \tag{12.87}$$

$$0 = \pi_{1,j-1}\mu_1 - \pi_{0,j}(\lambda + u_2) + \pi_{0,j+1}\mu_2, \quad 0 < j < N \tag{12.88}$$

$$0 = \mu_1\pi_{1,N-1} - (\lambda + \mu_2)\pi_{0,N}. \tag{12.89}$$

The interior equations are the equations of the states with $i > 0$.

$$0 = \pi_{i-1,0}\lambda - \pi_{i,0}(\lambda + \mu_1) + \pi_{i,1}u_2 \tag{12.90}$$

$$0 = \pi_{i-1,j}\lambda + \pi_{i+1,j-1}\mu_1 - \pi_{i,j}(\lambda + \mu_1 + \mu_2) + \pi_{i,j+1}\mu_2, \quad 0 < j < N \tag{12.91}$$

$$0 = \pi_{i-1,N}\lambda - \pi_{i,N}(\lambda + \mu_1 + \mu_2) + \pi_{i+1,N-1}\mu_1 + \pi_{i+1,N}\mu_1. \tag{12.92}$$

We now set

$$\pi_{i,j} = g_j x^i.$$

Replacing the $\pi_{i,j}$ in equations (12.90) to (12.92) by these values and dividing by x^{i-1} yields

$$0 = -g_0((\lambda + \mu_1)x - \lambda) - g_1\mu_2 x \tag{12.93}$$

$$0 = g_{n-1}\mu_1 x^2 - g_n((\lambda + u_1 + \mu_2)x - \lambda) + g_{n+1}\mu_2 x, \tag{12.94}$$

$$n = 1, 2, \ldots, N - 1 \tag{12.95}$$

$$0 = g_{N-1}\mu_1 x^2 - g_N(-\mu_1 x^2 + (\mu_1 + \mu_2 + \lambda)x - \lambda). \tag{12.96}$$

The solution $x = 0$ can be excluded because in this case, the vector g is zero, and only the trivial solution is possible. Similarly, if $g_0 = 0$ then $g_i = 0$ for all i. We can therefore set $g_0 = 1$, and since $x \neq 0$, (12.93) yields

$$g_1 = \frac{\lambda + \mu_1 - \lambda/x}{\mu_2}. \tag{12.97}$$

The remaining g_n can now be calculated by solving (12.94), which yields

$$g_{n+1} = g_n \frac{\lambda + \mu_1 + \mu_2 - \lambda/x}{\mu_2} - g_{n-1}\frac{\mu_1}{\mu_2}x, \quad n = 1, 2, \ldots, N - 1. \tag{12.98}$$

Given these g_n, any x satisfying (12.95) is obviously an eigenvalue. Now it is convenient to introduce g_{N+1} as follows:

$$g_{N+1} = g_N \frac{(\mu_1 + \mu_2 + \lambda) - \mu_1 x - \lambda/x}{\mu_2} - g_{N-1}\frac{\mu_1}{\mu_2}x. \tag{12.99}$$

With this, g_{N+1} is a function of x and every eigenvalue x must satisfy $g_{N+1}(x) = 0$.

Equation (12.98) is a difference equation, and to solve it, we set $g_n = y^n$. Substituting this value for g_n in (12.98) and simplifying leads to the following quadratic equation for y.

$$0 = y^2 - y\frac{\lambda + \mu_1 + \mu_2 - \lambda/x}{\mu_2} + \frac{\mu_1}{\mu_2}x. \tag{12.100}$$

This equation has two roots, y_1 and y_2. If we define

$$b(x) = \frac{\lambda + \mu_1 + \mu_2 - \lambda/x}{\mu_2} \tag{12.101}$$

$$d(x) = b(x)^2 - 4\frac{\mu_1}{\mu_2}x, \tag{12.102}$$

then we get

$$y_1 = \frac{1}{2}\left(b(x) - \sqrt{d(x)}\right), \quad y_2 = \frac{1}{2}\left(b(x) + \sqrt{d(x)}\right). \tag{12.103}$$

If $d(x) \neq 0$, there must be two distinct values y_1 and y_2, and g_n is given by

$$g_n = c_1 y_1^n + c_2 y_2^n.$$

Here, c_1 and c_2 are chosen such that one obtains the correct values for $g_0 = 1$ and g_1, which is, according to (12.97)

$$g_1 = \frac{\lambda + \mu_1 - \lambda/x}{\mu_2} = b(x) - 1.$$

With $g_n = c_1 y_1^n + g_2 y_2^n$, this leads to

$$c_1 + c_2 = g_0 = 1$$
$$c_1 y_1 + c_2 y_2 = g_1 = b(x) - 1$$

The solution of these equations is

$$c_1 = -\frac{y_2 - 1}{\sqrt{d(x)}}, \quad c_2 = \frac{y_1 - 1}{\sqrt{d(x)}}.$$

Therefore, we conclude that for $d(x) \neq 0$:

$$g_n = \frac{1}{\sqrt{d(x)}}((y_2 - 1)y_2^n - (y_1 - 1)y_1^n). \tag{12.104}$$

We now write (12.99) as

$$g_{N+1} = g_N\left(b(x) - \frac{\mu_1}{\mu_2}x\right) - g_{N-1}\frac{\mu_1}{\mu_2}x.$$

Since $b(x) = y_1 + y_2$ and $\frac{\mu_1}{\mu_2}x = y_1 y_2$, this yields

$$g_{N+1} = \frac{1}{\sqrt{d(x)}}\left[(y_1 - 1)(y_2 - 1)(y_1^{N+1} - y_2^{N+1})\right]. \tag{12.105}$$

To find the eigenvalues, we need the zeros of $g_{N+1} = 0$. Obviously, $g_{N+1} = 0$ if either $y_1 = 1$ or $y_2 = 1$ or $y_1^{N+1} = y_2^{N+1}$. Each alternative must be investigated, and once this is done, we can find the eigenvalues x_i resulting from these alternatives.

In the alternative $y_1 = 1$, equation (12.103) leads to

$$y_1 = \frac{1}{2}\left(b(x) - \sqrt{d(x)}\right) = 1.$$

We replace $b(x)$ and $d(x)$ by their definitions given in equations (12.101) and (12.102), respectively, and after some calculations, we find the following quadratic equation:

$$\mu_1 x^2 - (\lambda + \mu_1)x + \lambda = 0.$$

The solutions of this equation are $x = 1$ and $x = \frac{\lambda}{\mu_1}$.

We now calculate g_n, using $x = \frac{\lambda}{\mu_1}$. Since we are considering the solution with $y_1 = 1$, equation (12.104) becomes

$$g_n = \frac{(y_2 - 1)y_2^n}{\sqrt{d(x)}}.$$

Now, using also equations (12.104) and (12.103), we get

$$2y_2 = b(x) + \sqrt{d(x)} = 2(b(x) - 1),$$

and $y_2 = b(x) - 1$. For $x = \frac{\lambda}{\mu_1}$, $b(x) - 1$ is $\frac{\lambda}{\mu_2}$, consequently

$$y_2 = \frac{\lambda}{\mu_2}.$$

After some minor calculations, we get

$$g_n = \left(\frac{\lambda}{\mu_2}\right)^n, \quad n = 0, 1, 2, \ldots, N.$$

Therefore, a solution of equations (12.90) to (12.92) is given by

$$\pi_{i,j} = x^i y_2^j = \left(\frac{\lambda}{\mu_1}\right)^i \left(\frac{\lambda}{\mu_2}\right)^j.$$

We now have to verify that with these values, equations (12.87) and (12.88) are also satisfied, which indeed they are. As one sees, the solution obtained is really a *product-form solution* as they are discussed in Chapter 11.

The question is: are there other eigenvalues, and when should they be used? To explore this question, consider the alternatives $y_2 = 1$ and $y_1^{N+1} = y_2^{N+1}$. The alternative $y_2 = 1$ leads to the same eigenvalues $x_1 = 1$ and $x_2 = \frac{\lambda}{\mu_1}$, and we do not consider it further. Now, $y_1^{N+1} = y_2^{N+1}$ is true whenever we have

$$y_1 = z y_2,$$

where z is the solution of

$$z^{N+1} = 1.$$

In other words, whenever z is an $N + 1$rst root of 1, then x is an eigenvalue if it satisfies

$$2y_1 = b(x) - \sqrt{d(x)} = z(b(x) + \sqrt{d(x)}) = 2z y_2.$$

For instance, if N is odd, $z = -1$, one has

$$b(x) - \sqrt{d(x)} = -b(x) - \sqrt{d(x)}.$$

This leads to $b(x) = 0$, or

$$\frac{\lambda + \mu_1 + \mu_2 - \lambda/x}{\mu_2} = 0,$$

and the solution of this equation, $x = \frac{\lambda}{\lambda + \mu_1 + \mu_2}$, is an eigenvalue, as long as N is odd.

One obvious equation for determining an eigenvalue is $y_1 = y_2$ for x, but we deal with this case later. Except for $z = 1$ and $z = -1$ when N is odd, all other roots of $z^{N+1} = 1$ are complex, and they are given as

$$z = \cos\left(\frac{n\pi}{N+1}\right) + i \sin\left(\frac{n\pi}{N+1}\right) = a + i\alpha.$$

$y_1 = z y_2$ now becomes

$$b(x) - \sqrt{d(x)} = (a + i\alpha)\left(b(x) + \sqrt{d(x)}\right).$$

Since the right side of this equation is complex, the left side must also be complex, and this requires the $d(x) < 0$. Hence we have

$$b(x) - i\sqrt{-d(x)} = (a + i\alpha)\left(b(x) + i\sqrt{-d(x)}\right).$$

This equation can now be solved for x, which, on closer inspection, turns out to be an equation of third degree for x. According to [31], all eigenvalues of tridiagonal matrices are real. Therefore, only the real roots are of interest.

Now consider the case where $y_1 = y_2$, which happens if $d(x) = 0$. This case is only of minor interest because y_1 can only be equal to y_2 if $\lambda = \mu_2$. To prove this, note that if $y_1 = y_2$ then g_n is given, as is well known:

$$g_n = c_1 y^n + c_2 n y^{n-1}.$$

Hence

$$g_0 = c_1$$

$$g_1 = c_1 y + c_2.$$

Since $g_0 = 1$ and $g_1 = b(x) - 1$, this yields

$$c_1 = 1, \quad c_2 = b(x) - 1 - y = y - 1.$$

The last step follows from the fact that $y = b(x)/2$ according to equation (12.103). Consequently

$$g_n = y^n + n(y - 1)y^{n-1}.$$

Since $b(x) = 2y$ and $\frac{\mu_1}{\mu_2} x = y^2$, we have

$$
\begin{aligned}
g_{N+1} &= g_N \left(b(x) - \frac{\mu_1}{\mu_2} x \right) - g_{N-1} \frac{\mu_1}{\mu_2} x \\
&= g_N(2y - y^2) - g_{N-1}y^2 \\
&= (N+1)y^N - Ny^{N-1})(2y - y^2) - (y^{N+1} + (N-1)y^N(y-1)).
\end{aligned}
$$

After some calculation, this yields

$$g_{N+1} = -(N+1)(y-1)^2 y^N. \tag{12.106}$$

Hence, $d(x) = 0$ only if $y = y_1 = y_2 = 1$. When $y_1 = 1$, we showed earlier that $y_2 = \frac{\lambda}{\mu_2}$, and since this is 1, $\lambda = \mu_2$. Hence, only if $\lambda = \mu_2$, one has an eigenvalue such that $y_1 = y_2$.

Now we want to know when the eigenvalues that are different from $x = \lambda/\mu_1$ are needed? They will be required if equation (12.88) changes. To determine when this occurs, consider the following situation: Suppose that as long as the first line is empty, the first server will help the second server, increasing in this way the service rate of server 2 from μ_2 to μ_3. This implies that instead of (12.88), we have

$$0 = \pi_{1,j-1}\mu_1 - \pi_{0,j}(\lambda + \mu_2) + \pi_{0,j+1}\mu_3, \quad j > 0. \tag{12.107}$$

In this case, all eigensolutions $(x_k, g^{(k)})$ with $g^{(k)} = [g_j^{(k)}]$ will be needed. We thus have to set

$$\pi_{i,j} = \sum_{k=1}^{N} c_k g_j^{(k)} x_k^i,$$

where the c_k must be chosen such that the equations given by (12.107) are satisfied.

12.5 Conclusions

Solving queueing problems by using characteristic roots has a long tradition, beginning with Crommelin [16] who applied it to the $M/D/c$ queue and Kendall [58] who applied it to the $GI/M/1$ queue. It was used extensively by Morse [70] for queues with Erlang distributions, and Chaudhry and Templeton [10] to analyze bulk queues. The extrapolating Crout method was first described in [27].

Regarding transition matrices with repeating rows of block matrices, we have to name Neuts [72], [73]. Part of his work is based on the work of Wallace [93] who generalized the idea of characteristic roots in the scalar case to matrices, resulting in matrix polynomials. Initially, no connection to Gaussian elimination was realized. The first paper that showed this connection appeared in 1993 [35] and later in [99]. In a way, it was the simplest method that was discovered last. The eigenvalue method was used by several authors, including [4], [22] and [68]. In [18], the equilibrium probabilities were obtained from generating functions, which were analyzed by using eigenvalues.

The method that works best depends very much on the properties of the model. It also depends on how frequently a model is used. If the model is used only a few times, say to explore a specific situation, then the simplest method is best, and this would be the method discussed in Section 12.3.4.1, which is an extension of the state elimination method. Possible improvements can be obtained when the blocks in question are banded. If execution times are important, then one should attempt to generalize the scalar case given by Algorithm 32 to the matrix case. One should also look at the logarithmic reduction method as described in [6], [62] or [37, pages 170ff]. However, none of these methods reduces the order of the time complexity. They mainly reduce the number of iterations, but they all have essentially the same time complexity per iteration. In fact, the logarithmic reduction algorithm increases the density of the matrices involved, increasing in this way the number of flops per iteration.

When rows repeat in a scalar sense, then the execution times are typically not an issue, and it may not pay to spend much time to search for the best possible method. In general, we prefer the extrapolating state elimination method, but if routines for finding roots of polynomial equations are available, then one can consider methods based on characteristic roots. If the rows of matrix blocks repeat, then the use of generalized eigenvalues is recommended when the rank of the matrix with the rates that increase the level has only one non-zero row, because then, there is only one generalized eigenvalue that must be determined to find the equilibrium probabilities. Also, when the $\sum_k Q_k$ is tridiagonal, then all eigenvalues are real, and they can be found with relative ease [31].

Problems

12.1 Write a program to find the π_i for the $M/D/c$ queue by using the extrapolating Crout method. Test this program for $\lambda = 1.8$, $\mu = 1$ and $c = 2$.

12.2 For the model with repeating rows, use the generating function $P(z)$ to find the variance of X, provided the a_i as they are given by equation (12.19) are known.

12.3 Write a program to generate the matrices Q_{-1}, Q_0 and Q_1 as defined in Section 12.3 for the $E_2/H_2/2$ queue, and find its equilibrium probabilities, using the extrapolating state elimination method.

12.4 Consider the $GI/G/1$ model discussed in Section 12.2.4 with both inter-arrival and service time Pareto when the truncation points h and g increase. In particular, what happens to the probability that the waiting time is zero, and what happens with the expected line length?

12.5 Prove that in a QBD process, $P_1 G = RP_{-1}$.

12.6 Consider a shorter queue problem as follows: Arrivals are Poisson, and they always join the shorter line. If both lines have the same length, they join line 1. The service times are exponential, but the rates of the two servers are different. They are μ_1 for server 1 and μ_2 for server 2. Derive the formulas for this system. Use X_1 for the length of the waiting line in front of server 1, and X_2 for the difference in line length, which can be positive or negative. Derive the essential formulas if jockeying occurs as soon as the lines differ by more than m.

12.7 Discuss under which conditions the shorter queue model will require only one eigenvalue to find all equilibrium probabilities. Can probabilistic jockeying be accommodated, and how would this work in the case of two lines? Does this also work for 3 or more lines? In a two-line model, can arrivals join line 1 unless line 2 is shorter by a specific amount?

12.8 Equations (12.80) to (12.84) can be solved as difference equations, along the lines of the solution given in Section 12.4.4.1. Derive the formulas for this case.

12.9 Consider a QBD process and answer the following questions:

1. If P_1 has a rank of r, what is the rank of A_1? Can this be exploited to make Algorithm 33 more efficient?
2. What can you say about the number of non-zero eigenvalues of A_1?
3. If P_{-1} has a rank of r, what is the rank of G? Can this be exploited to make Algorithm 33 more efficient?

Appendix A
Language Conventions for Algorithms

The computer programs underlying the algorithms of this book were either written in MATLAB or VBA, two languages that were accessible to us and, we believe, accessible to almost everyone. However, today it seems that most researchers in the area of stochastic modeling use MATLAB, and many consider VBA, which comes with Microsoft Excel$^{©}$ as not very sophisticated, which we believe is wrong. Yes, MATLAB and its Gnu version Octave have powerful facilities for working with matrices, polynomials, and eigenvalues, and in many applications, these facilities are crucial. However, VBA has all facilities for higher computer languages. In particular, VBA has the facility to write recursive procedures, and to dimension arrays with lower bound other than 1 and arrays with arbitrary dimensions, facilities missing in MATLAB.

The algorithm in this book is neither in MATLAB nor VBA. There are many languages in use, and new languages are being developed, and though settling for one language may please some people, it may displease others. We thus write our languages in a pseudo-code, which also allows us to use informal constructs to increase clarity. A general code is justified because most computer languages have similar commands for program control, such as "for" and "while" loops, "if" statements, and so on. We essentially use the MATLAB conventions for this purpose. Also, the arithmetic operations are similar, and so are the names of the built-in functions, such as "exp" for the exponential function and "sqrt" for the square root, though names may vary. We also use the function sum(array) for the sum of an array. For the relation operators, we use "=" rather than "==" for equality, and the mathematical symbols "$\leq$" and "$\geq$". The use of the operators "and" "or" and "not" in logical expressions should be self-explanatory.

One convention seldom used outside MATLAB is the notion of a sequence of integers. Thus, the sequence of integers from a to b is denoted as a:b, where a and b are integers. Also, if c is an integer, a:c:b will be the sequence with the sequence a, a+c, a+2c, and so on, as long as a+k*c<b. For instance, 2:5 will produce 2,3,4,5, and 3:2:8 will produce 3,5,7. Hence, the loop

W. Grassmann and J. Tavakoli, *Numerical Methods for Solving Discrete Event Systems*.
CMS/CAIMS Books in Mathematics 5, https://doi.org/10.1007/978-3-031-10082-6

```
for i = 2:5
    statement
end for
```

will execute "statement" with i running from 1 to 5. If "for i = 5:-1:2" is used in place of "for i = 2:5", then "statement" is executed with i running from 5 to 2. The colon operator is also used for arrays. Thus, if "myarray" is a one-dimensional array (something that does not exist in MATLAB), then myarray(3:5) is the array consisting of the third, fourth, and fifth entry of "myarray". Note that, in MATLAB, one uses "endfor" instead of "end for". Similarly, we write "end if" instead of "endif", and use a similar notation elsewhere.

Clearly, the conventions for writing symbols in mathematics are different from the conventions of writing symbols in computer languages. In particular, in mathematics, symbols are written in italic, but in computer programs, they are roman. Also, in mathematics, symbols typically only have one letter, whereas, in computer languages, you are encouraged to use meaningful variable names which consist of several letters. Finally, in mathematics, one uses subscripts and superscripts. To do the conversion, we use the following conventions.

- If a letter is used in both a mathematical formula and a computer code, then we use the same letter, but write it roman in the computer code.
- Greek letters are written in full.
- Subscripts are converted to array arguments. The order of the arguments is as follows: first comes the subscript, then the superscript, and finally the argument. For instance, $f_k(X)$ is written as "f(k,X)", and $p_i^{(k)}$ is written as "p(i,k)"..
- For clarity, we often enclose the variables used in programs by quotes when they are defined in the text.

In some programs, we use sentences for explanations. In these sentences, we feel free to either use mathematical or computer conventions.

References

1. D. Aldous and L. Shepp. The least variable phase type distribution is Erlang. *Stochastic Models*, 3:467–473, 1987.
2. A. S. Alfa. Markov chain representation of discrete distributions applied to queueing models. *Computers and Operations Research*, 31:2365–2385, 2004.
3. J. Banks, J. S. Carson, B. L. Nelson, and D. M. Nicol. *Discrete-Event Systems Simulation*. Prentice Hall, Upper Sassle River, NJ, 2005.
4. D. Bertsimas. An analytic approach to a general class of G/G/s queueing systems. *Operations Research*, 38:139–155, 1990.
5. D. Bertsimas and D. Nakasato. The distributional Little's law and its applications. *Operations Research*, 43:298–310, 1995.
6. D. A. Bini, G. Latouche, and B. Meini. *Numerical Methods for Structured Markov Chains*. Oxford University Press, Oxford, 2005.
7. G. Bolch, S. Greiner, H. de Meer, and K.S. Trivedi. *Queueing Networks and Markov Chains*. IIE Transactions. John Wiley & Sons (Electronic copy), Hoboken, NJ, 2nd edition, 2006.
8. J. P. Buzen. Computational algorithms for closed queueing networks with exponential servers. *Communications of the ACM*, 16(9):527–531, 1973.
9. C. G. Cassandras and S. Lafortune. *Introduction to Discrete Event Systems*. Springer Verlag, New York, second edition, 2008.
10. M. Chaudhry and J. Templeton. *A first Course on Bulk queues*. John Wiley, New York, 1983.
11. M. L. Chaudhry. Alternative solutions of stationary queueing-time distributions in discrete-time queues: GI/G/1. *The Journal of the Operational Research Society*, 44(10):1035–1051, 1993.
12. H. Chen and D. Yao. *Fundamentals of Queueing Networks*. Springer Verlag, New York, 2001.
13. E. Çinlar. *Introduction to Stochastic Processes*. Prentice-Hall, Englewood Cliffs, N.J., 1975.
14. T. H. Cormen, C. E. Leiserson, and R. L. Rivest. *Algorithms*. MIT Press, Cambridge, Massachusetts, 1990.
15. D. R. Cox. *Reenwal Theory*. Methuen, London, 1962.
16. C. D. Crommelin. Delay probability formulae when the holding times are constant. *Post Office Electrical Engineers Journal*, 26:266—274, 1934. reference found in Wikipedia, M/D/c queue.
17. M. El-Taha and S. Stidham Jr. *Sample-Path Analysis of Queueing Systems*. Springer Verlag, New York, 1999.

© The Author(s), under exclusive license to Springer Nature Switzerland AG 2022
W. Grassmann and J. Tavakoli, *Numerical Methods for Solving Discrete Event Systems*,
CMS/CAIMS Books in Mathematics 5, https://doi.org/10.1007/978-3-031-10082-6

18. A. Elvalid and D. Mitra. Markovian arrival and service communication systems: Spectral expansion, separability and Kronecker product forms. In W. J. Stewart, editor, *Computations with Markov Chains*, pages 507–546, Boston, MA, 1995. Kluwer Academic Publishers.
19. Agner Krarup Erlang. Wikipedia.org. Accessed March 11, 2020.
20. W. Feller. *Introduction to Probability Theory and Its Applications*, volume 2. Wiley, New York, second edition, 1971.
21. B. L. Fox and P. W. Glynn. Computing Poisson probabilities. *Communications of the ACM*, 31(4):441–445, 1988.
22. H.R. Gail, S. L. Hantler, and B. A. Taylor. Spectral analysis of M/G/1 and G/M/1 type Markov chains. *Adv. Appl. Prob.*, 28:114–165, 1996.
23. I. Gohberg, P. Lancaster, and L. Rodman. *Matrix Polynomials*. Academic Press, New York, 1982.
24. W. Grassmann and J Tavakoli. Comparing some algorithms for solving QBD processes exhibiting special structures. *INFOR*, 48:133–141, 2010.
25. W. K. Grassmann. Transient solutions in Markovian queueing systems. *Computers and Operations Research*, 4:47–56, 1977.
26. W. K. Grassmann. The $GI/PH/1$ queue: A method to find the transition matrix. *INFOR*, 20:144–156, 1982.
27. W. K. Grassmann. The factorization of queueing equations and their interpretation. *Journal of the Operational Research society*, 36(11):1041–1050, 1985.
28. W. K. Grassmann. Means and variances of time averages in Markovian environments. *EJOR*, 31:132–139, 1987.
29. W. K. Grassmann. Means and variance in Markov reward systems. In C. D. Meyer and R. J. Plemmons, editors, *Linear Algebra, Markov Chains, and Queueing Models*, chapter 18, pages 193–204. Springer-Verlag, New York, 1993.
30. W. K. Grassmann. Rounding errors in certain algorithms involving Markov chains. *ACM Transactions in Mathematical Software*, 19:496–508, 1993.
31. W. K. Grassmann. Real eigenvalues of certain tridiagonal matrix polynomials, with queueing applications. *Journal of Linear Algebra and Its Applications*, 342:93–106, 2002.
32. W. K. Grassmann. The use of eigenvalues for finding equilibrium probabilities of certain Markovian two-dimensional queueing problems. *INFORMS Journal on Computing*, 15(4):412–421, 2003.
33. W. K. Grassmann. Warm-up periods in simulation can be detrimental. *Probabilitiy in Engineering and Informational Sciences*, 22:415–429, 2008.
34. W. K. Grassmann. Efficient methods to find the equilibrium distribution of the number of customers in $GI/M/c$ queues. *INFOR*, 52:197–205, 2014.
35. W. K. Grassmann and D. P. Heyman. Equilibrium distribution of block-structured Markov chains with repeating rows. *Journal of Applied Probability*, 27:557–576, 1990.
36. W. K. Grassmann and J. L. Jain. Numerical solutions of the waiting time distribution and the idle time distribution of the arithmetic $GI/G/1$ queue. *Operations Research*, 37(1):141–150, 1989.
37. W. K. Grassmann and David A. Stanford. Matrix analytic methods. In W. K. Grassmann, editor, *Computational Probability*, International Series in Operations Research and Management Science, chapter 7, pages 153–202. Kluwer Academic Publishers, Boston, 2000.
38. W. K. Grassmann, Michael Taksar, and Daniel P. Heyman. Regenerative analysis and steady state distributions for Markov chains. *Operations Research*, 33:1107–1117, 1993.
39. W. K. Grassmann and J. Tavakoli. Stochastic and substochastic solutions for infinite-state Markov chains with applications to matrix analytic methods. *Adv. Appl. Prob.*, 40:1157–1173, 2008.
40. W. K. Grassmann and J. Tavakoli. Transient solutions for multi-server queues with finite buffers. *Queueing Systems*, 62:35–49, 2009.

41. W. K. Grassmann and J. Tavakoli. The distribution of the line length in a discrete GI/G/1 queue. *Performance Evaluation*, 131:43–53, 2019

42. W. K. Grassmann and J. Tavakoli. A numerical method to determine the waiting time and the idle time distributions of a GI/G/1 queue when inter-arrival times and / or service times have geometric tails. *INFOR*, 57:286–295, 2019

43. D. Gross and D. R. Miller. The randomization technique as a modeling tool and solution procedure for transient Markov processes. *Operations Research*, 32:343–361, 1984.

44. D. Gross, J. F. Shortle, J.M. Thompson, and C. M. Harris. *Fundamentals of Queueing Theory*. Wiley, New York, fourth edition, 2008.

45. G. Haßlinger. A polynomial factorization approach to the discrete time $GI/G/1/\langle N\rangle$ queue size distribution. *Performance Evaluation*, 23:217–240, 1995.

46. Qi Ming He. *Fundamentals of Matrix-Analytic Methods*. Springer, New York, 2014.

47. D. P. Heyman and M. J. Sobel. *Stochastic Models in Operations Research*, volume 1 McGraw-Hill, New York, 1982.

48. R. A. Horn and C. R. Johnson. *Matrix Analysis*. Cambridge University Press, New York, 1985.

49. Horner's method. Wikipedia.org. Accessed Dec. 1, 2021.

50. R. A. Howard. *Dynamic Programming and Markov Processes*. John Wiley, New York, 1960.

51. B.F. Adan I.J. *A Compensation Approach to Queueing Problems*. PhD thesis, Technische Universiteit Eindhoven, Eindhoven, The Netherlands, 1991.

52. N. K. Jaiswal. *Priority Queues*. Academic Press, New York, 1968.

53. A. Jensen. Markoff chains as an aide to study Markoff processes. *Skand Aktuarietidskrift*, 36, 1953.

54. S. Karlin and H. M. Taylor. *A First course in Stochastic Processes*. Academic Press, New York, NY, second edition, 1975.

55. J. Keilson. *Markov Chain Models - Rarity and Exponentiality*. Springer, New York. 1979.

56. J. G. Kemeny and J. L. Snell. *Finite Markov Chains*. Springer, New York, 1976.

57. J. G. Kemeny, J.L. Snell, and A. W. Knapp. *Denumerable Markov Chains*. Van Nostrand, Princeton, NJ, 1966.

58. D. G. Kendall. Stochastic processes occurring in the theory of queues and their analysis by the method of imbedded Markov chains. *Annals of Mathematical Statistics*, 24(3):338–354, 1953.

59. N. K. Kim and M. L. Chaudhry. The use of the distributional Little's law in the computational analysis of discrete-time $GI/G/1$ and $GI/D/c$ queues. *Performance Evaluation*, 65:3–9, 2008.

60. P. Lancaster. *Lambda-matrices and Vibrating Systems*. Pergamon Press, Oxford, 1966.

61. G. Latouche. Algorithms for infinite Markov chains with repeating columns. In *Linear Algebra, Markov Chains, and Queueing Models*, volume 48 of *IMA Volumes in Mathematics and its Applications*, pages 231–265. Springer-Verlag, Heidelberg, Germany, 1992.

62. G. Latouche and V. Ramaswami. A logarithmic reduction algorithm for quasi-birth-death processes. *J. Appl. Prob.*, 30:650–674, 1993.

63. G. Latouche and V. Ramaswami. *Introduction to Matrix Analytic Methods in Stochastic Modelling*. ASA-SIAM Series on Applied Probability. SIAM, 1999.

64. D. A. Levin and Y. Peres. Markov chains and mixing times. downloaded April 2, 2022 from https://pages.uoregon.edu/dlevin/MARKOV/markovmixing.pdf.

65. M. A. Marsan, G. Balbo, G. Conte, S. Donatelli, and G. Franceschinis. *Modelling with Generalized Stochastic Petri Nets*. Wiley series in parallel computing. John Wiley, New York, 1995.

66. Matrix pencil. Wikipedia.org. Accessed August 21, 2021.

67. B. Melamed and M. Yadin. Numerical computation of soujourn-time distributions in queueing networks. *Journal of the ACM*, 31:839–854, 1984.

68. I. Mitrani and R. Chakka. Spectral expansion solution for a class of Markov models: Application and comparison with the matrix-geometric method. *Performance Evaluation*, 23:241–260, 1995.

69. A. Moorsel, L. Kant, and W. Sanders. Computation of the asymptotic bias and variance for simulation of Markov reward models. *Proceedings of the 29th Annual Simulation Symposium*, pages 173–182, 1996.

70. P. M. Morse. *Queues, Inventories and Maintenance*. John Wiley, New York, 1958.

71. M. F. Neuts. Pobability distributions of phase type. In *Liber Amicorum Prof. emeritus H. Florin*, pages 173–206. University of Louvain, Belgium, 1973.

72. M. F. Neuts. *Matrix-Geometric Solutions in Stochastic Models*. Johns Hopkins University Press, Baltimore, 1981.

73. M. F. Neuts. *Structured Stochastic Matrices of M/G/1 Type and Their Applications*. Marcel Dekker, New York, 1989.

74. C. A. O'Cinneide. Entrywise perturbation theory and error analysis for Markov chains. *Numerische Mathematik*, 65:109–120, 1993.

75. N. U. Prabhu. *Queues and Inventories*. John Wiley, New York, 1965.

76. W. H. Press, B. P. Flannery, S. A. Teukolsky, and W. T. Vetterling. *Numerical Recipes*. Cambridge University Press, Cambridge, 1986.

77. J. F. Reynolds. The covariance structure of queues and related processes–a survey of recent work. *Advance of Applied Probability*, 7:383–415, 1975.

78. S. Ross. *Introduction to Probability Models*. Academic Press, twelfth edition, 2019.

79. W. H. Sanders et al. Möbius, model-based environment for validation of system reliability, availability, security and performance. https://www.mobius.illinois.edu, accessed June 21, 2016.

80. R. Schassberger. *Warteschlangen*. Springer Verlag, 1973.

81. J. G. Shantikumar. Bilateral phase-type distributions. *Naval Research Logistic Quarterly*, 32:119–136, 1985.

82. E. A. Silver, D. Pyke, and R. Peterson. *Inventory Management and Production Planning and Scheduling*. John Wiley and Sons, New York, 1998.

83. D.A. Stanford and W.K. Grassmann. The bilingual server system: A queueing model featuring fully and partially qualified servers. *INFOR*, 31:261–277, 1993.

84. G. W. Stewart. *Matrix Algorithms. Volume I: Basic Decompositions*. SIAM, Philadelphia, 1998.

85. G. W. Stewart. *Matrix Algorithms. Volume II: Eigensystems*. SIAM, Philadelphia, 1998.

86. W. J. Stewart. *An Introduction to the Numerical Solution of Markov Chains*. Princeton University Press, Princeton, New Jersey, 1994.

87. W. J. Stewart. *Probability, Markov Chains, Queues, and Simulation*. Princeton University Press, Princeton, New Jersey, 2009.

88. S. Stidham. A last word on $L = \lambda W$. *Operations Research*, 22(2):417—421, 1974.

89. J. C. François Sturm. Mémoire sur la résolution des équations numériques. *Mém Savans Etrang.*, pages 271–318, 1835.

90. H. Takagi. *Analysis of Polling Systems*. MIT Press Computer Systems Series. MIT Press, Cambridge, MA 02142, 1986.

91. H. C. Tijms. *A First Course in Stochastic Models*. Wiley & Sons Ltd, West Sussex, 2003.

92. K. S. Trivedi. *Probability and Statistics with Reliability, Queueing and Computer Science Application*. John Wiley, second edition, 2002.

93. V. Wallace. *The Solution of Quasi Birth and Death Processes Arising from Multiple Access Computer Systems*. PhD thesis, University of Michigan, 1969.

94. W. Whitt. *Stochastic-Process Limits*. Springer, New York, 2002.

95. J. S. Wilkinson. *The Algebraic Eigenvalue Problem*. Clarendon Press, Oxford, 1965.

96. T. Yang and M. L. Chaudhry. On steady-state queue size distributions of the discrete-time *GI/G/*1 queue. *Advances in Applied Probability*, 28(4), 1996.

97. Y. Zhao and W. K. Grassmann. The shortest queue model with jockeying. *Naval Research Logistics*, 37:773–787, 1990.

98. Y. Zhao and W. K.Grassmann. Queueing analysis of the jockeying model. *Operations Research*, 43:520–529, 1995.

99. Yiqiang Q. Zhao, W. Li, and W. J. Brown Infinite block-structured transition matrices and their properties. *Adv. Appl. Prob.*, 30:365–384, 1998.

Index

Printed by Printforce, United Kingdom